Source Separation and Decentralization for Wastewater Management

Source Separation and Decentralization for Wastewater Management

Tove A. Larsen, Kai M. Udert
and Judit Lienert

Publishing
London • New York

Published by **IWA Publishing**
Alliance House
12 Caxton Street
London SW1H 0QS, UK
Telephone: +44 (0)20 7654 5500
Fax: +44 (0)20 7654 5555
Email: publications@iwap.co.uk
Web: www.iwapublishing.com

First published 2013

Cover Image: New technologies for the treatment of source-separated waste streams will improve the recovery of resources such as nutrients and water. Copyright EOOS Design.

British Library Cataloguing in Publication Data
A CIP catalogue record for this book is available from the British Library

ISBN 9781843393481 (Paperback)
ISBN 9781780401072 (eBook)

Printed and bound in Great Britain by Bell and Bain Ltd., Glasgow

FSC www.fsc.org MIX Paper from responsible sources FSC® C007785

Contents

Chapter 3

Chapter 4

Chapter 5

Chapter 8
The rationale for decentralization of wastewater infrastructure 101

George Tchobanoglous and Harold Leverenz

Chapter 9
Cities of the global South – is decentralized sanitation a solution? 117

Barbara Evans

Chapter 17

Chapter 18

Chapter 24

Chapter 25

Authors' List

Yuval Alfiya

Technion – Israel Institute of Technology
Faculty of Civil and Environmental Engineering
Department of Environmental, Water and Agricultural Engineering
Technion City, Haifa 32000, Israel
alfiya@tx.technion.ac.il

Prof. Dr. M. Bruce Beck

University of Georgia
Warnell School of Forestry and Natural Resources
Athens, Georgia 30602, USA
mbbeck@uga.edu

Dr. Christian Binz

Eawag, Swiss Federal Institute of Aquatic Science and Technology
Innovation Research in Utility Sectors (Cirus)
8600 Dübendorf, Switzerland
christian.binz@eawag.ch

Prof. em. Dr. Markus Boller

Eawag, Swiss Federal Institute of Aquatic Science and Technology
Department of Urban Water Management (SWW)
8600 Dübendorf, Switzerland
markus.boller@emeriti.eawag.ch

Zenah Bradford-Hartke

The University of New South Wales
School of Chemical Engineering *and* UNESCO Centre for Membrane Science and Technology
Sydney, New South Wales 2052, Australia
zenah@unsw.edu.au

Dr. Shelley Brown-Malker

Very Small Particle Company Ltd
Brisbane, Queensland 4076, Australia
shelleyb@vspc.com

Bert Bundervoet

Ghent University
Laboratory Microbial Ecology and Technology (LabMET)
9000 Ghent, Belgium
b.bundervoet@colsen.nl

Prof. Dr. David Butler

University of Exeter
Centre for Water Systems
Exeter, EX4 4QF, United Kingdom
d.butler@exeter.ac.uk

Dr. Christopher A. Buzie

Hamburg University of Technology
Institute of Wastewater Management and Water Protection
21072 Hamburg, Germany
christopher.buzie-fru@tu-harburg.de

Dr. Dana Cordell

University of Technology, Sydney (UTS)
Institute for Sustainable Futures (ISF)
Sydney, New South Wales 2007, Australia
dana.cordell@uts.edu.au

Dr. Vasileios Diamantis

Democritus University of Thrace
Department of Environmental Engineering
Laboratory of Wastewater Management and Treatment Technologies
67100 Xanthi, Greece
bdiamant@env.duth.gr

Prof. Dr. Jan Willem Erisman

Louis Bolk Institute
3972 LA Driebergen, The Netherlands
j.erisman@louisbolk.nl

and

VU University Amsterdam
Department of Earth Sciences
1081 HV Amsterdam, The Netherlands

Barbara Evans

University of Leeds
School of Civil Engineering
Water Engineering and Environment
Leeds, LS2 9JT, United Kingdom
b.e.evans@leeds.ac.uk

Prof. Dr. Malin Falkenmark

Stockholm International Water Institute
111 51 Stockholm, Sweden
malin.falkenmark@siwi.org

Prof. Dr. Eran Friedler

Technion – Israel Institute of Technology
Faculty of Civil and Environmental Engineering
Department of Environmental, Water and Agricultural Engineering
Technion City, Haifa 32000, Israel
eranf@techunix.technion.ac.il

Dr. Ted Gardner

Central Queensland University
Institute for Resource Industries and Sustainability
Rockhampton, Queensland 4702, Australia
e.gardner@cqu.edu.au

Dr. Heiko Gebauer

Eawag, Swiss Federal Institute of Aquatic Science and Technology
Innovation Research in Utility Sectors (Cirus)
8600 Dübendorf, Switzerland
heiko.gebauer@eawag.ch

Prof. em. Dr. Willi Gujer

Swiss Federal Institute of Technology Zürich (ETHZ)
Department of Civil, Environmental and Geomatic Engineering (BAUG)
Institute of Environmental Engineering
8093 Zürich, Switzerland
gujer@ifu.baug.ethz.ch

Prof. Dr. Bruce Jefferson

Cranfield University
Cranfield Water Science Institute
Cranfield, MK43 0AL, United Kingdom
b.jefferson@cranfield.ac.uk

Prof. Dr. Paul Jeffrey

Cranfield University
Cranfield Water Science Institute
Cranfield, MK43 0AL, United Kingdom
p.j.jeffrey@cranfield.ac.uk

Sarina Jenni

Eawag, Swiss Federal Institute of Aquatic Science and Technology
Process Engineering Department (Eng)
8600 Dübendorf, Switzerland
sarina.jenni@eawag.ch

Prof. Dr. Håkan Jönsson

SLU - Swedish University of Agricultural Sciences
Department of Energy and Technology
750 07 Uppsala, Sweden
hakan.jonsson@slu.se

Prof. Dr. Işık Kabdaşlı

İstanbul Technical University
Civil Engineering Faculty
Environmental Engineering Department
34469 İstanbul, Turkey
kabdasli@itu.edu.tr

Prof. Dr. Jürg Keller

The University of Queensland
Advanced Water Management Centre (AWMC)
Brisbane, Queensland 4072, Australia
j.keller@uq.edu.au

Prof. Dr. Klaus Kümmerer

Leuphana Universität Lüneburg
Institute of Sustainable and Environmental Chemistry
21335 Lüneburg, Germany
klaus.kuemmerer@uni.leuphana.de

Dr. Katarzyna Kujawa-Roeleveld

Wageningen University
Agrotechnology and Food Sciences Group
Sub-department of Environmental Technology *and* Lettinga Associates Foundation (LeAF)
6700 AA Wageningen, The Netherlands
katarzyna.kujawa@wur.nl

Dr. Tove A. Larsen

Eawag, Swiss Federal Institute of Aquatic Science and Technology
Department of Urban Water Management (SWW)
8600 Dübendorf, Switzerland
tove.larsen@eawag.ch

Michele Laureni

Eawag, Swiss Federal Institute of Aquatic Science and Technology
Process Engineering Department (Eng)
8600 Dübendorf, Switzerland
michele.laureni@eawag.ch

Prof. Dr. Gregory Leslie

The University of New South Wales
School of Chemical Engineering *and* UNESCO Centre for Membrane Science and Technology
Sydney, New South Wales 2052, Australia
g.leslie@unsw.edu.au

Dr. Harold Leverenz

University of California at Davis
Department of Civil and Environmental Engineering
Davis, California 95616, USA
hlleverenz@ucdavis.edu

Dr. Judit Lienert

Eawag, Swiss Federal Institute of Aquatic Science and Technology
Department of Environmental Social Sciences (ESS)
8600 Dübendorf, Switzerland
judit.lienert@eawag.ch

Prof. Dr. Jörg Londong

Bauhaus-Universität Weimar
Department of Urban Water Management and Sanitation
99423 Weimar, Germany

Dr. Christoph Lüthi

Eawag, Swiss Federal Institute of Aquatic Science and Technology
Water and Sanitation in Developing Countries (Sandec)
8600 Dübendorf, Switzerland
christoph.luethi@eawag.ch

Prof. Dr. Max Maurer

Eawag, Swiss Federal Institute of Aquatic Science and Technology
Department of Urban Water Management (SWW)
8600 Dübendorf, Switzerland
max.maurer@eawag.ch

and

Swiss Federal Institute of Technology Zürich (ETHZ)
Department of Civil, Environmental and Geomatic Engineering
Institute of Environmental Engineering
8093 Zürich, Switzerland

Prof. em. Dr. Gustaf Olsson

Lund University
Department of Measurement Technology and Industrial Electrical Engineering (MIE)
Division of Industrial Electrical Engineering and Automation (IEA)
221 00 Lund, Sweden
gustaf.olsson@iea.lth.se

Prof. Dr. Ralf Otterpohl

Hamburg University of Technology
Institute of Wastewater Management and Water Protection
21072 Hamburg, Germany
otterpohl@tu-harburg.de

Dr. Bert Palsma

STOWA, Dutch Foundation for Applied Water Research
3800 CD, Amersfoort, The Netherlands
palsma@stowa.nl

Dr. Arne R. Panesar

Deutsche Gesellschaft für Internationale Zusammenarbeit (GIZ) GmbH
Department Asia and Latin America
65760 Eschborn, Germany
arne.panesar@giz.de

Prof. Dr. Bruce E. Rittmann

Arizona State University
Swette Center for Environmental Biotechnology
Biodesign Institute
Tempe, Arizona 85287, USA
rittmann@asu.edu

Prof. Dr. Hansruedi Siegrist

Eawag, Swiss Federal Institute of Aquatic Science and Technology
Process Engineering Department (Eng)
8600 Dübendorf, Switzerland
hansruedi.siegrist@eawag.ch

Dr. Ashok Sharma

Commonwealth Scientific and Industrial Research Organisation, Australia
Land and Water Division
Highett, Victoria 3190, Australia
ashok.sharma@csiro.au

Prof. Dr. Thor Axel Stenström

Stockholm Environment Institute
Bioresources Group
106 91 Stockholm, Sweden
thor.axel.stenstrom@sei.se

and

Norwegian University of Life Sciences
Department of Mathematical Science and Technology
1432 Ås, Norway

Dr. Eckhard Störmer

Eawag, Swiss Federal Institute of Aquatic Science and Technology
Innovation Research in Utility Sectors (Cirus)
8600 Dübendorf, Switzerland
stoermer@z-punkt.de

Bjartur Swart

STOWA, Dutch Foundation for Applied Water Research
3800 CD Amersfoort, The Netherlands

and

MWH North Europe
2612 PA Delft, The Netherlands
bjartur.swart@nl.mwhglobal.com

Prof. em. Dr. George Tchobanoglous

University of California at Davis
Department of Civil and Environmental Engineering
Davis, California 95616, USA
gtchobanoglous@ucdavis.edu

Elizabeth Tilley

Eawag, Swiss Federal Institute of Aquatic Science and Technology
Department of Water and Sanitation in Developing Countries (Sandec)
8600 Dübendorf, Switzerland
elizabeth.tilley@eawag.ch

and

Swiss Federal Institute of Technology Zürich (ETHZ)
Centre for Development and Cooperation (NADEL)
8092 Zürich, Switzerland

Prof. Dr. Bernhard Truffer

Eawag, Swiss Federal Institute of Aquatic Science and Technology
Innovation Research in Utility Sectors (Cirus)
8600 Dübendorf, Switzerland
bernhard.truffer@eawag.ch

Prof. Dr. Olcay Tünay

İstanbul Technical University
Civil Engineering Faculty
Environmental Engineering Department
34469 İstanbul, Turkey
tunayol@itu.edu.tr

Dr. Kai M. Udert

Eawag, Swiss Federal Institute of Aquatic Science and Technology
Process Engineering Department (Eng)
8600 Dübendorf, Switzerland
kai.udert@eawag.ch

Prof. em. Dr. Willy Verstraete

Ghent University
Laboratory Microbial Ecology and Technology (LabMET)
9000 Gent, Belgium
willy.verstraete@ugent.be

Prof. Dr. Björn Vinnerås

SLU - Swedish University of Agricultural Sciences
Department of Energy and Technology
750 07, Uppsala, Sweden
bjorn.vinneras@slu.se

Prof. Dr. Urs von Gunten

Eawag, Swiss Federal Institute of Aquatic Science and Technology
Department of Water Resources and Drinking Water (W+T)
8600 Dübendorf, Switzerland
urs.vongunten@eawag.ch

and

Ecole Polytechnique Fédérale de Lausanne (EPFL)
School of Architecture, Civil and Environmental Engineering (ENAC)
Institute of Environmental Engineering
1015 Lausanne, Switzerland

Prof. em. Dr. Peter A. Wilderer

Technische Universität München
Institute for Advanced Study
85748 Garching, Germany
peter.wilderer@mytum.de

Prof. Dr. Jun Xia

Chinese Academy of Sciences (CAS)
Center for Water Resources Research *and* Key Laboratory of Water Cycle and Related Surface Processes
100101, Beijing, China
xiaj@igsnrr.ac.cn

Prof. Dr. Grietje Zeeman

Wageningen University
Agrotechnology and Food Sciences Group
Sub-department of Environmental Technology *and* Lettinga Associates Foundation (LeAF)
6700 AA Wageningen, The Netherlands
grietje.zeeman@wur.nl

Preface

Until 20 years ago, the combination of the term "source separation" and "wastewater" was hardly found in the scientific literature. Source separation was mainly an issue for those interested in the treatment of solid waste. Today, source separation is one of the most exciting developments in the area of wastewater treatment, driven by the increasingly visible resource restrictions of the 21st century.

The book, which you now hold in your hands, is a landmark in this development. It summarizes in a systematic and thorough way the advantages and challenges of source separation and shows that this notion is inherently connected to decentralization, another of today's important issues. The book, however, does not stay conceptual. As a process engineer by education and by heart, I have enjoyed immensely the technology part, where a whole new world of processes and techniques opens up. Most of the new technologies not only ensure safe discharge, but also allow for the recovery of precious resources.

From the chapters on the international experience, I realized how many different drivers and challenges all lead in the direction of source separation and decentralization. Although I was aware of many of the pilot projects from the literature, I am impressed by this condensed presentation of more than a decade of experience in a number of different countries. And I must agree with my colleagues of a life-time, Peter Wilderer and Bruce Beck, who in the two last chapters both dare to pronounce the possibility of a paradigm shift in wastewater treatment: It certainly looks as if a new field has evolved.

I wish you an inspiring reading and learning experience and hope that the next generation of wastewater professionals will develop this innovative and exciting approach into maturity. The world desperately needs a new approach to wastewater management and based on my experience, I predict that this approach will include source separation and decentralization.

Willi Gujer

ideas to use the nutrients in decentralized wastewater systems for the growth of biomass and energy production.

Our inadequate approach to the resources captured in the wastewater system becomes especially obvious in the case of phosphorus. Phosphorus recycling was one of the early arguments for source separation of wastewater, and Dana Cordell shows that phosphorus belongs to the most important elements for humanity as it is crucial for global food security (Chapter 3). Cordell demonstrates that phosphorus scarcity cannot simply be expressed by the depletion of the phosphate rock reserves. Rather, five interrelated dimensions must be considered, namely physical, managerial, institutional, economic and geopolitical scarcity. The central importance of phosphorus recovery from excreta and other wastes for a sustainable future is of special importance for the readers of this book.

Like BOD, the nutrients phosphorus and nitrogen are also "misplaced resources." Especially in developing and fast-industrializing countries, the nutrients are desperately needed in agriculture, but at the same time they are polluting water resources worldwide. In Chapter 4, Jan Willem Erisman and Tove Larsen focus on the dramatic increase of environmental pollution due to excess nitrogen. While nitrogen shortage in some areas has severe consequences for human nutrition, the production of harmful reactive nitrogen is expected to increase dramatically because of population growth, increased protein consumption and biofuel production. Only with a better understanding how different nitrogen sources give rise to environmental effects will it be possible to develop policies that are effective in tackling these problems. As described in the chapter, source separation could be an important measure amongst other policy means.

Water scarcity as discussed by Malin Falkenmark and Jun Xia in Chapter 5 is probably the clearest example of the importance of source separation and decentralization for resource efficiency. Water sets distinct limits for population growth and human welfare, and water efficiency will help extract more welfare per drop of water. Water efficiency and especially also water recovery is greatly enhanced by separating less polluted water from toilet waste, but as discussed later in the book, the entire concept of sewer-based urban water management is challenged by water scarcity.

Our approach in this book is clearly resource-oriented. However, some substances contained in wastewater are so highly dispersed that an efficient recovery is hardly feasible. This applies especially to micropollutants, as discussed by Klaus Kümmerer in Chapter 6. Source separation in the case of micropollutants mainly aims at removing these potentially harmful substances as near to the source as possible. The "benign-by-design" principle goes a step further, to the real beginning of the pipe, the industrial production of chemicals.

In our daily life as engineers, we are confronted with physical problems concerning the functioning of the system and are asked to find the most cost-effective solutions for the requested services. However, the prevailing urban water management system is hardly set up to deal with the increasingly complex challenges of the 21st century. Max Maurer introduces us to the undisputed advantages of the conventional centralized system and explains its success. However, Chapter 7 also helps to understand the problems of sewer-based wastewater management: high capital-intensity and extreme inflexibility. Based on in-depth analyses, Maurer identifies entry markets for on-site treatment systems with better abilities to adapt to growing demands, especially in rapidly expanding cities.

In Chapter 8, George Tchobanoglous and Harold Leverenz illustrate similar problems from a more practical point of view. Not only in fast-growing cities, but especially also in regions that have to cope with water scarcity, sewer-based wastewater management comes to its limits. In the USA, the increasing introduction of water saving devices such as low-flush toilets, conservation programs and water extraction from sewers have led to reduced wastewater flow. This has a number of negative effects including increased corrosion rates. The authors also introduce a typology of wastewater infrastructures, which is highly useful to structure discussions about decentralized technologies.

For developing countries, such problems may seem trivial. Developing countries are facing dramatic urban water management challenges, as introduced by Barbara Evans in Chapter 9. In the cities of the global South, access to basic sanitary services is low, with severe public health consequences. Although the boundary conditions are very different to those in industrialized countries, decentralized wastewater systems are becoming prevalent for similar reasons. As others, Evans also introduces tools to systematize the field, namely the advantages of “vertical” and “horizontal” unbundling of wastewater services. These tools enable us to tackle wastewater management problems efficiently, with more flexibility, and better adapted to different realities. “Vertical unbundling”, for instance, allows creating different incentives along the value chain of urban sanitation, thus increasing the chance to develop a functioning system in a city.

Although there are many potential advantages of source separation and decentralization, which are extensively discussed and referenced in Part I of this book, the challenges for a paradigm change are huge (Part II). This is particularly evident for cities, where the central paradigm is deep-rooted. However, as discussed by Tove Larsen and Willi Gujer in Chapter 10, there are many chances for technology development and technology learning in various niches of the system. Depending on the socio-economic environment, these niches look different. There are many possibilities for cost-effective improvements of the present system, which may eventually lead to the development of viable alternatives to sewer-based sanitation, also in an urban environment.

Changes to the existing urban water management system inevitably invoke fears that urban hygiene could be jeopardized. Of course, this is not trivial, and the problem must be tackled with due respect. However, as Thor Axel Stenström demonstrates in Chapter 11, the risk that exposed humans are infected by pathogens is not inherently larger in decentralized wastewater systems than in sewer-based ones. To quantify the risk, an integral assessment must include the reduction of pathogens, the transmission routes and the exposure. It is also of vital importance to follow the entire "flow" of wastewater, which originates in the household and passes through the collection and treatment part to the point of re-use or disposal. Also the downstream populations must be included.

If the end-products from decentralized wastewater systems are re-used in agriculture, the risk of contamination by pathogens must be minimized, but additionally they must reach the agricultural land efficiently and in the appropriate form. Håkan Jönsson and Björn Vinnerås provide the fundamentals to understand the agricultural perspective (Chapter 12), hereby supporting engineers in developing technologies that transform human excreta to a marketable product.

Farmers must accept fertilizers from human excreta, as well as the consumers who buy agricultural produce. In Chapter 13, Judit Lienert reviews social science studies on the acceptance of urine source separation, including the re-use of human urine in agriculture. The results are generally positive, but based on the questioning of more than 2700 users of NoMix toilets in seven European countries, the weak points of the technology also become clear. Additionally, Lienert gives some guidance on how to explore aspects of social acceptance of source-separating technologies. This illustrates how essential the involvement of social scientists becomes the closer wastewater treatment gets to the consumers.

Gustaf Olsson (Chapter 14) provides an approach based on modular build-up and standardization to increase the acceptance of source-separating technologies and make them work in practice. Similar to cars or computers, the complexity of the machinery does not mean that only specialists can use it. However, the responsibility for proper operation and the handling of failures should be delegated to professional service enterprises. Olsson suggests using remote sensing, a proven technology in other areas and simple sensors along the whole chain of "smart water grids." To allow for mass production and the economic viability of decentralized systems, we must think in terms of "plug and play" of the components.

Also from a socio-economic perspective, Bernhard Truffer, Christian Binz, Heiko Gebauer and Eckhard Störmer arrive at the conclusion that the success of source separation and decentralized technologies will depend on reliable and effective components, but even more so on integrating these into working systems (Chapter 15). These authors understand source-separating, decentralized wastewater technologies as a "systemic innovation" problem. They draw on experience from other domains to understand what it takes to develop such a field.

In Part III a wide range of technologies for the treatment of source-separated waste streams are presented. Some commercially manufactured reactors already exist, especially for greywater treatment. Many other treatment processes have been investigated in the laboratory or on a pilot scale. Nevertheless, source separation and decentralization still offer extensive research opportunities for engineers and urban planners, because the reliable operation of small reactors with concentrated source-separated waste streams poses new challenges.

Compared to conventional wastewater management, a sanitation system based on source separation and decentralization consists of many different waste streams, dispersed treatment units and involved stakeholders. This new system is more complex, but has the key advantage of allowing for wider range of technologies and business models. In Chapter 16, Elizabeth Tilley presents a conceptual approach to describe the functional groups (e.g., user interface, collection and storage) and product flows (e.g., brownwater) in any kind of sanitation system. This approach helps to identify the treatment steps and linkages that provide reliable and cost-effective sanitation. The chapter clearly shows that not only do novel technologies have to be developed, but business relationships also have to be enabled.

A thorough understanding of the composition of wastewater streams is needed to choose effective treatment reactors and identify business opportunities. On the basis of a comprehensive summary of literature data, Eran Friedler, David Butler and Yuval Alfiya (Chapter 17) discuss how socio-economic conditions and the technological development of appliances determine the composition and variability of wastewater streams. The analysis shows that treating the source-separated waste streams according to their composition allows for efficient recovery of water and nutrients. Additionally, targeted treatment improves the removal of pathogens and emerging pollutants such as personal care products.

Most pathogens are excreted via faeces. In Chapter 18, Ralf Otterpohl and Christopher Buzie present various processes to treat faecal solids. They focus on technologies with a low degree of mechanization which can be easily applied on-site in locations without infrastructures. These are mainly biological processes such as composting, but also simple chemico-physical processes such as dehydration. The authors emphasize the need for further development of the technology. Future research should focus not only on minimizing energy demand: simple operation and low maintenance are at least equally important.

Extensive experience with decentralized treatment is available for two source-separated waste streams, for faecal solids as well as for greywater. In fact, Bruce Jefferson and Paul Jeffrey (Chapter 19) argue that aerobic biological treatment of greywater is the most successful application of decentralized treatment of any source-separated wastewater. Intensive systems such as membrane bioreactors (MBRs) as well as extensive technologies such as reed beds (constructed wetlands) exhibit a similarly high performance of almost 90% BOD removal. The key challenges for greywater treatment are the high variability of the BOD load and concentration and the high fraction of xenobiotic compounds such as personal care products.

Besides organic compounds and pathogens, nitrogen is another major target for wastewater treatment. Most of the nitrogen is excreted via urine, so that nitrogen treatment technologies are mostly needed for urine or blackwater. Since nitrogen is a valuable nutrient in agriculture, the target of the treatment is recovery instead of removal if the costs are comparable to those of local fertilizers. In Chapter 20,

Kai Udert and Sarina Jenni argue that autotrophic denitrification is the most energy-efficient process for removing nitrogen from urine, but diligent process control is required to ensure stable process performance. Nitrogen can be recovered by biologically oxidizing a part or all of the ammonia in urine to nitrate. The resulting solution can be used directly as fertilizer, or a concentrated fertilizer can be produced if the water is removed, for example by distillation.

While urine contains most of the excreted nutrients, faeces have the highest chemical energy content in the form of organic substances. Gretjie Zeeman and Katarzyna Kujawa-Roeleveld (Chapter 21) discuss the use of anaerobic digestion for recovering this energy as methane gas. The most suitable waste streams for anaerobic digestion are brownwater and blackwater, due to their high faecal content. The authors show that anaerobic digestion of blackwater in an upflow anaerobic sludge blanket reactor (UASB) is a proven technology: in pilot studies nearly 90% of the chemical oxygen demand (COD) can be degraded and 60% can be recovered as methane. Further research is required to develop effective post-treatment options to remove pharmaceutical residues and hormones.

Energy recovery is also one of several applications of electrochemical processes. In Chapter 22, Kai Udert, Shelley Brown-Malker and Jürg Keller give an overview of a variety of processes which have been tested on a laboratory scale. Some of them have high potential for the treatment of source-separated waste streams. The main advantage of electrochemical processes for decentralized reactors is their direct use of electric current and voltage for process control and automation. In electrolysis, electricity is applied to remove substances such as ammonia, urea, organics and pathogens, while fuel-cell applications allow the direct conversion of chemical energy into electrical energy. The authors also discuss the use of electroactive bacteria in bioelectrochemical systems, a new technology which has lately received considerable attention.

Ammonia stripping from urine has already been tested successfully on a pilot scale by several research groups. In Chapter 23, Hansruedi Siegrist, Michele Laureni and Kai Udert present the basic principles of ammonia stripping with air and the subsequent ammonia recovery in sulfuric acid. They also discuss literature data on steam stripping and report about their own experience with passive ammonia stripping in urine-collecting systems. The ammonia stripping technology is most suitable for medium sized reactors, since corrosive chemicals or steam at high pressure and temperature are needed. An interesting combined process is struvite precipitation and ammonia stripping, which allows for the recovery of phosphorus and nitrogen as two different products.

Struvite precipitation by magnesium dosage is the most intensively researched nutrient recovery process from urine. Struvite precipitation is therefore at the core of Chapter 24, in which Işık Kabdaşlı, Olcay Tünay and Kai Udert discuss treatment processes based on the transfer of nutrients into or onto a solid phase. Again, urine and blackwater are the main substreams for this type of treatment process, because they contain the most nutrients. Aluminium, iron or calcium can be used to precipitate the phosphate as alternatives to magnesium. Other processes that recover nutrients at or in a solid phase are adsorption, ion exchange and water removal (e.g., distillation).

In contrast to struvite precipitation reactors, which are currently tested on a pilot scale, membrane bioreactors (MBRs) are an established technology for decentralized wastewater treatment. In Chapter 25, Gregory Leslie and Zenah Bradford-Hartke report on the use of MBRs on full, pilot and laboratory scales for combined wastewater, blackwater, greywater and urine. MBRs have some ideal properties for on-site reactors: they can be compact, modular, scalable and provide consistent product quality. However, further research is needed to ensure consistent throughput capacity and to reduce their energy consumption.

Insufficient micropollutant removal is a common shortcoming of MBRs and other established technologies. Urs von Gunten (Chapter 26) discusses a wide range of chemical oxidation processes which can be used to oxidize micropollutants and remove pathogens in a post-treatment step. He argues that ozone, •OH radicals and ferrate exhibit the highest overall performance based on their reaction kinetics, oxidant stability and by-product formation. Further research is required to better understand the conditions under which chemical oxidation of micropollutants can produce toxic degradation products or unwanted by-products such as bromo-organic compounds.

Willy Verstraete, Vasileios Diamantis and Bert Bundervoet conclude the technology part of this book by presenting a concept for enhanced energy recovery from existing sewer-based sanitation systems (Chapter 27). Instead of establishing a new sanitation system based on separating the wastewater streams at the source, the authors suggest that the solids could be intercepted and concentrated in the sewer to recover as much as possible of the chemical energy of the organic solids by anaerobic digestion. Solid recovery (fractionation) can be increased by several processes such as chemically enhanced sedimentation, dissolved air flotation, bioflocculation and direct sewage filtration. The authors also present technologies to obtain high-quality effluents which can be used for irrigation or disposed of to sensitive water bodies.

Part III of this book shows that the basic idea behind source separation, that is, the efficient management of the resources contained in wastewater, can be approached with a variety of technologies and concepts depending on local socio-economic conditions and the existing infrastructure. We hope that continuous development will result in a wide range of technologies so that engineers and urban planners of the future will have more flexibility in implementing appropriate sanitation systems. Besides the development of technologies and management schemes, pioneers are needed who have the courage and the confidence to implement new concepts in pilot projects which can later serve as references. Pioneering projects and their initiators are presented in Part IV of this book.

The very early pioneers of source separation are found in Scandinavia and especially in Sweden. These actors developed the first (modern) NoMix toilet, which was a key precondition for the implementation of decentralized urine source-separating solutions also in other European countries. Björn Vinnerås and Håkan Jönsson introduce the Swedish story of source separation, which started in the early 1990s (Chapter 28). The reasons for an increased interest in urine-diverting technologies were growing environmental concerns and the ambition of the Swedish government to create closed loops. This environmental concern was reflected in the building of eco-villages, which often included on-site treatment of wastewater. Today, Sweden has around 700,000 on-site sanitation systems in a variety of settings and system configurations.

Also in Germany, diverse projects with on-site wastewater treatment have been implemented in the last two decades. In Chapter 29, Jörg Londong describes the German development of source separation. Initially, these projects were driven by universities, but relatively rapidly the field was structured via a working group within the German Association for Water, Wastewater and Waste (DWA). In a variety of German pilot projects, mainly blackwater and urine source-separating systems were tested. In the newest large-scale project in Hamburg, black and greywater are treated separately to create a fully decentralized treatment system. DWA still views the implementation of new sanitation strategies as a major task and is currently working on a German standard. Recently, the German Environmental Ministry has indicated interest in the topic, which is certainly a positive signal.

Rather similarly, but at an even greater pace, decentralized and source-separating technologies were introduced in The Netherlands. Bjartur Swart and Bert Palsma present this success story in Chapter 30. The driving force was certainly STOWA, the Dutch Foundation for Applied Water Research, which coordinates research on behalf of 26 Water Boards, the Provinces and the Ministry of Infrastructure and the Environment. Until STOWA picked up the topic of "New Sanitation" in the year 2000, source-separating technologies were of minor importance. STOWA made two crucial decisions: to take responsibility *and* to place wastewater treatment into a wider social context. Within a few years, more than half of all water boards in the Netherlands were involved in one of the 40 pilot or research projects. New Sanitation has reached a transition phase in 2011. It is now time to move from research to implementation, and several of the most promising initiatives will undergo up-scaling.

Markus Boller shows that the experience in Switzerland is rather different (Chapter 31). Switzerland took a lead in Europe by introducing a phosphate ban for textile detergents in 1986. From 1991, on-site infiltration of stormwater and separate sewer systems for all new or renovated buildings and roads were required by law. This made studies to control hazards from construction materials and traffic vehicles possible, and source control with respect to stormwater management can currently be seen as a system change in a transition phase. In contrast, in-house installations, especially of urine source separation, have much larger consequences and are far from being accepted as state of the art technology. In Switzerland, these initiatives were strongly driven by research, with only smaller implementation projects. Markus Boller introduces the three main types: on-site wastewater treatment and re-use, separate collection and processing of urine and small-scale autarkic material and water cycles. Several of the projects were motivated by the topography in the Swiss mountains, where connection to sewers is not possible. From a technical point of view, these decentralized water schemes performed satisfactorily, but also in Switzerland, stakeholder participation and acceptance will be key to their widespread implementation.

Australia faces fundamentally different challenges than these European countries. Ted Gardner and Ashok Sharma illustrate that the primary driver for the uptake of decentralized wastewater systems in Australia is water scarcity (Chapter 32). The Australian projects are also rather different to those in Switzerland, for instance, because often the private sector has taken the lead, and not research. However, similar to Swedish holiday homes along the Baltic Sea, or cable car stations in the Swiss Alps, the original driver in Australia, which is still important today, was the provision of on-site sanitation systems in non-sewered urban and peri-urban communities. Later important drivers included sustainability aspects, which inevitably lead to the consideration of source-separating technologies.

In Chapter 33, Christoph Lüthi and Arne Panesar argue that the drivers and constraints for source-separating wastewater technologies are drastically different between industrialized and middle- or low-income countries. The solutions, however, are in principle similar to those that provide sustainable answers for the problems in industrialized countries. For different reasons, the vast majority of households in the global South will remain to be served by on-site sanitation. Often, only rudimentary pit latrines or cesspits are common. The authors understand this deplorable situation also as an opportunity to leapfrog into a fundamental system change. They demonstrate this with two interesting "case studies" with more than impressive numbers. In eThekwini Metropolitan Municipality, South Africa, more than 90,000 urine-diverting dry toilets were introduced. A program in the Shaanxi province in China by an NGO aims at introducing 27,000 source-separating sanitation systems. Overall in China, more than two million urine-diverting dry toilets have been built between 2000 and 2010. It is likely that industrialized countries will strongly profit from the experience gained in low- and middle-income countries. Possibly, we will install source-separating, decentralized wastewater technologies in the global North, which were essentially developed, tested and improved in the global South.

Finally, in Part V, we have asked two experienced scientists to give their personal opinions on a possible paradigm shift in the area of wastewater treatment. In Chapter 34, Bruce Beck discusses the historical development from his perspective and argues for the beneficial co-existence of different wastewater paradigms. In the last chapter of this book, Peter Wilderer reflects on the nature of innovation and paradigm change. Based on decades of experience in the area of conventional urban water management and the paradigm shift discussed in this book, Wilderer encourages us to continue on the path of source separation and decentralization. As the editors of this book, we can only agree.

Part I

The advantages of source separation and decentralization

Chapter 2

The energy issue in urban water management

Bruce E. Rittmann

2.1 INTRODUCTION – THINK GLOBALLY AND ACT LOCALLY

A long-standing good policy for environmental action is to "think globally and act locally." Global thinking ensures that the policies are directed towards achieving the right long-term goal. Acting locally makes the steps to implement the proper policy practical and achievable.

Source separation and decentralization – the topics of this book – would seem to capture the "act locally" part perfectly. The question is then about how well these localized actions fit with good global thinking about energy. In this chapter, I lay out the global context for viewing energy as a resource to be recovered from wastewater. Then, I describe several ways in which we can capture the energy value from wastewater (Figure 2.1).

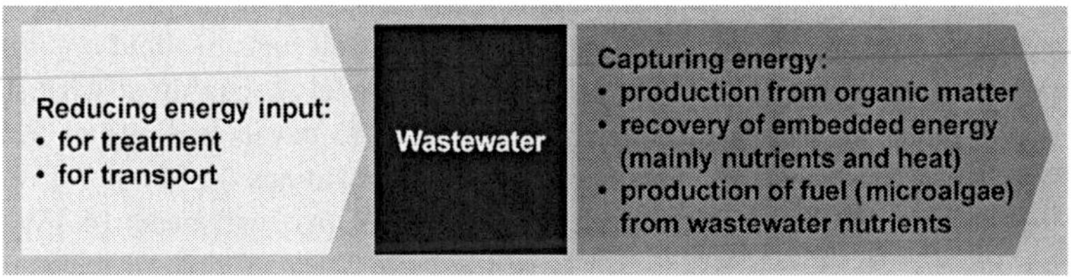

Figure 2.1 Energy recovery in wastewater treatment.

Capture of energy can be accomplished directly by producing renewable energy in useful forms, such as methane. It also can be achieved indirectly, for example, by lowering the energy cost of treatment or by capturing nutrients or heat, which already contain an energy investment. Rather new is the idea of producing fuel based on wastewater nutrients, which I will discuss in detail. Finally, I place source separation and decentralization in this overall context to see how these local actions can serve the global goals.

2.2 GLOBAL ENERGY GOAL

Human society today consumes energy at a rate of approximately 13 TW (primary energy; Rittmann 2008, Goldemberg and Johannson 2004). When divided by the global population of around 6.5 billion people, this gives an average of nearly 2,000 $W \cdot p^{-1}$, which has been suggested in Switzerland as a reasonable goal to balance lifestyle and sustainability needs (http://www.novatlantis.ch/en/2000-watt-society.html). Of course, the range of actual energy use is huge, ranging from <400 $W \cdot p^{-1}$ in the poorest countries (e.g., Bangladesh, Haiti, and Congo), to about 5,000 $W \cdot p^{-1}$ for the typical Western developed country, and to as high as 10,000 $W \cdot p^{-1}$ in the United States of America and 14,000 $W \cdot p^{-1}$ in the United Arab Emirates (Meda *et al.* 2010; http://en.wikipedia.org/List_of_countries_by_energy_consumption_per_capita). Attaining the 2,000 $W \cdot p^{-1}$ goal would lead to a total energy demand of 18 TW if the population stabilizes at 9 billion, an average number from global demographic scenarios.

Of the 13 TW used today, about 11 TW (or ~84%) comes from combusting fossil fuels. This high rate of fossil-fuel use is the main cause of the rapidly increasing concentrations of CO_2 in the Earth's atmosphere and the consequent changes in the global climate (IPCC 2007). Global climate change will put very large stresses on the world's food supply and is likely to lead to more severe weather events. Society's large demand for petroleum and natural gas also are fostering geopolitical and economic stresses. All of these stresses challenge the ability of our political institutions to maintain a world free from war, hunger, and disease.

Clearly, human society would be wise to move away from its overwhelming dependence on fossil fuels to provide its energy needs. However, it needs to do this by finding renewable, carbon-neutral substitutes, not by a catastrophic decline in energy use worldwide. Looking at the challenge from the point of view of global climate change, human society needs to cut its use of fossil fuels to about one-third of today's use rate if it is to hold the atmospheric CO_2 level roughly at what it is today (~390 ppm). In simple numbers, society needs to cut its use rate of fossil fuels from ~11 TW to ~4 TW. This means finding substitutes for about 7 TW of fossil fuel energy. If the demand increases to 18 TW, we will need 14 TW of substitute energy. Although we do not have the technological tools to make this substitution right now, this is the right policy goal to set for 2 to 3 decades to the future.

Our society needs to take actions today that will make it more realistic to achieve the 4-TW goal in 2 or 3 decades. At the same time, we need to reduce our energy demand by eliminating unnecessary energy use and improving the efficiency of the ways we use energy. While Western societies must drastically reduce their total energy demand to 2,000 $W \cdot p^{-1}$, emerging and developing countries need to improve their standard of living without surpassing this same level. As this chapter shows, source separation and decentralization can have an important role in improving our society's energy efficiency, and it also can help produce some of the substitute energy.

2.3 RENEWABLE ENERGY SOURCES

All major forms of renewable energy come ultimately from sunlight, which is plentiful on the Earth's surface, but not in a form useful for most of the work the society wants to do. Solar energy reaches the Earth at a rate of 173,000 TW (Goldemberg and Johannson 2004), or roughly 16,000 times greater than fossil-fuel use. However, solar energy is in the form of ultraviolet, visible, and near-infrared radiation, and we need to convert it to other forms if the energy is to be used to power airplanes, engines, pumps, computers, light bulbs, and so on. Thus, the challenge is conversion, not lacking an ample source.

We are able to capture sunlight energy in numerous ways: falling water (traditional water wheels and hydroelectric turbines), wind, ocean currents, photovoltaics, solar thermal, and biomass. Wastewater treatment fits into the last category, biomass. Therefore, I focus on photosynthetic biomass and how it finds its way to wastewater from its ultimate starting point, which is photosynthesis.

2.4 PHOTOSYNTHESIS, BIOMASS, AND BOD

Plants, algae, and bacteria are capable of photosynthesis, in which they capture visible and near-infrared radiation using a series of molecular antennae and photosystems (Madigan *et al.* 2003). They use the energy to synthesize and maintain themselves. While details of the photosynthesis strategies differ somewhat among the photosynthetic species, the basic pattern is the same. The most common type of photosynthesis is oxygenic photosynthesis, and it provides a good example of the basic pattern.

All life depends on the transfer of electrons from a high-energy state to a low-energy state. The organisms harvest energy from the transfer of the electrons, and they use that energy to grow. For oxygenic phototrophs, their electron donor is H_2O, which provides electrons at a very low-energy state. Thus, phototrophs utilize the energy captured in sunlight to boost the energy level of the electrons (e^-) as they separate them from protons (H^+) and generate oxygen (O_2):

$$H_2O + \text{ radiation energy } \longrightarrow 2H^+ + \mathbf{2e^-} + 0.5O_2 \tag{2.1}$$

The production of O_2 is the reason for the name oxygenic photosynthesis, and this reaction is responsible for Earth's atmosphere having 21% O_2.

This production of high-energy e^- and H^+ allows the photosynthetic organisms to make the internal energy carriers (e.g., ATP) and electron carriers (e.g., NADH and NADPH) that they utilize as they reduce environmental nutrients (mainly CO_2) to the right oxidation state and assemble the macromolecules that they build into new cells (Rittmann and McCarty 2001, Madigan *et al.* 2003). Here is an example of a half reaction in which biomass (represented very simply by $C_5H_7O_2NP_{0.1}$) is constructed from environmental nutrients and electrons derived originally from H_2O:

$$5HCO_3^- + NO_3^- + 0.1H_2PO_4^- + \mathbf{28.5e^-} + 34.6H^+ \longrightarrow C_5H_7O_2NP_{0.1} + 16.4H_20 \quad (2.2)$$

The 28.5 e^- needed in this synthesis reaction come from 14.25 equivalents of the oxygenic reaction shown in Eqn. 1. Then, the overall photosynthesis reaction is

$$5HCO_3^- + NO_3^- + 0.1H_2PO_4^- + 6.1H^+ + \text{energy} \longrightarrow C_5H_7O_2NP_{0.1} + 2.15H_20 + 14.25O_2 \quad (2.3)$$

When the photosynthetic biomass is a food crop, it enters the human food system as either a direct food input to humans or as feed to animals that eventually become food for humans. Some of the food consumed by humans finds its way into domestic wastewater. The food industry also generates a large amount of wastes: crop residues, animal manures, and food- and beverage-manufacturing wastes. While the food-industry wastes often are not introduced to the domestic wastewater, they constitute a large portion of the biomass originally generated by photosynthesis (Energy Information Administration 2005).

Whether the biomass enters domestic wastewater or is in a separate waste stream, it contains electrons (and their energy) that were fixed into it during photosynthesis. Instead of being valued as food, these “waste electrons” now become pollution that we call biological oxygen demand, or BOD. Discharged to receiving water, BOD causes depletion of dissolved oxygen and a myriad of related bad consequences (Masters and Ela 2008). Indeed, BOD is simply “misplaced electrons,” a valuable resource “gone bad.”

One thing we need to do is recapture the energy value of the misplaced electrons. Then, BOD pollution returns to its original state of a BOD energy resource. However, BOD in wastewater, waste sludge, animal manure, and food-processing wastewater cannot be used by society in these forms. The goal is to redirect the electrons in the BOD into chemical forms that society can use as an energy resource.

2.5 MICROBIAL ENERGY CONVERSION

Fortunately, different groups of microorganisms are able to convert the energy and electrons in biomass into forms readily used by human society (Rittmann 2006, 2008). Among the useful forms are methane gas (CH_4), hydrogen gas (H_2), ethanol (C_2H_6O), butanol (C_4H_8O), and electricity. For two reasons, I focus on CH_4, H_2, and electricity. The first reason is that these energy forms are easily harvested from water. CH_4 and H_2 are very low-solubility gases that naturally evolve from the water. Electricity is simply the movement of electrons, and electrons do not dissolve in water at all. This ready harvesting eliminates the large energy costs and technology complications for separating ethanol and butanol from water (Shapouri *et al.* 2001, Rittmann 2008). Second, CH_4, H_2, and electrons can be formed from almost any biomass component. They do not require a purified sugar feedstock, as do ethanol and butanol. The biomass contained in organic wastes and in photosynthetic organisms is comprised of mixtures of carbohydrate, protein, and lipids, often intertwined. Being able to produce an energy product from "real biomass" simplifies the technology and results in a much higher net-energy output.

For all the biomass-conversion approaches, the first step is to have the microorganisms oxidize the carbon in biomass to release the electrons. The oxidation reaction for the biomass formula shown above is:

$$C_5H_7O_2NP_{0.1} + 13.4H_2O \longrightarrow 5HCO_3^- + NH_4^+ + 0.1H_2PO_4^- + \mathbf{20.5e^-} + 24.6H^+ \quad (2.4)$$

The release of 20.5 e^- equivalents per mole of $C_5H_7O_2NP_{0.1}$ provides the energy input to the microorganisms for making CH_4, H_2, or electricity. The half reactions for using the electrons to make CH_4 and H_2 are:

$$CO_2 + 8H^+ + \mathbf{8e^-} \longrightarrow CH_4 + 2H_2O \quad (2.5)$$

$$2H^+ + \mathbf{2e^-} \longrightarrow \mathbf{H_2} \quad (2.6)$$

Electrical power comes from the movement of electrons through a potential difference. In practice, the electrons extracted from biomass are moved to where they can react with O_2, which establishes a large potential difference:

$$\mathbf{O_2} + 4H^+ + \mathbf{4e^-} \longrightarrow 2H_2O \quad (2.7)$$

Methanogenesis is a long-standing, mature technology for converting complex organics into CH_4, or natural gas (Speece 1996, McCarty and Smith 1986). The energy value of natural gas is widely recognized, and the developed world has extensive infrastructure in place to distribute and utilize natural gas for heating and electricity generation. Likewise, the wastewater-treatment field has extensive experience with sludge digestion for the purpose of methane production (McCarty 1964, Speece 1996, McCarty and Smith 1996, Rittmann and McCarty 2001). For practical purposes, methane has to be combusted to gain its energy value:

$$CH_4 + 2O_2 \longrightarrow CO_2 + 2H_2O \quad (2.8)$$

The conversion efficiency when generating electricity or motive force normally is about 35% (Rittmann *et al.* 2008).

H_2 is an alternative energy output from biomass. Similar to CH_4, H_2 can be combusted to generate electricity or motive force, and its conversion efficiency is about the same, ~35% (Rittmann *et al.* 2008):

$$2H_2 + O_2 \longrightarrow 2H_2O \tag{2.9}$$

However, H_2 can be oxidized in a chemical fuel cell to produce combustionless electricity with higher conversion efficiency, around 55% (Rittmann *et al.* 2008). Thus, H_2 can be a superior fuel over CH_4 in terms of energy conversion and by avoiding pollution associated with combustion. Furthermore, H_2 also is a widely used feedstock in the chemical industry (Lee *et al.* 2010a) and can be used to reduce a wide range of oxidized contaminants in water (Rittmann 2007).

H_2 can be produced from biomass in two ways that were reviewed by Lee *et al.* (2010a). The more traditional approach is fermentation, usually at a low pH to suppress methanogens. While fermentation is able to produce H_2 at a high volumetric rate, its conversion efficiency to H_2 is low, typically less than 17%. Not only does this reduce the energy capture to H_2, it produces a liquid effluent that is high in organic acids and alcohols, making the effluent a high-BOD stream.

The second approach is much newer; it is the microbial electrolysis cell (MEC). In an MEC, simple organic compounds are oxidized by bacteria that are able to transfer the electrons directly to an anode in a novel form of respiration. The electrons move through an electrical circuit to the cathode, where they reduce H^+ to H_2. These bacteria have been given several names: anode-respiring bacteria (ARB) and exoelectricigens seem to be the most popular ones. The most efficient ARB produce an electrically conductive biofilm matrix that transfers the electrons to the anode with minimal potential loss, while also firmly attaching them to the anode (Torres *et al.* 2010). By exploiting respiration instead of fermentation, an MEC can achieve high conversion efficiency to H_2. Using an MEC following H_2 fermentation seems to be an ideal approach for maximizing H_2 production, since the organic products from fermentation are excellent substrates for the anode of an MEC (Lee *et al.* 2010a).

The final conversion approach produces electrical power directly from oxidation of the organic matter in biomass. This approach is the microbial fuel cell (MFC), which is a variation from the MEC (Rittmann *et al.* 2008, Logan 2004, Logan *et al.* 2006a, b). Like with an MEC, ARB oxidize organic compounds at the anode and transfer the electrons to the anode. In an MFC, when the electrons reach the cathode, they reduce O_2 to H_2O, much the same as in a chemical fuel cell. Thus, the output is combustionless electricity produced with renewable fuel, the biomass.

How much energy is contained in the BOD of wastewater? A simple computation can be based on the typical BOD production rate of an adult human being,[1] which is

[1]Please note that this is a typical value in the US, but considerably higher than the European numbers referred to by Friedler *et al.* (2013).

about 0.135 $kg \cdot p^{-1} \cdot d^{-1}$. If all the energy could be captured, it would be approximately 23 $W \cdot p^{-1}$; this is about 20% of the energy intake of the food that an adult human eats (Meda *et al.* 2010). For several reasons, we cannot capture all the energy. One reason is that the microorganisms need to take some of the energy to maintain themselves. All of the microbial energy-recovery systems are anaerobic, which means that the microorganisms do not need to take a "big cut." Thus, about 90% of the energy could be sent along to the energy output, or about 21 $W \cdot p^{-1}$. A second reason is that typical wastewater treatment is aerobic, which means that the majority of the energy is lost to oxidation of the C to CO_2 with reduction of O_2 to H_2O. In a typical sewage treatment plant that uses anaerobic digestion, only about 40% of the BOD makes it to the digester, and it is in a form that is only about 50% digested; thus, only about 20% of the original BOD is captured as methane: 4.6 $W \cdot p^{-1}$. Another loss is if the desired energy output is electricity. Production of electricity via combustion of CH_4 or H_2 is only about 35% efficient, and this lowers the energy capture still further, to about 1.6 $W \cdot p^{-1}$. Conversion of H_2 to electricity in a chemical fuel cell is as much as 55% efficient, a noticeable improvement over 35%. Direct production of electricity in an MFC may be as much as 65% efficient (Rittmann *et al.* 2008).

Table 2.1 summarizes the amount of energy output that can be expected for different approaches. The calculations assume 0.135 $kgBOD \cdot p^{-1} \cdot d^{-1}$, an energy value of 14.7 $kJ \cdot gBOD^{-1}$, 90% conversion to the energy output in the anaerobic process, 35% efficiency of electricity production by combustion of CH_4 or H_2, 55% efficiency of electricity production by oxidation of H_2 in a conventional fuel cell, 65% efficiency of electricity production by oxidation of BOD in an MEC, and 20% transfer of influent BOD to waste-sludge degradable BOD in aerobic treatment.

One thing that is clear from Table 2.1 is that conventional aerobic treatment of wastewater needs to be avoided as much as possible to prevent the loss of BOD to oxidation without energy recovery; all of the approaches that produce the least energy involve aerobic treatment of the wastewater. If electricity is the desired output, then the best approaches are using an MFC for direct production or using an MEC to make H_2 to fuel a chemical fuel cell. With conventional technology, only highly concentrated wastewater (e.g., source-separated faeces) are suitable for anaerobic degradation (Zeeman and Kujawa-Roeleveld 2013). The innovative technologies, MEC and MFC, may be applicable only for relatively diluted wastewater, while the high efficiencies cited in Table 2.1 may be achievable only with concentrated solutions (Udert *et al.* 2013). Finally, Table 2.1 highlights that the amount of energy present in wastewater is far too little to provide the energy needed to run the modern human society: for example, 2,000 or 5,000 $W \cdot p^{-1}$. Nevertheless, capturing the energy in wastewater will have a large positive impact on the energy, greenhouse-gas, and economic bottom lines of the city or utility managing the wastewater.

do not produce nearly enough biomass to meet our society's demands for energy. For example, capturing all residuals from agriculture and the food system could supply as much as 25% of the world's energy demand (Rittmann 2008, Goldemberg and Johannson 2004). While 25% is significant, it does not approach the ~67% needed to arrest global warming. Besides, we cannot capture anywhere near all 25%. Therefore, we need to develop sources that can produce "new biomass" at very large rates (i.e. containing TW of energy), but without damaging the environment or our food-supply system.

By far the best option for making large amounts of "new biomass" is with photosynthetic microorganisms: algae and cyanobacteria, which are grouped together as "microalgae." Most people think of plants when they contemplate photosynthesis and biofuels. However, the photosynthetic microorganisms are vastly superior to plants in these ways (Rittmann 2008, Chisti 2007):

- The biomass yield per hectare is 10 to 100 times greater with microorganisms than with plants. Therefore, the amount of new biomass needed to supplant fossil fuel can be produced on a reasonable land surface area.
- Microorganisms grow in a slurry, not in soil. Therefore, using photosynthetic microorganisms for bioenergy does not compete with the food-supply system, since the microorganisms do not grow in arable soil.
- The growth of microorganisms need not place a large demand on water resources (Harto *et al.* 2010), since microorganisms do not transpire H_2O (as do plants) and do not need to have uncontrolled evaporation. Furthermore, the water can be recycled after the biomass is harvested, making the system nearly "closed loop" for water.
- The nutrients can be captured and recycled from the harvested biomass (Rittmann *et al.* 2011), making the system nearly "closed loop" for nutrients. In addition to lowering the demand for nutrients, this closed-loop approach eliminate nutrient run-off, which is a serious environmental hazard of plant-based approaches (NRC 2007).

Since wastewaters contain nutrients, they can serve as the nutrient sources for generating new microbial biomass from photosynthesis. The uptake of the nutrients into photosynthetic biomass also brings about nutrient removal, one of the major tools for fighting eutrophication worldwide. Phototrophic lagoons are a long-standing method to treat wastewaters in rural areas (Tchobanoglous *et al.* 2003). When the lagoon is working properly, the phototrophic microorganisms directly remove N and P through their own growth and also stimulate the aerobic biodegradation of BOD by generating O_2. The concept of using the nutrients in the wastewater to grow phototrophic biomass that is harvested for energy changes the main objective from improving water quality to producing renewable energy. These two goals ought to be compatible, since good nutrient removal also means that biomass production is at its maximum. It may be advantageous to utilize the off-gas CO_2 from anaerobic biomass conversion processes (recall sections 2.5 and 2.6) to stimulate biomass production and nutrient uptake together.

2.8 LOWER ENERGY USE

One goal for wastewater treatment is to reduce its CO_2 footprint by lowering its demand for fossil-fuel energy. Capturing energy from BOD and nutrients is an obvious way to achieve the goal, but it is not the only tool. Another tool is to lower the total energy-use rate by making the entire treatment system more energy efficient. Some examples are:

- As illustrated in Table 2.1, the biggest loss of energy in conventional wastewater treatment is aerobic treatment. This loss is exacerbated by the reality that aeration to provide O_2 is highly energy-intensive, normally comprising the majority of the energy demand for a conventional treatment facility (Tchobanoglous *et al.* 2003, Meda *et al.* 2010). Therefore, the best way to save energy in wastewater treatment is to maximize energy recovery by eliminating aerobic treatment as much as possible.
- When aeration is needed, such as for nitrification, the aeration system should be made as efficient as possible by avoiding an excessively high dissolved-oxygen concentration, using high-efficiency air diffusers and compressors, and matching the oxygen supply rate to the demand (Tchobanoglous *et al.* 2003).
- When denitrification is necessary, the electron donor should be from an internal (e.g., influent BOD or waste sludge) instead of an external donor, such as methanol (Rittmann and McCarty 2001, Lee *et al.* 2010b).
- Because urine contains about 80% of the N in wastewater, urine separation eliminates the need for nitrification and denitrification, and it lowers the energy demand for aerobic degradation of organic matter (Larsen and Gujer 1996). Wilsenach and van Loosdrecht (2006) showed that, with 85% urine separation, a simple BOD-removing treatment plant with a very low sludge age (and no nitrification/denitrification) will produce an effluent with 2 $gN_{total} \cdot m^{-3}$ and result in net energy production from wastewater. Urine separation also makes N and P recovery more feasible (Maurer *et al.* 2006, Udert and Jenni 2013).
- Normal human activities increase the temperature of the wastewater over that of its source water. This is another form of embedded energy. A gain of 5°C represents energy of 60 $W \cdot p^{-1}$ (at a total wastewater volume of 240 $L \cdot p^{-1} \cdot d^{-1}$). Capturing the added heat from wastewater effluent provides a heat-energy source and also protects receiving waters from thermal pollution.
- Minimizing volumes for and distances of transport of the wastewater (in sewers for discharge and pipes for reuse) and sludge (in trucks) lowers

energy costs. As shown by Kenway *et al.* (2011), not only desalination and (centralized) wastewater recovery technologies, but also the transport of these "new" water resources may cause a dramatic increase in water-related energy consumption (see also Falkenmark and Xia 2013 for a discussion of demand management versus supply-oriented measures for water provision).

- Energy for operation can be obtained from renewable sources, such as photovoltaics for electricity and renewable CH_4 or H_2 for trucks and electricity.

2.9 THE IMPACT OF SOURCE SEPARATION AND DECENTRALIZATION

Source separation and decentralization can be ideal manifestations of acting locally, as long as the actions are consistent with a global outlook. In principle, they should be well aligned with long-term global goals: shifting from >80% reliance of fossil fuels to <30% reliance, advancing microbial photobioenergy, recovering valuable nutrients instead of accelerating eutrophication, lowering our society's greenhouse-gas footprint, and minimizing environmental pollution.

It is important to keep our field's contributions in perspective. In terms of energy recovery, the energy value of BOD in domestic wastewater (Table 2.1) is only a small fraction of our society's energy demand (Rittmann 2008, Energy Information Administration 2005). This means that outstanding achievements by the domestic-wastewater sector are not going to come close to "solving the global energy problem," although they will contribute to a global solution. However, the wastewater sector can put into practice most of its sustainability improvements relatively rapidly. This will be an important leading-edge role for wastewater treatment.

Capturing the energy value of the organic wastes from the food-supply system will greatly expand the impacts and probably should be embraced by the wastewater field. Being part of energy capture from photosynthetic microorganisms can expand the impact to an even larger realm. The wastewater field is well positioned to embrace these new directions and take a leadership role in developing the microbiological tools to produce massive amounts of renewable energy from many sources, not just from typical wastewater.

In terms of traditional wastewater treatment, the most profound impact of energy recovery (and corollary nutrient recovery) should be on the economic "bottom line" of cities and utilities performing wastewater treatment. Reaping the benefits of the energy in BOD can dramatically improve the economic viability of wastewater treatment. Supplying renewable N and P for agriculture also can add a complementary income stream that also lowers the energy demand for fertilizer. Being more economically viable will allow cities and utilities to be more effective toward protecting the water environment and advancing the wide range of sustainability goals that require social investments.

Decentralization may have a unique positive role if it accentuates the advantages of some of the strategies to lower energy use. Among the possibilities, decentralization clearly is well suited to minimizing transport distances and volumes. Decentralized systems also may be able to link effectively to other forms of renewable energy, such as wind farms, photovoltaics, and microbial photosynthesis, creating what is called a "renewable energy park" (Subhadra and Edwards 2010). Adding nutrient recovery will be an especially good fit for rural areas, where the recovered N and P can be used in agriculture or to grow photosynthetic microorganisms. This advantage can be extended to urban areas as the concept of growing food in the city takes hold. Furthermore, the technologies for recovering nutrients in a concentrated, and therefore transportable, form discussed in this book open up for nutrient recovery on a large scale.

In order to reap the benefits of decentralization, it will be necessary to avoid problems frequently associated with small systems. The first problem is that small systems often have inadequate financial and technical support, which renders them inefficient and unreliable (NRC 1997). Society must understand that decentralized systems require the same kind of investment as do large systems. Second, small plants tend to experience "diseconomies of scale," particularly for energy-use efficiency (Tchobanoglous *et al.* 2003). Again, society must realize that high-efficiency operation must be extended to small facilities.

REFERENCES

Chisti Y. (2007). Biodiesel from microalgae. *Biotechnol. Adv.*, **25**, 294–306.

Cordell D. (2013). Peak phosphorus and the role of P recovery in achieving food security. In: Source Separation and Decentralization for Wastewater Treatment, T. A. Larsen, K. M. Udert and J. Lienert (eds.), IWA Publishing, London, UK, pp. 29–44.

Elser J. and White S. (2010). Peak phosphorus, and why it matters. *Foreign Policy*, http://www.foreignpolicy.com/articles/2010/04/20/peak_phosphorus.

Energy Information Administration (2005). *Annual Energy Outlook: with Projection to 2030*. U.S. Department of Energy, Washington, DC.

Falkenmark M., Xia J. (2013). Urban water supply under expanding water scarcity. In: Source Separation and Decentralization for Wastewater Treatment, T. A. Larsen, K. M. Udert and J. Lienert (eds.), IWA Publishing, London, UK, pp. 59–70.

Friedler E., Butler D., Alfiya Y. (2013). Wastewater composition. In: Source Separation and Decentralization for Wastewater Treatment, T. A. Larsen, K. M. Udert and J. Lienert (eds.), IWA Publishing, London, UK, pp. 241–257.

Goldemberg J. and Johansson T. B. (2004). *World Energy Assessment Overview: 2004 Update*. United Nations Development Programme, New York.

Harto C., Meyers R. and Williams E. (2010). Life cycle water use of low carbon transport fuels. *Energy Policy*, **38**, 4933–4944.

Intergovernmental Panel on Climate Change (IPCC) (2007). *Climate Change 2007: Synthesis Report*. United Nations, New York.

Kenway S. J., Lant P. A., Priestley A. and Daniels P. (2011).The connection between water and energy in cities, a review. *Water Sci. Technol.*, **63**(9), 1983–1990.

Larsen T. A. and Gujer W. (1996). Separate management of anthropogenic nutrient solutions (Human urine). *Water Sci. Technol.*, **34**(3–4), 87–94.

Lee H.-S., Vermaas W. and Rittmann B. E. (2010a). Biological hydrogen production: perspectives and challenges. *Trends Biotechnol.*, **28**(5), 262–271.

Lee I., Parameswaran P., Alder J. and Rittmann B. E. (2010b). Feasibility of Focused-Pulsed-treated WAS as a supplemental electron donor for denitrification. *Water Environ. Res.*, **83**, 2316–2324.

Logan B. E. (2004). Extracting hydrogen and electricity from renewable resources: a road map for developing sustainable processes. *Environ. Sci. Technol.*, **38**, 161A–167A.

Logan B. E., Hamelers B., Rozendal R., Schröder U., Keller J., Freguia S., Aelterman P., Verstraete W. and Rabaey K. (2006a). Microbial fuel cells: methodology and technology. *Environ. Sci. Technol.*, **40**, 5181–5192.

Logan B. E. and Regan J. M. (2006b). Microbial fuel cells: challenges and applications. *Environ. Sci. Technol.*, **41**, 5172–5180.

Madigan M. T., Martinko J. M. and Parker J. (2003). *Brock Biology of Microorganisms*, 10th edn, Prentice Hall, Saddle Creek, NJ.

Masters G. M. and Ela W. P. (2008). *Introduction to Environmental Engineering and Science*, 3rd edn. Prentice–Hall, Upper Saddle River, NJ.

Maurer M., Pronk W. and Larsen T. A. (2006). Treatment processes for source-separated urine. *Water Res.*, **40**, 3151–3166.

Maurer M., Schwegler P. and Larsen T. A. (2003). Nutrients in urine: energetic aspects of removal and recovery. *Water Sci. Technol.*, **48**(1), 37–46.

McCarty P. L. (1964). Anaerobic waste treatment fundamentals: Part IV, Process design. *Public Works*, **95**(Dec.), 3–22.

McCarty P. L. and Smith D. P. (1986). Anaerobic wastewater treatment. *Environ. Sci. Technol.*, **20**, 1200–1206.

Meda A., Cornel P. and Henkel J. (2010). Wastewater as a source of energy – can wastewater treatment plants be operated energetically self-sufficient? In: Proc. of the IWA Leading Edge Conference on Water and Waste Water Technologies, Phoenix, AZ, USA, 2–4 June.

National Research Council (NRC) (1997). *Safe Water from Every Tap*. National Academy Press, Washington, DC.

National Research Council (NRC) (2007). *Water Implications of Biofuels Production in the United States*. National Academy Press, Washington, DC.

Rittmann B. E. (2006). Microbial ecology to manage processes in environmental biotechnology. *Trends Biotechnol.*, April 28, 2006, pp. 261–268.

Rittmann B. E. (2007). The membrane biofilm reactor is a versatile platform for water and wastewater treatment. *Environmental Engineering Research*, **12**(4), 157–175.

Rittmann B. E. (2008). Opportunities for renewable bioenergy using microorganisms. *Biotechnol. Bioeng.*, **100**, 203–212.

Rittmann B. E., Mayer B., Westerhoff P. and Edwards M. (2011). Capturing the lost phosphorus. *Chemosphere*, **84**, 846–853.

Rittmann B. E. and McCarty P. L. (2001). *Environmental Biotechnology: Principles and Applications*. McGraw–Hill Book Co., New York.

Rittmann B. E., Torres C. I. and Marcus A. K. (2008). Understanding the distinguishing features of a microbial fuel cell as a biomass-based renewable energy technology. *Emerging Environmental Technologies*, V. Shah (ed.), Springer, pp. 1–28.

Shapouri H., Duffield J. A. and Wang M. (2001). *The Energy Balance of Corn Ethanol: an Update*. U. S. Department of Agriculture, Washington, DC.

Speece R. E. (1996). *Anaerobic Biotechnology for Industrial Wastewater*. Archae Press, Nashville, TN.

Subhadra B. and Edwards M. (2010). An integrated renewable energy park approach for algal biofuels production in the United States. *Energy Policy* **38**, 4897–4902.

Tchobanoglous G., Burton F. L. and Stensel H. D. (2003). *Wastewater Engineering: Treatment and Reuse*. McGraw–Hill Book Co., New York.

Torres C. I., Marcus A. K., Lee H.-S., Parameswaran P., Krajmalnik-Brown R. and Rittmann B. E. (2010). A kinetic perspective on extracellular electron transfer by anode-respiring bacteria. *FEMS Microbiol. Rev.*, **34**, 3–17.

Udert K. M., Brown-Malker S., Keller J. (2013). Electrochemical systems. In: Source Separation and Decentralization for Wastewater Treatment, T. A. Larsen, K. M. Udert and J. Lienert (eds.), IWA Publishing, London, UK, pp. 321–335.

Udert K. M, Jenni S. (2013). Biological nitrogen conversion processes. In: Source Separation and Decentralization for Wastewater Treatment, T. A. Larsen, K. M. Udert and J. Lienert (eds.), IWA Publishing, London, UK, pp. 291–305.

Wilsenach J. A. and van Loosdrecht M. C. M. (2006). Integration of processes to treat wastewater and source-separated urine. *J. Environ. Eng.-ASCE*, **132**, 331–341.

Zeeman G., Kujawa-Roeleveld K. (2013). Anaerobic treatment of source-separated domestic wastewater. In: Source Separation and Decentralization for Wastewater Treatment, T. A. Larsen, K. M. Udert and J. Lienert (eds.), IWA Publishing, London, UK, pp. 307–319.

Chapter 3

Peak phosphorus and the role of P recovery in achieving food security

Dana Cordell

3.1 INTRODUCTION

Global phosphorus scarcity, alongside scarcity of other critical resources, such as water and energy, is an emerging global challenge. Increasing scarcity and cost of phosphate rock will likely mean phosphorus will need to be recovered from excreta and other wastes and reused in food production. Key drivers for the recovery of phosphorus from wastewater today include: prevention of phosphorus from polluting waterways, improved maintenance of wastewater treatment plants (to prevent phosphate scale clogging pipes), and resource recovery of phosphorus for its fertilizer or industrial value. This chapter first demonstrates the significance of phosphorus to global food security, followed by the nature of phosphorus scarcity and finally demonstrates that averting a crisis and achieving a sustainable future situation of "phosphorus security" for global food production will require an integrated, multi-scale and multi-sectoral response, including a high recovery and reuse rate of all sources of phosphorus.

3.2 PHOSPHORUS AND GLOBAL FOOD SECURITY

Phosphorus is one of the most important elements for humanity, because it underpins our ability to produce food. Phosphorus is an essential element for all living organisms. In humans and animals, phosphorus is a building block of DNA and RNA, required for storing and transporting energy as ATP and the structure of cell walls (Emsley 2000). Humans and animals obtain phosphorus from food and to a lesser extent through food and feed additives. Plants in turn obtain phosphorus from soil solution and use the nutrients for growth, fruit/seed development and ripening (Johnston 2000)[1]. Because there is no substitute for phosphorus in crop growth, ensuring availability and accessibility of phosphorus in both the short and long term is critical to global food production. Prominent science writers such as chemist Isaac Asimov described phosphorus as "life's bottleneck": "*We may be able to substitute nuclear power for coal, and plastics for wood, and yeast for meat, and friendliness for isolation – but for phosphorus there is neither substitute nor replacement*" (Asimov 1974).

Historically farmers relied on natural soil-phosphorus to grow crops (with addition of some local manures and human excreta). However increased famine and soil degradation led to a search for external sources of phosphorus fertilizers. The discovery and consumption of phosphorus-rich guano and phosphate rock in certain geographical corners of the world contributed to a dramatic increase in global crop yields, allowed the population to rapidly expand and saved many from starvation over the past half-century (IFPRI 2002). Today, humanity is effectively dependent on mined phosphate rock to maintain high crop yields to meet increasing food and fibre demand (Cordell 2009a).

Since the 1950s, humans have mobilized around 900 million tonnes of phosphorus from non-renewable phosphate rock, most of which has ended up in rivers and lakes and ocean sediments, unavailable for recovery (USGS 2004, Rosemarin 2004). The importance of phosphorus and the need to raise soil fertility in nutrient deficient areas like Sub-Saharan Africa is relatively well understood by the food security community (Blair 2008). However, until recently, few food security discussions have explicitly addressed the emerging challenge of *where* and *how* phosphorus will be obtained in the future to ensure continuous food availability for a growing world population (Cordell 2010).

3.3 GLOBAL PHOSPHORUS SCARCITY AND POLLUTION

The global phosphorus problem is one of both scarcity and pollution. Current global phosphorus usage practices are threatening the world's future ability to produce food and are simultaneously responsible for widespread eutrophication and "dead zones" in some regions (Neset and Andersson 2008). The dual challenge of scarcity and pollution occurs at multiple socio-ecological scales, including at the global, regional, catchment, farm and human body levels (Table 3.1). For example, at the regional level, while phosphorus pollution from manure and excessive fertilizer use is a major problem in parts of Europe and North America, phosphorus scarcity is a wide-spread problem in Sub-Saharan African agriculture

[1]Mineral sources of soil phosphorus originally come from rock containing phosphorus-rich apatite that has taken around 10–15 million years to form (White 2000). These sources started their life as remains of aquatic life (such as shells) which were eventually buried on the sea floor, and transferred to the lithosphere via mineralization and tectonic uplift over millions of years and eventually weathered down via wind and rain erosion.

where most soils are phosphorus deficient, market access to fertilizers is low and phosphate fertilizers can cost 2–6 times more than in Europe due to higher logistic costs (Runge-Metzger 1995, Cordell *et al.* 2009a).

Table 3.1. Phosphorus in excess and scarcity – a double-edged sword. The table indicates examples of phosphorus excess and scarcity at multiple scales.

Scale	P Excess	P Scarcity
GLOBAL:	Excess phosphorus use can lead to major water pollution of global significance (eutrophication and dead zones).	Depletion of high-grade non-renewable phosphate reserves can limit future food production and therefore global food security.
WATER CATCHMENT:	Excess nutrients (eutrophication) can damage aquatic ecosystems.	Lack of nutrients (oligotrophication) can also reduce the population of aquatic organisms.
FARM:	Excess manure can leak from livestock production facilities and pollute water bodies.	Lack of access to fertilizers can limit crop growth and adversely affect farmer livelihoods.
SOILS:	Excess phosphorus beyond critical soil levels or imbalanced N-P-K nutrient ratios can both pollute and reduce farmer profits.	Phosphorus deficiency in soils can limit crop growth.
HUMAN BODY:	Excess phosphorus consumption can act as a deadly poison (in extreme cases).	Phosphorus deficiency can lead to physical illnesses (in extreme cases).

Source: Cordell (2010).

Globally, the phosphorus in fertilizers is predominantly sourced from phosphate rock – a non-renewable resource that is becoming increasingly scarce (Cordell *et al.* 2009a, see section 3.4 for further details). At the same time, eutrophication of inland and coastal waters caused by phosphorus (and nitrogen) pollution is considered a major global environmental challenge of the century (Millennium Ecosystem Assessment 2005, Steffen *et al.* 2004). Dead zones or serious algal blooms are occurring from Chesapeake Bay in the US to the Baltic Sea to Australia's Great Barrier Reef (Chudleigh and Simpson 2000, Commonwealth of Australia 2001, HELCOM 2005, World Resources Institute 2008). Removing phosphorus from wastewater has been the most effective means of water pollution control especially for lakes (Larsen *et al.* 2007), but on a global scale, only a fraction of wastewater systems currently remove nutrients.

The phosphorus pollution challenge extends beyond eutrophication. It is often argued that phosphate fertilizers contribute very little to greenhouse gas production, especially compared to nitrogen fertilizers (which require large amounts of natural gas to produce, and result in GHGs leaking from a fertilized field) (Prud'Homme 2010, Cordell 2010). However, a substantial amount of energy is required to transport phosphate rock and phosphate fertilizer to the end user (typically the farm gate) in the order of 20 $MJ \cdot kgP^{-1}$ (Schröder *et al.* 2010). Phosphate rock is one of the most highly traded commodities on the international market; approximately 30 million tonnes are traded globally each year (IFA 2006). This largely relies on cheap fossil fuel energy for transport.

Another important pollution aspect of phosphate fertilizer production is the generation of the radioactive waste by-product phosphogypsum. For every tonne of phosphate (P_2O_5) produced, five tonnes of phosphogysum is generated, which is considered too radioactive for reuse, and hence must be stockpiled (USGS, 1999). Uranium and cadmium (among other heavy metals) are naturally geochemically associated with phosphate rock and if not removed as phosphogysum during processing, these elements pose a risk to ecosystems and humans once released into the environment or transferred to soils (Saueia *et al.* 2005). Direct application of crushed phosphate rock also results in the transfer of radionuclides of the decay series of uranium and thorium to agricultural soils. There are concerns that in some cases this could result in soil radiation levels that are above acceptable limits, risking exposure to workers and crops and ultimately the food chain and consumers (Saueia *et al.* 2005). While radiation levels can vary above and below acceptable limits, there are no standard procedures in place for ensuring measurement of soil radioactivity due to applied phosphate rock (or mineral phosphate fertilizers).

3.4 FIVE DIMENSIONS OF PHOSPHORUS SCARCITY

Whilst many of the recent debates on phosphorus scarcity have focused on physical or market availability, such as how many million tonnes of phosphate rock remain (Gilbert 2009, IFA 2008, FAO 2008), the picture is much more complicated. This section highlights five important and interrelated dimensions of phosphorus scarcity (physical, managerial, institutional, economic and geopolitical scarcity) that together limit the availability of phosphorus for productive use by humans for fertilizers and hence food production (Cordell 2010).

While the element phosphorus is relatively abundant in the earth's upper crust, the amount available to humans for productive use in society is many order of magnitudes smaller (Cordell 2010). This is because only a small and finite amount of phosphate is naturally present in high concentration. Further, a substantial proportion of these high-concentrated deposits are not readily accessible – such as nodes of phosphate on the deep sea bed and continental shelves – or the deposits contain too many contaminants/impurities or cannot be legally extracted because they are in environmentally or culturally sensitive areas. Of the reserves considered economically and technically exploitable and subsequently extracted and processed into fertilizers, only a smaller percentage will ever be taken up by plants in fertilized agricultural lands due to the naturally low bioavailability of phosphorus in soil solution (Johnston 2000). The UN's Food and Agricultural Organization (FAO) estimate that plants only take up approximately 20–30% of P in applied fertilizers each year (Syers *et al.* 2008). The remainder will accumulate in the soil and is potentially available for uptake by crops in subsequent years. However in practice, much of this phosphorus can be temporarily unavailable to plants if the phosphorus is not present in solution, or permanently lost via wind or water erosion.

In some parts of the world, such as North America, Northern and Western Europe, Japan and Australia, many agricultural soils have an over-accumulation of phosphorus due to decades of over-application of fertilizers and high livestock densities leading to excessive phosphorus-rich manure loads on surrounding farmland. In these regions, phosphorus could be better managed via more efficient use of phosphorus and by replacing only what is removed in harvest (see Schröder *et al.* 2011). In other parts of the world, the situation is quite different. For example, in Sub-Saharan Africa where soils are naturally phosphorus deficient and fertilizer application rates have been extremely low, leading to deficiencies in phosphorus. In these soils, phosphorus needs to be boosted to reach optimum or critical phosphorus levels (after which point they can be managed via compensating what is removed in harvest).

While fertilizer demand has been stabilizing in the developed world due to previous decades of over-application (and hence built up phosphorus in soils), the demand in emerging and developing economies like China, India and Brazil is anticipated to rise over the coming decades (IFA 2008), resulting in a net global increased demand for phosphorus. This is not only due to increasing population, but also changing diets and the need to boost soil fertility of the world's agricultural soils. When mining and processing of phosphate rock became commonplace after the Green Revolution in the 1960's, there were few supply concerns as phosphate rock was a cheap and easy source of highly concentrated phosphorus that was thought to be plentiful. However the most highly accessible and high quality sources were mined first, and today's reserves are lower grade, more expensive, harder to access, contain lower concentrations of phosphorus and more impurities (Prud'Homme 2010). This means that while the element phosphorus is not "running out," the world's supply of cheap phosphorus is being depleted.

Further, like oil and other non-renewable resources, the global production of phosphate rock will eventually reach a peak due to the non-homogeneous nature of the reserve and the economic and energy constraints of accessing the lower quality and more difficult to reach rock. A fundamental notion behind the peak theory is that the growing demand for the resource will outstrip economically available supply (annual production) at some point, despite some advances in technology and efficiency. Important here is that the critical point (the "peak") will occur much sooner than depletion of the entire reserve. Peak phosphorus is estimated to occur by 2033 (Figure 3.1), after which demand will continue to increase while annual supply decreases (Cordell *et al.* 2009a). Yet there are currently no alternative sources of phosphorus on the market that could replace the large demand for phosphate rock at any significant scale. It would take decades to develop infrastructure at such a scale. Whilst there is a vigorous debate today around the lifetime of phosphate rock reserves and the timeline of peak phosphorus, the underlying problem remains the same. Extracting the same nutrient value from rock will increasingly require more inputs of energy, resources and costs, while resulting in increased volumes of waste and pollution.

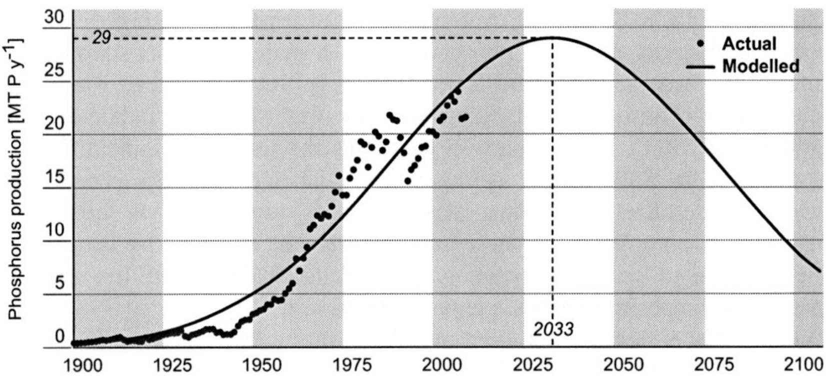

Figure 3.1 Peak phosphorus: global annual production of current phosphate rock reserves is estimated to peak by 2033 at 29 MtP·a^{-1} (equivalent to approximately 220 million tonnes of phosphate rock per year) while demand will continue to increase (based on best available industry and US Geological Survey data). *Source*: Cordell *et al.* (2009a).

The recent short-term tightness between supply and demand (and the associated 800% price spike in 2008) resulted in increased investment in exploration of new phosphate rock deposits and commissioning of new mines, most notably in Saudi Arabia, Australia and seafloor sediments off the coast of Namibia (USGS 2009, 2010; Jung 2010; Drummond 2010). While these may increase the overall tonnages of world phosphate rock reserves in the coming years, the quality (e.g., % P_2O_5) and accessibility of these reserves are markedly lower than current reserves.

While physical scarcity of high quality phosphate rock is a serious concern, there is also a scarcity of management of phosphorus throughout the food production and consumption system. For example, only 20% of the phosphorus mined for food production actually reaches the food the global population eats due to substantial inefficiencies in the entire food production and consumption system (Cordell *et al.* 2009a). Phosphorus is lost or wasted during mining and processing, transport, fertilizer application, food processing and retail, food preparation and consumption (see Section 3.5 for more detail).

Institutional scarcity is also inhibiting the productive use of phosphorus by humans. That is, there is a substantial lack of effective policies and actors explicitly governing global phosphorus resources to ensure availability and accessibility of phosphorus for food security, both in the short and long term (Cordell 2010). Further, there are no structures for monitoring and evaluating the long-term global situation and no effective feedback loop designed to correct the system. The lack of effective global governance is compounded by a lack of stakeholder consensus on the issues and institutional fragmentation. While phosphorus is relevant to numerous different sectors (e.g., a commodity in the mining sector, a pollutant in the wastewater sector) phosphorus *scarcity* is not a priority within any sector, and hence long-term phosphorus security has no obvious institutional home. Phosphorus is by default governed by the market system, which is only sufficient for a very narrowly defined system, but is not sufficient to adequately address the much broader sustainability implications, such as the need for access to phosphorus for all farmers, the finite nature of phosphate rock resources and the long-term situation (Cordell 2010).

Phosphorus is also scarce in an economic sense, for example, when farmers cannot access phosphorus, due to a lack of purchasing power. The current demand for phosphorus only represents those users who have the capital to procure phosphate rock or fertilizers. In order to feed 9 billion mouths by 2050, soil fertility must be increased, particularly in areas with phosphorus-deficient soils and a high rate of food insecurity like Sub-Saharan Africa. This means ensuring farmer access to phosphorus (Cordell 2010). Indeed, the recent short-term price spike resulted in many farmers around the world not buying fertilizers which in turn will affect current crop yields, farmer livelihoods and soil fertility for subsequent crop production. Financial scarcity on the supply side can occur when investments in new capacity (such as phosphate rock mines) and commercial production do not keep up with market demand for the resource (time lags can be 5–10 years). This was thought to be a significant factor leading to the 2008 short-term phosphate rock scarcity situation (IFA 2008), shown in Figure 3.2.

Finally, geopolitical scarcity can restrict the availability of phosphorus resources in the short or long term. For example, while all farmers need phosphorus, approximately 85% of the world's remaining phosphate rock reserves are controlled by five countries, the main players being Morocco, China and the US (Figure 3.3). In 2008 China imposed a 135% export tariff to secure domestic supply for food production; a move which essentially halted exports from the region overnight and contributed to the 2008 price spike (Fertilizer Week 2008, Euronews 2009). The US is expected to deplete its own high-grade reserves in the coming decades and increasingly imports rock phosphate from Morocco.

However, Morocco currently occupies Western Sahara and controls that region's reserves in defiance of UN resolutions. According to Hagen (2008), Morocco's claim is not recognized by any other country and "numerous UN resolutions support the conclusion that extracting and trading with phosphates from Western Sahara are contrary to international law."

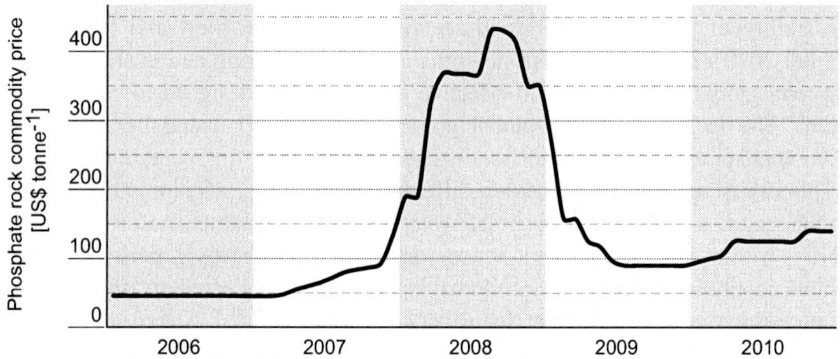

Figure 3.2 Price of phosphate rock – January 2006 to January 2011, indicating 800% price spike in mid-2008. *Source*: World Bank Commodity Price Data.

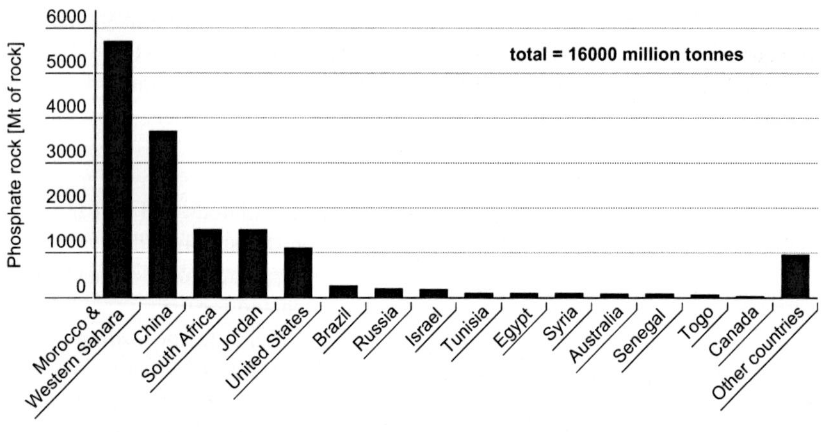

Figure 3.3 World phosphate rock reserves by country. *Source*: USGS (2010).

3.5 PHOSPHORUS USE IN THE GLOBAL FOOD SYSTEM

The anthropogenic phosphorus cycle, that is, the flow of phosphorus through the food system, extends from mine to field to fork to excreta (Figure 3.4). As identified in section 3.4, there are substantial avoidable and unavoidable losses

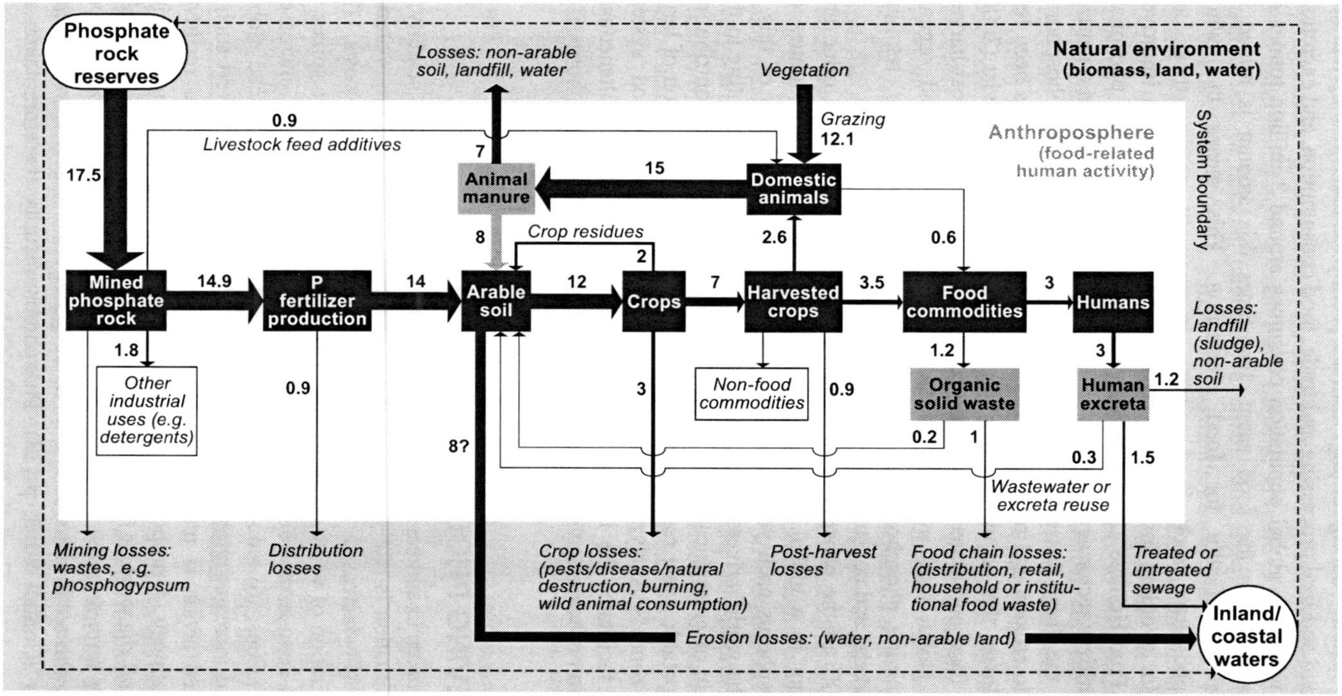

Figure 3.4 Global phosphorus flows through the food production and consumption system (in million tonnes P·a^{-1}). The figure indicates substantial avoidable and unavoidable phosphorus losses from mine to field to fork.
Source: redrawn from Cordell *et al.* (2009a).

3.6.2 The role of decentralized sanitation systems

As indicated above, achieving phosphorus security will likely require a high recovery rate of the 3 million tonnes P per year generated in human excreta. While excreta reuse alone will not be sufficient to meet the world's growing phosphorus demand, phosphorus security cannot be achieved without it. There are numerous processes by which phosphorus can be recovered from excreta, as discussed in this book (see also Cordell *et al.* 2011 for a framework of phosphorus recovery and reuse options). While no single recovery solution will fit all contexts, sustainable sanitation systems must be designed in a way that facilitates not only pollution prevention, but also the reuse of phosphorus as fertilizer in food production, thereby addressing the following important aspects:

(1) Quality and effectiveness of the final product as a fertilizer for crop production
(2) The percentage of phosphorus that can be efficiently recovered
(3) Life-cycle costs (including energy) of recovering, transporting and reusing the nutrients from excreta
(4) New policies, partnerships and institutional arrangements that link the sanitation sector with the food, agriculture and fertilizer sectors

Source-separating and decentralized sanitation systems can thus play a critical role in addressing phosphorus security, by:

- Facilitating the *efficient* recovery of phosphorus from excreta
- Providing a *local*, *renewable* and potentially cost-competitive substitute to phosphate rock in the future
- Reducing global society's *dependence* on a resource subject to increasing price fluctuations and from geopolitically unstable regions
- Facilitating local communities' "*phosphorus sovereignty,*" particularly in regions of low farmer access to fertilizers
- Reducing phosphorus *pollution* in receiving waterways, thereby reducing eutrophication and algal bloom potential.

3.6.3 Key challenges and opportunities

A major challenge in achieving phosphorus security is the lack of effective governance, including institutional fragmentation where phosphorus scarcity is not a priority within any sector. Indeed, there is no single organization or policy that are explicitly designed to ensure that phosphorus is accessible for the world's farmers. However, this also presents an opportunity for new partnerships between different sectors. Because phosphorus scarcity means that nutrient recovery is not only about environmental protection, but also about reuse, the sanitation and the food production sectors need to be more explicitly linked.

More people are living in urban areas than rural areas globally and the trend will increase over the coming half-century. Therefore cities are becoming "phosphorus hotspots" (Cordell 2010) of phosphorus excreted by city-dwellers. Substantial changes in urban infrastructure will be required to collect, store and transport urban nutrients (such as excreta, greywater, food waste) back to agriculture. In many instances, retrofitting existing infrastructure (such as household toilets) is likely to be costly and resource-intensive. Future urban planning, including both sanitation and food production planning could take phosphorus security into account. For example, urban agriculture fertilized by urban phosphorus "wastes" may present an opportunity due to the challenge of the often large distances between phosphorus "wastes" in cities and agricultural fields.

A short-term challenge will be the demonstration of the cost-effectiveness of recovered phosphorus (compared to phosphate rock) to farmers, politicians and industry. However, accurately estimating costs-effectiveness is difficult because a) the price of phosphate rock and fertilizer commodities has fluctuated as much as 800% in the past few years, making a comparative costs highly variable; b) due to the substantial uncertainty and lack of public data associated with costs, such as life-cycle energy costs, especially for new sources, for example, continental shelves; c) the cost-effectiveness is highly dependent on where the boundary around the system is drawn, and what type of costs are included; and d) such analyses will be largely context specific (e.g., transport cost for recovered urban nutrients). Despite these difficulties, there is general consensus that phosphate rock prices as well as the already high environmental cost of mining, processing and transporting are likely to increase over the long term.

3.7 CONCLUSIONS

The phosphorus challenge is a serious global problem threatening the world's fundamental ability to produce food in the long-term in addition to the health of the world's inland and coastal waters. Yet it requires multi-level and multi-sectoral responses. At the international level, effective global governance of phosphorus is required, including clear stakeholder roles and responsibilities. While there is no single solution, a high recovery and reuse rate of phosphorus from all sources is likely to be required, in addition to substantially increasing the efficiency by which phosphorus is used to meet the global food needs. Because the situation differs between regions, national policy-makers can facilitate the assessment of a region's phosphorus vulnerability to scarcity and pollution, *and* the prioritization, development and implementation of cost-effective, socially robust and environmentally sound, context-specific responses for the recovery and efficient use of phosphorus. All key stakeholders, the fertilizer industry, water service providers, farmers, and so on must actively be involved in the solutions.

REFERENCES

Asimov I. (1974). Asimov on Chemistry, Doubleday Publishing, New York, USA.

Blair D. (2008). Green revolution needed to feed the poor: UN, Sydney Morning Herald, New York, USA. http://www.smh.com.au/articles/2008/05/19/1211182703408.html (accessed 14th March 2012).

Chudleigh P. and Simpson S. (2000). The national eutrophication management program – a review, Land & water resources research & Development corporation, Canberra, Australia.

Commonwealth of Australia, (2001). Australian Agriculture Assessment 2001. National Land and Water Resources Audit, Canberra, Australia.

Cordell D., Rosemarin A., Schröder J. J. and Smit A. L. (2011). Towards global phosphorus security: A systemic framework for phosphorus recovery and reuse options. *Chemosphere*, **84**(6), 747–758.

Cordell D. (2010). The Story of Phosphorus: Sustainability implications of global phosphorus scarcity for food security. PhD thesis. Institute for Sustainable Futures, University of Technology, Sydney and Linköping University, Linköping, Sweden.

Cordell D., Drangert J.-O. and White S. (2009a). The story of phosphorus: Global food security and food for thought. *Global Environmental Change*, **19**(2), 292–305.

Cordell D., White S., Drangert J. -O. and Neset T. S. S. (2009b). Preferred future phosphorus scenarios: A framework for meeting long-term phosphorus needs for global food demand. In: International Conference on Nutrient Recovery from Wastewater Streams Vancouver, 10–13th May, 2009, D. Mavinic, K. Ashley and F. Koch (eds.), IWA Publishing, London, UK.

Corell H. (2002). Letter dated 29 January 2002 from the Under-Secretary-General for Legal Affairs, the Legal Counsel, addressed to the President of the Security Council, S/2002/161, United Nations Security Council.

Correll D. L. (1998). The role of phosphorus in the eutrophication of receiving waters: a Review. *Journal of Environmental Quality*, **27**(2), 261–266.

Drummond A. (2010). Minemakers: targeting phosphate production from two continents. In: Phosphates 2010 International Conference, 22–24 March, Brussels, Belgium.

Emsley J. (2000). The 13th Element: The Sordid Tale of Murder, Fire, and Phosphorus, John Wiley & Sons, New York, USA.

Euronews (2009). US and EU go to WTO over Chinese export curbs, http://www.euronews.com/2009/06/23/us-and-eu-go-to-wto-over-chinese-export-curbs/ (accessed 23 June 2009).

FAO (2008). Global fertilizer supply expected to outstrip demand – New FAO fertilizer outlook to 2011/12 published, FAO Newsroom, UN Food & Agriculture Organisation, 26 February 2008, Rome, Italy.

FAO (2006). Plant nutrition for food security: A guide for integrated nutrient management, FAO Fertilizer And Plant Nutrition Bulletin 16, Food and Agriculture Organization of the United Nations, Rome, Italy.

Fertilizer Week (2008). Industry ponders the impact of China's trade policy, In: Thursday Markets Report, 24th April 2008, British Sulphur Consultants, CRU.

Gilbert N. (2009). The Disappearing Nutrient, Nature News Feature, **461**(8), 716–718.

Hagen E. (2008). The role of natural resources in the Western Sahara conflict, and the interests involved, International conference on multilateralism and international law, with Western Sahara as a case study, 4–5 Dec 2008, Pretoria, South Africa.

HELCOM (2005). Nutrient pollution to the Baltic Sea in 2000, Baltic Sea Environment Proceedings No.100, Helsinki Commission, Baltic Marine Environment Protection Commission, Helsinki, Finland.

IFA (2006). Production and International Trade Statistics, International Fertilizer Industry Association, Paris, France.

IFA (2008). Feeding the Earth: Fertilizers and Global Food Security, Market Drivers and Fertilizer Economics International Fertilizer Industry Association, Paris, France.

IFPRI (2002). Green Revolution: Curse or Blessing? International Food Policy Research Institute, Washington DC, USA.

Jasinski S. M. (2009). Phosphate Rock, Mineral Commodity Summaries, US Geological Survey, Washington DC, USA.

Jasinski S. M. (2010). Phosphate Rock, Mineral Commodity Summaries, US Geological Survey, Washington DC, USA.

Johnston A. E. (2000). Soil and plant phosphate, International Fertilizer Industry Association (IFA), Paris, France.

Jung A. (2010). Phosphates Fertilizer Outlook, British Sulphur Consultants, Phosphates 2010 International Conference, 22–24 March 2010, Brussels, Belgium.

Lundqvist J., Fraiture C. d. and Molden D. (2008). Saving Water: From Field to Fork – Curbing Losses and Wastage in the Food Chain, SIWI Policy Brief, Stockholm International Water Institute, Stockholm, Sweden.

Millennium Ecosystem Assessment (2005). Global Assessment Reports, Volume 1: Current State and Trends, Chapter 12: Nutrient Cycling, Island Press, Washington DC, USA.

Neset T-S S. and Andersson A. (2008). Environmental impact of food production and consumption – from phosphorus leakage and resource depletion to recycling. In: Water for Food, The Swedish Research Council Formas, Stockholm, Sweden, pp. 99–108.

Prud'Homme M. (2010). World phosphate rock flows, losses and uses, International Fertilizer Industry Association, Phosphates 2010 International Conference, 22–24 March 2010, Brussels, Belgium.

Rosmarin A. (2004). The precarious geopolitics of phosphorous, Down to Earth, Science and Environment Fortnightly, 2004(June 30), 27–31.

Runge-Metzger A. (1995). Closing the cycle: obstacles to efficient P management for improved global food security. In: SCOPE 54 – Phosphorus in the global environment – transfers, cycles and management John Wiley & Sons, Chichester, UK.

Saueia C. H., Mazzilli B. P. and Fávaro D. I. T. (2005). Natural radioactivity in phosphate rock, phosphogypsum and phosphate fertilizers in Brazil. *Journal of Radioanalytical and Nuclear Chemistry*, **264**(2), 445–448.

Schröder J. J., Smit A. L., Cordell D. and Rosemarin A. (2011). Improved phosphorus use efficiency in agriculture: a key requirement for its sustainable use. *Chemosphere*, **84** (6), 822–831.

Schröder J. J., Cordell D., Smit A. L. and Rosemarin A. (2010). Sustainable Use of Phosphorus. European Union tender project ENV.B.1/ETU/2009/0025. Report 357, Plant Research International, Wageningen University and Research Centre, Wageningen, The Netherlands.

Smil V. (2000). Feeding the World: A Challenge for the 21st Century, MIT Press, Cambridge, USA.

Steffen W., Sanderson A., Tyson P.D., Jäger J., Matson P.A., Moore III B., Oldfield F., Richardson K., Schellnhuber H.J., Turner II B.L. and Wasson R.J. (2004). Global change and the earth system: a planet under pressure, executive summary, IGBP Secretariat, Royal Swedish Academy of Sciences, Stockholm, Sweden.

Syers K., Johnston A. E. and Curtin D. (2008). Efficiency of soil and fertilizer phosphorus use: Reconciling changing concepts of soils phosphorus behaviour with agronomic information. *FAO Fertilizer and Plant Nutrition Bulletin 18*, Food and Agriculture Organization of the United Nations, Rome, Italy.

USGS (1999). Fertilizers – Sustaining global food supplies, USGS Fact Sheet FS–155–99. US Geological Survey, Reston, USA.

USGS (2004). Phosphate rock statistics: 1900–2002, David A. Buckingham and Stephen M. Jasinski, US Geological Survey, Washington DC, USA.

World Resources Institute (2008). Agriculture and "Dead Zones", Jonathan Lash, Washington DC, USA.

Chapter 4

Nitrogen economy of the 21st Century

Jan Willem Erisman and Tove A. Larsen

4.1 INTRODUCTION

In the past 50–100 years, the world's nitrogen cycle has been radically altered by global food and energy production systems, which have caused an increase in the total production rate of reactive nitrogen by a factor of 1.5 to 2 compared to its natural production. Reactive nitrogen, Nr, encompasses all forms of nitrogen occurring on earth except for atmospheric N_2. Reactive nitrogen both promotes and endangers human welfare. Humanity benefits from energy use and particularly from the food and feed produced through the availability of mineral fertilizer. These activities bring enormous benefits, but losses of nitrogen from agriculture, human waste and combustion processes also have detrimental effects on the environment (see e.g., Galloway *et al.* 2008). These effects include the exceeding of critical loads resulting in loss of biodiversity, groundwater pollution, eutrophication of inland ecosystems and coastal areas causing algae blooms and fish kills, elevated atmospheric levels of ammonia, nitrogen oxide, nitrous oxide and aerosols resulting in climate change and/or human health impacts (Cowling *et al.* 2002; Galloway *et al.* 2003, 2008; Townsend *et al.* 2003; Reid *et al.* 2005). Due to the many actors, interactions and effects involved, the nitrogen issue is very complex. Each of the losses to the environment as part of the Nr cycle can give rise to a number of different sequential effects, known as the nitrogen cascade (Galloway *et al.* 2003). Thus reactive Nr released to the atmosphere from fuel combustion can lead to the following sequence: an increase in tropospheric ozone levels, a decrease of visibility and an increase of atmospheric acidity. Upon deposition from the atmosphere, Nr can acidify water and soil over-fertilize grassland, forests and coastal ecosystems, followed by re-emission to the atmosphere as nitrous oxide, again contributing to climate change and the depletion of stratospheric ozone. As long as Nr remains in the environment, this cascading effect will continue to be active. As it moves down the cascade, the original source of Nr becomes less important because of the many transformations and interactions (Dumont *et al.* 2004). Although all the issues highlighted above are global in scale, they affect various parts of the globe very differently. The most striking example concerns the different degrees of fertilizer application: whereas over-fertilization is causing severe environmental problems in some parts of the world, other parts suffer substantially from a lack of nitrogen fertilizer. Obviously, the magnitude of these problems varies. Consequently, local and regional perspectives are of vital importance for understanding the differing nature and priority of nitrogen issues in various places. This regional context is also important for the design of solutions due to differing cultural, social and economic factors, which must also be considered to ensure sound policy implementation.

In areas with a nitrogen shortage, human beings suffer from malnutrition, and it is a huge challenge to solve this pressing issue without causing even further acceleration of the nitrogen cascade. There is thus an urgent need for understanding how the different Nr sources translate into Nr effects. Quantitative knowledge of the sources and effects of Nr in areas with excessive amounts, combined with information from regions struggling with deficiencies, will help develop policies to minimize the negative effects of surplus Nr while maximizing its benefits. Some effective policies in the area of wastewater treatment will be discussed later on in this chapter. The production of Nr is expected to grow dramatically due to population growth, increased protein consumption and biofuel production. It is unrealistic to suppose that increased efficiency of Nr use will be able to offset this growth, so other interventions are needed to counteract these impacts, which are already clearly visible. In this chapter, we summarize the global consequences of an increase in reactive nitrogen production. We first describe the sources, followed by the major emission pathways to the environment. We then discuss the resulting effects on the nitrogen cascade and finally draw up some options for policy implementation.

4.2 NITROGEN SOURCES

Most of the nitrogen on earth is available in the form of dinitrogen gas, N_2, which makes up 78% of the gases in the atmosphere. Only very few organisms can exploit this form of nitrogen directly, but it can be used to create Nr in both natural and industrial processes. Many forms of Nr exist. Ammonia is the most important gas used in the chemical industry, urea and ammonium nitrate are widely applied as fertilizers, and nitrogen oxides are mainly found in combustion gases. Globally, around 75% of anthropogenic Nr production stems from industrial nitrogen fixation and 25% from fossil fuel burning (Galloway *et al.* 2003).

The *natural sources* of Nr are volcanic eruptions, lightning and biological nitrogen fixation (Galloway *et al.* 2003). Volcanic eruptions release stored Nr from the earth's crust in the form of ammonia, whereas lightning produces NO_x in the troposphere, followed by deposition onto the earth's surface. Biological fixation, the transformation of N_2 into Nr, is the main source of Nr in natural (non-agricultural) landscapes. This process is performed by microorganisms living in symbiosis with higher organisms, most often plants from the legume family. The natural contribution to Nr is estimated at around 125 TgN per year, biological nitrogen fixation being the dominant source with 120 $TgN{\cdot}a^{-1}$ and lightning producing only around 5 $TgN{\cdot}a^{-1}$ (Fig. 4.1; Galloway *et al.* 2008; Schlesinger 2009).

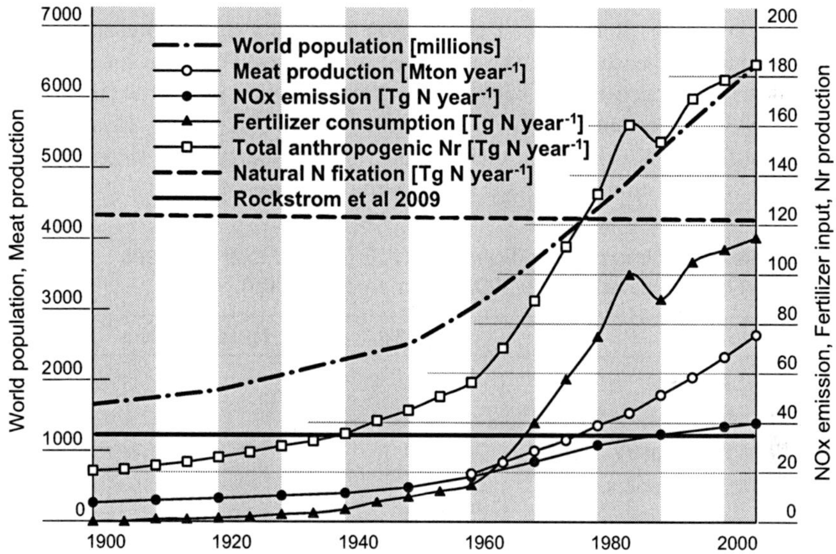

Figure 4.1 Changes in total anthropogenic and natural Nr production, population growth and meat production.

The main sources of anthropogenic Nr are fossil fuel combustion (producing 40 $TgN{\cdot}a^{-1}$ in the form of NO and NO_2), industrial ammonia production (122 $TgN{\cdot}a^{-1}$ from N_2), and biological nitrogen fixation in agriculture (30 $TgN{\cdot}a^{-1}$).

Industrial ammonia is mainly used for fertilizer production (82%), but also for a range of products including plastics, explosives and cleaning agents (Smil 2001; Erisman *et al.* 2008). The industrial production of ammonia has dramatically increased the global productivity of agriculture. From 1908 to 2008 the average number of persons fed per hectare of arable land increased from 1.9 to 4.3.

4.3 RELEASE OF NITROGEN TO THE ENVIRONMENT

Nr is retained in products, soil and vegetation. When in excess, Nr cascades through the different environmental compartments. It is returned to atmospheric nitrogen mainly through denitrification processes, either in the environment, in agricultural soil or through sewage treatment. A recent article in the journal *Nature* ranked the top ten global issues exceeding the earth's planetary boundaries. The preliminary estimates reported in this article showed nitrogen to be in the top three, way beyond any reasonable boundaries of sustainability (Rockström *et al.* 2009). According to these authors, anthropogenic nitrogen production should be reduced by 75% to 35 $TgN{\cdot}a^{-1}$ (Fig. 4.1).

Losses of reactive nitrogen to the environment occur mainly due to a generally low nitrogen use efficiency (NUE). Erisman *et al.* (2005) calculated the NUE for a number of sectors in the Netherlands (Table 4.1) and showed that the overall NUE is below 50%. It is zero in combustion processes (traffic, transport and energy production) because Nr is an unwanted by-product. The NUE for industry and fertilizer production is high because only the production process is taken into account, without accounting for losses further down the chain. Households are quite efficient N-users, but their annual loss is the second largest, with 125 $ktN{\cdot}a^{-1}$ ending up in household waste and wastewater.

Table 4.1. Nitrogen use efficiencies, NUE (%) and total Nr losses (ktN) in the Netherlands in 2000 (Erisman *et al.* 2005).

Sector	N_{eff} (%)	Nr loss ($ktN{\cdot}a^{-1}$)
Agriculture	34	571
Industry	91	89
Fertilizer industry	98	54
Traffic and transport	0	88
Energy production	0	62
Households	71	125
Wastewater treatment	40	52

Fossil fuel combustion results in the emission of NO_x to the atmosphere, from where these compounds are distributed on a large scale to managed and unmanaged ecosystems. NUE will always be zero in this process, where Nr is created as an unwanted side effect.

In contrast, the notion of efficiency is central for our understanding of the mechanisms of nitrogen loss in agriculture. NUE, defined as Nr content in agricultural products divided by the total input, generally decreases as the input increases. With a few exceptions, NUE in European agriculture in the mid 90s ranged from 30 to 75%, with actual Nr losses ranging from close to zero to more than 120 $kgN{\cdot}ha^{-1}{\cdot}a^{-1}$ (Erisman *et al.* 2005). Generally, higher productivity is linked to lower nitrogen efficiency, but management strategies are obviously also important. Additionally, meat consumption plays a central role, because of the low nitrogen efficiency of meat production. Even if perfect recirculation of nitrogen from manure were possible, this internal cycle will always reduce the resulting NUE in agriculture due to the inevitable losses of nitrogen during the period of plant growth. Moreover, meat production entails more intensive agriculture in order to feed the world, which again tends to lead to higher nitrogen losses.

In reality, manure is generated on different farm types, ranging from extensive to intensive breeding situations, giving rise to different hot spots around the globe (Steinfeld *et al.* 2006). During the production, management and application of manure to agricultural fields, the nitrogen efficiency is low and contributes to Nr emissions to air, groundwater and surface water.

A major but uncertain fraction of the reactive nitrogen emissions from agriculture is lost through denitrification to atmospheric N_2. Recent estimates range between 30–80% of the total losses from agriculture (e.g., Schlesinger 2009). In principle, this loss does not directly harm the environment, but it represents a loss of energy spent for industrial fertilizer production. More seriously, however, the rest of the nitrogen emitted from agriculture produces a number of harmful environmental effects (Galloway *et al.* 2003; 2008; Gruber and Galloway 2008; Erisman *et al.* 2008). These losses mainly consist of ammonia (NH_3), nitric oxide (NO) and nitrous oxide (N_2O) emitted to the atmosphere, and nitrate (NO_3^-) as well as different organic nitrogen compounds emitted to water.

Recently, van Drecht *et al.* (2009) modelled the fate of Nr in urban wastewater on a country basis, using historical data as well as forecasts for the years 2030 and 2050 (Figure 4.2). On the basis of this work, Figure 4.2 summarizes the global pathways of wastewater nitrogen in the years 2000 and 2050 (forecast) for four different future scenarios developed in the Millennium Environmental Assessment (MEA, Reid *et al.* 2005). The four scenarios describe different pathways of societal development, but the main trends for nitrogen emissions from wastewater are similar.

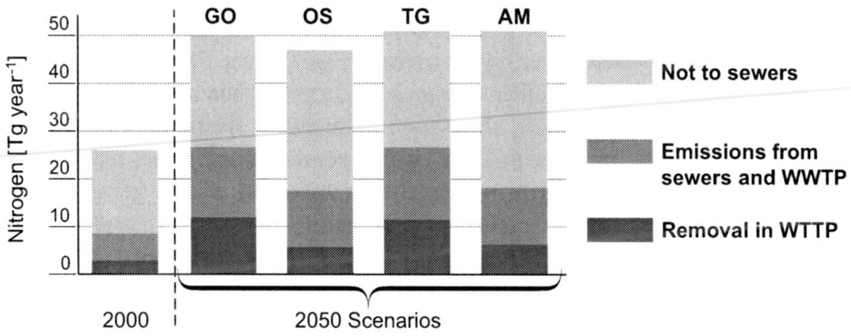

Figure 4.2 Fate of domestic wastewater nitrogen on a global scale. Based on data from Van Drecht *et al.* (2009). The 2050 predictions are based on the storylines of the four Millenium Environmental Assessment (MA) scenarios: Global Orchestration (GO), Order of Strength (OS), Technogarden (TG), and Adapting Mosaik (AM). For a summary of the storylines, see Cork *et al.* (2006).

The different prognoses differ by a factor of two in their predictions of nitrogen removal by wastewater treatment in 2050 (from around 6 to 12 $TgN{\cdot}a^{-1}$). However, the predictions of sewer emissions (11–15 $TgN{\cdot}a^{-1}$) and nitrogen emitted from households not connected to the sewer system (23–33 $TgN{\cdot}a^{-1}$) are in all cases higher that the predicted removal. Nitrogen emitted through a sewer system as NH_4^+, NO_3^- or organic nitrogen compounds will rapidly reach the riverine system and thus contribute to eutrophication. In all scenarios, this sewer-borne nitrogen is expected to increase by a factor of 2–2.5 from 2000 to 2050. The fate of wastewater-related nitrogen that never reaches a sewer system is unclear, but for concentrated inputs to the environment we expect different pathways than those found in agriculture. Urbanization without proper sanitation may thus have similar effects on the nitrogen cycle as the introduction of sewer systems: more nitrogen will be released into the environment. Without a sewer system, the route will be via the ground water and atmosphere rather than via the riverine system, but it is unrealistic to assume a similarly high nitrification/denitrification rate as in agriculture. The total removal of wastewater-related nitrogen is expected to increase from today's 10% to somewhere between 12 and 23%, and the total emissions from around 23 $TgN{\cdot}a^{-1}$ in 2000 to 38–45 $TgN{\cdot}a^{-1}$ in 2050.

On the basis of these three major sources of anthropogenic Nr emissions to the environment, global inputs of NO_x and NH_3 to the atmosphere have probably risen by a factor of more than three compared to the pre-industrial era. Regionally, there have been even more substantial increases, and emissions from large parts of North America, Europe and Asia rose by a factor of more than ten during the past century (van Aardenne *et al.* 2001). Large areas of the world now show average Nr deposition rates exceeding 10 $kgN{\cdot}ha^{-1}{\cdot}a^{-1}$. This represents an order of magnitude increase above natural rates.

Groundwater sources are threatened by NO_3^- and organic nitrogen compounds from agriculture, wastewater and natural sources, and the limits for nitrate and other nitrogen contaminants have now been exceeded in many groundwater bodies (UNEP 2007). Moreover, river transport and direct discharges transfer nitrogen to coastal waters and other marine ecosystems. Whereas the open ocean is mainly affected by atmospheric deposition of nitrogen, rivers are the primary contributors to coastal areas, with large and increasing impacts on ecosystems and people. These are not only growing in magnitude, but also in geographical extent (Selman *et al.* 2008). To understand the causes of coastal eutrophication, we need models describing quantitatively how nutrient inputs and the bio-geophysical properties of a watershed translate into nutrient export to a given coastal ecosystem. Recently, a new type of generally applicable model has been developed, tested and applied at both global and regional scales. Of particular interest for this chapter is the Global NEWS (Global Nutrient Export from WaterSheds) model. This global watershed nutrient export model predicts the annual average nitrogen exports from individual river basins to coastal systems. It makes spatially explicit predictions for the various Nr compounds as a function of nitrogen inputs to the watershed, land use and bio-geophysical characteristics (Seitzinger *et al.* 2010). More than 4,500 watersheds are included in the NEWS model, which can quantitatively evaluate the contribution of various nitrogen sources in a watershed. The results of implementing the model indicate a large spatial variation of nutrient exports to coastal ecosystems as well as a large range of nutrient sources in watersheds causing those exports.

When these contributions are summarized on a global level, however, the contributions from wastewater are smaller than other sources of Nr, but still significant (Figure 4.3). Furthermore, local input can be relatively high leading to local N-excess situations and hypoxic conditions. However, once nitrogen is in the river, it does not matter from where it comes, and as shown by Mulholland *et al.* (2008), the proportion of nitrate denitrified in rivers strongly decreases with increased concentration of nitrate.

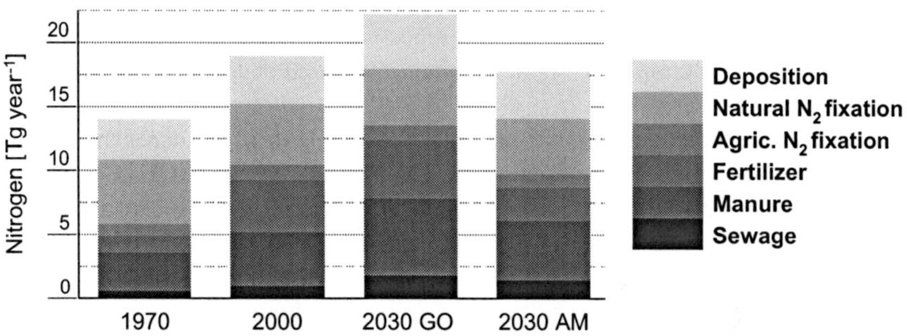

Figure 4.3 Model predicted global contribution of nitrogen sources in watersheds to river export of dissolved inorganic nitrogen (based on Seitzinger *et al.* 2010). For a definition of the 2030 scenarios, see Figure 4.2.

4.4 ENVIRONMENTAL CONSEQUENCES

A large part of the anthropogenic Nr reaches N-limited aquatic and terrestrial ecosystems, resulting in unintentional fertilization and loss of biodiversity. Increased Nr loads lead to algal blooming in coastal areas and a loss of drinking-water quality of ground and surface water. In coastal areas, there is a rapid increase of oxygen-depleted zones which cannot sustain many forms of marine life. Furthermore, three quarters of global fish stocks are being exploited at levels that threaten their long-term sustainability. Greater biodiversity can increase the resilience of the earth's ecosystems towards such anthropogenic impacts, but we are unfortunately reducing biodiversity instead.

The increased anthropogenic production of nitrogen not only causes eutrophication of terrestrial and freshwater ecosystems, but also affects atmospheric quality through direct emissions, particulate matter formation and the nitrogen cascade. Its effects include global warming, changes in the ozone concentration in the troposphere and stratosphere, soil acidification and the formation of secondary organic particulate matter: these have a negative impact on both ecosystems and human health (Cowling *et al.* 2002; Galloway *et al.* 2003).

The complex role of the cascading nitrogen cycle is well understood in some areas. Thus the role of N_2O and tropospheric O_3 (increasing due to rising NO_x emissions) is known to be a contributing factor to greenhouse gas emissions. However, there is less understanding of the role played by the nitrogen cycle for other issues, including its importance for carbon sequestration and other interactions between the carbon, nitrogen and phosphorus cycles (see Gruber and Galloway 2008).

Even where the contributing factors are well known, the cascading effects are not always taken into account. While denitrification in managed (wastewater treatment) and unmanaged (wetland) ecosystems certainly plays an important role in combating eutrophication, increased denitrification may also boost N_2O emissions, especially under less than optimal conditions (Larsen 2011). Thus, solving one part of the problem may not necessarily solve the whole, in particular if end-of-pipe solutions are applied. Often, the solution to one problem will just create new challenges.

Nitrogen indirectly affects the amount of available drinking water by decreasing the quality of ground and surface water. The most common contaminant identified in ground water is nitrate (NO_3^-), which is particularly widespread in shallow aquifers. Its main sources are manmade activities, and the problem is found in both rural and urban areas. Fertilizer application in agriculture is neither equable nor optimal (Townsend *et al.* 2003). In many developed countries, N-intensive agriculture contributes to unhealthy diets and promotes excess food intakes, while in many developing and rapidly industrializing countries, especially in sub-Saharan Africa, a lack of fertilizers is a major cause of malnutrition (Sanchez and Swaminathan 2005). Additionally, the environmental effects of excess nitrogen can affect human health and welfare both directly and indirectly. For instance, the nitrogen-driven increase in tropospheric O_3 poses a direct health threat to humans (Levy *et al.* 2005) and is the cause of substantial productivity losses in agriculture (Reilly *et al.* 2007). Atmospheric reactive nitrogen is also involved in the formation of fine particulate matter, a substantial health threat in heavily polluted regions (Wolfe and Patz 2002), and it is speculated that excess nitrate in drinking water may constitute a risk factor for cancer and reproductive problems (Ward *et al.* 2005). These effects combined are likely to incur multi-billion dollar costs.

Other potentially important and costly effects are less well known. They include the possible consequences of eutrophication for certain human infectious and parasitic diseases (Townsend *et al.* 2003), such as malaria, cholera, West Nile virus and schistosomiasis (McKenzie and Townsend 2007). Unfortunately, it is currently impossible to make a comprehensive assessment of these nutrient-disease connections due to a lack of experimental data. However, the greatest diversity of human infectious and parasitic diseases is found in tropical regions, which will face a marked increase in nutrient loading in the coming decades, highlighting the importance of understanding these connections (McKenzie and Townsend 2007). Finally, adequate nutrition is a necessary requirement for a healthy immune system, and the supply of fertilizer may consequently be one of the most critical links between tropical diseases and nitrogen fixation (Sanchez and Swaminathan 2005). All these impacts are multiplied by the "nitrogen cascade" as described above.

4.5 THE FUTURE AND POSSIBLE INTERVENTIONS

Several current trends indicate that the "nitrogen economy" of the 21st century will become increasingly dependent on industrially produced Nr (Figure 4.4). Erisman *et al.* (2008) considered five parameters (population, NUE, change in protein consumption, protein form and production of biofuels) as the most important factors for trends up to 2100. They applied the philosophy of the four storylines from the IPCC SRES scenarios (Nakicenovic *et al.* 2000) that is "global" vs. "regional" and "technology-oriented" vs. "environmentally conscious" to predict future global fertilizer demand (Figure 4.4). Despite improvements in NUE, limited population growth, optimized diets and the banning of biofuels, even the most optimistic scenario leads to a similar annual Nr production as today, which is far above the planetary limit of 35 $TgN{\cdot}a^{-1}$ defined as safe by Rockström *et al.* (2009). Several interventions are consequently required.

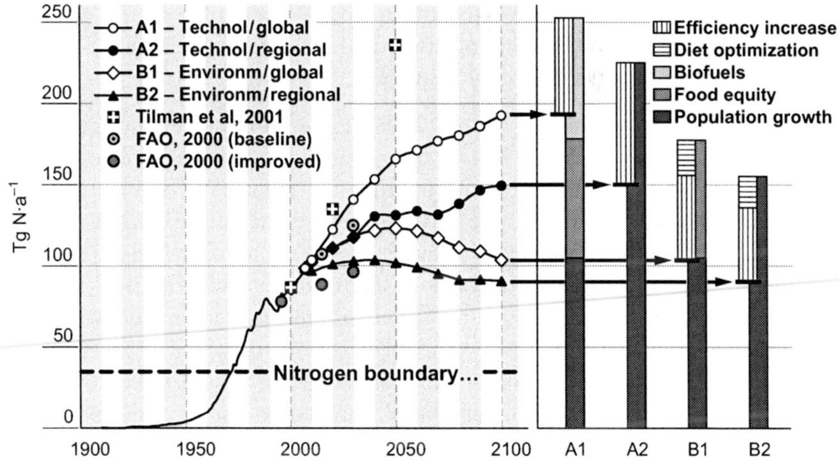

Figure 4.4 Global scenarios for nitrogen fertilizer consumption (left) and the impact of specific drivers on 2100 consumption (right). The resulting consumption ($TgN{\cdot}a^{-1}$) is always the sum (indicated with an arrow on the right graph) of the drivers increasing and decreasing nitrogen fertilizer consumption. Other estimates are presented for comparison. The A1, B1, A2 and B2 scenarios are based on the IPCC emission scenario storylines. The nitrogen boundary has been derived by Röckstrom *et al.* 2009. *Source*: Erisman *et al.* (2008).

The input to the nitrogen cascade can be reduced by a number of interventions, such as optimizing the fossil-fuel combustion process, increasing the uptake efficiency of agricultural nitrogen, changing human diets and improving wastewater and animal-waste treatment (Galloway *et al.* 2008). These authors list four options for Nr management on a global scale:

- Using the best available technologies, optimization of fossil fuel combustion can lead to a one third decrease in Nr creation from the current level of ~24 $TgN{\cdot}a^{-1}$, thus reducing the global rate of Nr creation by 8 $TgN{\cdot}a^{-1}$.
- Increasing the Nr use efficiency in crop production. The authors consider that the 38% increase in global cereal demand expected by 2025 can realistically be met with a 25% decrease in Nr fertilizer application, thus reducing the global rate of Nr creation by ~15 $TgN{\cdot}a^{-1}$.
- Improved feeding strategies and manure management on farms. A combination of barn adaptations, low-protein animal feeding, covered manure storage, exhaust air treatment and more efficient manure application can prevent a worldwide loss of ~17 $TgN{\cdot}a^{-1}$ to the environment. Additional losses could be prevented if the fraction of dairy products and meat in human diets were reduced, a measure which would also have potential health benefits.
- The global population emits 23 $TgN{\cdot}a^{-1}$ in domestic wastewater, of which only around 10% is removed (Fig. 4.2). The authors estimate a realistic improvement of wastewater treatment in the order of 5 $TgN{\cdot}a^{-1}$ by denitrification in treatment plants (if half the urban population of 3.2 billion people had access to tertiary treatment).

In the estimations of Galloway *et al.* (2008), these four interventions contribute a potential decrease of ~50 $TgN{\cdot}a^{-1}$, or ~25% of the total Nr created in 2005, representing a cost-effective way to improve the climate, environmental quality and human health. However, we have to realize that the expected growth in especially agricultural and wastewater associated nitrogen releases will be enormous (Fig. 4.4). More interventions are consequently needed to prevent dramatic increases in the nitrogen-related environmental and human-health effects described in this chapter.

Innovative sanitation approaches that simultaneously help to solve sanitation problems in developing countries, control nitrogen emissions and possibly even recycle nitrogen for agriculture offer a high potential for influencing the nitrogen cycle. Large amounts of Nr – in the order of 50 $TgN{\cdot}a^{-1}$ – are expected to be emitted by the world population in 2050 (Figure 4.2). In the "equity" scenarios depicted in Figure 4.4, it is assumed that around 70 $TgN{\cdot}a^{-1}$ of additional fertilizer nitrogen will be required to achieve this "equity." Since around 80% of the nitrogen from domestic wastewater stems from concentrated urine, which holds the potential of being transformed into fertilizer, there is an immense potential for "short circuiting." As shown in this book, the approach to this problem can very well be based on engineering solutions, which will also allow the nutrients from cities and not only from rural areas to be tapped.

4.6 CONCLUSIONS

The nitrogen challenge is complex and involves many interactions. Nitrogen pollution has a multitude of sources and cascading effects affecting different environmental compartments, locally and globally. In areas with a shortage, human beings suffer malnutrition, and it will be a huge challenge to find solutions to this problem without further accelerating the nitrogen cascade. Policies should be developed on the basis of the experience gained from regions with excess Nr combined with a synthesis of information from N-deficient regions. One possible policy would be to short-circuit the nitrogen from human waste as discussed in this book, which can help solve the problem of N-deficiency without leading to increased production of Nr.

The nitrogen effects in coastal areas, in aquatic and terrestrial ecosystems, on biodiversity, and in the atmosphere and stratosphere are well documented. Much less is known about the impact on human health from nitrogen compounds in the different environmental compartments. However, effective policy development urgently requires increased knowledge of the quantitative relationships between nitrogen sources and health effects.

Although there is great spatial (and temporal) variability between nitrogen fluxes, excesses, shortages and effects, a strong coupling exists due to the cascading of nitrogen compounds. For these reasons, the issue of scale is important, and better tools should be developed for the quantification of the cascading effects. Local sources contribute primarily to local effects through run-offs and/or air emissions, and yet these local sources also contribute to the regional, national, continental, and sometimes even global effects, such as in the occurrence of the greenhouse gas N_2O. Due to this link between the different scales, zooming models are required to treat the trans-boundary between water and air fluxes.

In many cases, nitrogen emission itself is the key driver of these effects (such as coastal and terrestrial eutrophication or nitrous oxide emissions), while being only a key contributor to others, thus aggravating a wider problem. The transport of nutrients (food, fertilizer) also couples in scales on the input side. These complicating aspects mean that the central role of nitrogen is not always evident, even though it is actually involved in many trans-boundary pollution problems. At present, no single study addresses this complex nitrogen issue in an integrated way.

Such an integrated approach to nitrogen that specifically takes the nitrogen cascade into account is needed. Finally, all reactive nitrogen produced and not denitrified into N_2 (without side products) leads to such effects, regardless of its origin. Although the qualitative cause-effect relationships of nitrogen are quite well known, more widely applicable quantitative relationships have not yet been

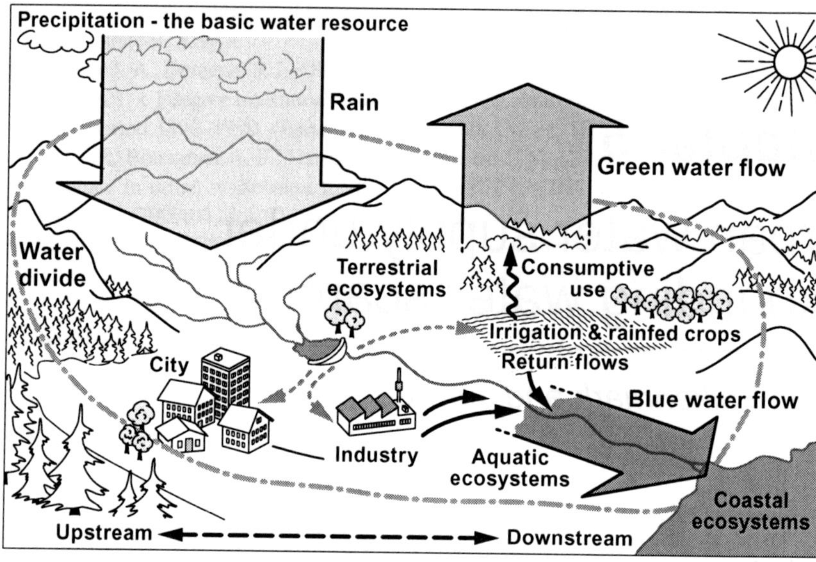

Figure 5.1 All water-dependent activities in a river basin depend on sharing the water resource of the basin. Drawing based on SIWI (2007).

The blue water in rivers and aquifers moves towards the river mouth and constitutes the principal water resource available to meet societal water requirements: the water supply for local residents and industry, irrigation, hydropower and other uses – both upstream and downstream. Some water uses are throughflow-based and produce a more or less polluted return flow; others are consumptive, such as irrigation, the water evaporating during use and being returned to the atmosphere, where the water vapour is caught by the wind system to form precipitation somewhere downwind.

Besides the urban water supply, therefore, many other water-related activities rely on the same water system from which the city residents have to compete for water. A major competitor, especially in dry climate regions, is irrigation-dependent agriculture; another one is the river itself, as some water has to remain in the river for aquatic ecosystems and fluvial processes and uses (known as environmental flow).

Today, an increasing number of rapidly growing cities are located in water-scarcity regions, a situation that calls for guidance on how limited blue-water availability in a river system may in principle be allocated between the most important water needs. It is quite thought-provoking to realize that practically all the population growth during the next few decades is expected to take place in urban areas, and that this is happening most rapidly in dry-climate, low-income regions (see also Larsen and Gujer 2013).

Therefore, in developing and rapidly industrializing regions in the developing world, the water supply to urban areas is now becoming a major challenge. At the same time, additional water may be required for agriculture, since in dry-climate regions food production tends to depend on extra water for irrigation in order to produce good yields. There is, however, also a third high-priority water need, namely the water that has to remain in the river, known as the environmental flow, that is, the flow reserved for aquatic ecosystems, for fish stocks, for flushing away silt so that it does not clog the river channel, for assuring protection against salinity intrusion and for the water balance of the estuary.

A basin is considered to be closing when all its commitments, including environmental flow, cannot be met for some part of the year, and closed when the same is true over the entire year (Molle *et al.* 2007). If a basin passes the point of closure, the question of sustainability arises. The river basin water would then be unable to support its many functions, and ecosystem services would consequently be lost unless a way is found to reduce the consumptive use of water or increase its supply.

5.2.1 On the verge of a new water scarcity

Although water scarcity is a grim reality for millions of people, this is still neither properly understood nor acknowledged in many front-line discussions. A regrettable confusion with respect to growing water scarcity distorts the formulation of adequate policies and the implementation of effective action programmes. The sheer magnitude of this task implies that "water scarcity should be everybody's business."

In the 2006 Human Development Report "Beyond Scarcity: Power, Poverty and the Global Crisis" (United Nations Development Programme 2006), water scarcity is considered from two points of view: (1) as a crisis caused by a lack of services for the provision of safe water, and (2) as a crisis arising from a scarcity of water resources. While governance will remain a key challenge, there is also a need to increase our understanding of the issue of "water crowding" amid growing pressure on finite, vulnerable, and erratically available water resources.

Physical water scarcity will cause immense problems for developing countries in arid and semi-arid regions, as they have to cope concurrently with rapidly expanding populations and the requirement to fight poverty and improve their quality of life. While water availability is mainly a product of climate, demand is determined by population size, the sectors of society competing for water, and the degree of water productivity achieved.

The effects of blue-water scarcity differ from those of green-water scarcity. The former may lead to a collapse of the water supply, failure of crops in irrigated fields, closure of river basins, and higher infrastructure costs for water provision. Other effects include disputes amongst stakeholders and increased water pollution due to a lack of dilution water in the river to dilute contaminants. Water-pollution control for the conservation of biodiversity requires constraints on acceptable upstream pollution loads as well as on consumptive use.

Instead of addressing management and governance problems, many countries instinctively prefer supply-side solutions such as desalination or the construction of reservoirs for storage of flood flow and other large-scale infrastructure. Obviously, there may be situations where supply-oriented approaches are needed. However, a wiser approach to increasing water scarcity is to implement demand management before investing in supply-side solutions (see also the discussion on energy issues in supply-side solutions by Rittmann 2013).

5.3 IMPLICATIONS OF INCREASING COMPETITION?

This section aims at clarifying the implications of sharper competition for water between its different uses, especially in water-scarce economies. We will analyze the situation as we seek reasonable allocation principles. What will be the implications for future water management? We will also comment briefly on the implications of source separation and decentralization, the topic of this book.

In terms of water demand for municipal and industrial uses (M/I) in water-scarce basins, our quantification in this chapter builds on the background reports to the Commission of Sustainable Development (CSD) study “Comprehensive Assessment of Freshwater Resources of the World” (Stockholm Environment Institute). In one of these reports (Lundqvist and Gleick 1997), Israeli expert Hillel Shuval noted that in a country where water availability is only 100–200 $m^3 \cdot p^{-1} \cdot a^{-1}$, after subtracting water for basic human needs and high-value economic demands, “little if any water is left for agricultural use.” He more specifically claimed that in an arid country like Israel, 190 $m^3 \cdot p^{-1} \cdot a^{-1}$ is in fact enough to meet domestic, urban and industrial requirements as well as to produce vegetables for local consumption, assuming large imports of staple food products in exchange for exports of industrial products. However, in many developing countries, agriculture is the foundation for development and general efforts to reduce poverty. In green-water-scarce countries, such agriculture is a large-scale water consumer.

On the basis of such statements, we have assumed a standard value for M/I of 200 $m^3 \cdot p^{-1} \cdot a^{-1}$ in our analysis, equivalent to 550 $L \cdot p^{-1} \cdot d^{-1}$. It should be stressed that this is a quite conservative level: it is only half of what Falkenmark and Lundqvist (1997) assumed to be the basic water requirements for households and industry in both temperate and tropical zones, whether humid or arid. In comparison, Shiklomanov (1998) noted that actual water withdrawals by many large cities – that is, only the municipal water supply component itself – amounted to 300 – 600 $L \cdot p^{-1} \cdot d^{-1}$ (108 – 216 $m^3 \cdot p^{-1} \cdot a^{-1}$).

For environmental flow, in terms of the water that has to be reserved as remaining river flow for the benefit of the aquatic ecosystems, river flushing needs and so on, we have assumed a water requirement of the order of 30% of the natural river flow. This fits well with earlier experience by South African scientists (Falkenmark 2003).

In a recent study for the Hai river basin in China, the better ecological situation in the 1970s was used as the baseline when analyzing a reasonable percentage to be reserved for "ecological water use" (Xia *et al.* 2006). In a similar way to the Australian "working river" approach (Whittington 2002), Xia and his colleagues stressed that the ecological water requirement has to be coordinated with the requirement for socio-economic development. Beyond paying attention to the river ecosystem, their analysis included the water requirement for the area's wetlands, lakes and water outflow to the sea to maintain the estuary's water balance. Comparing theoretical ecological flow requirement to actual river flow they concluded that ecological water requirement corresponds to some 35–50% of the flow to avoid a water deficit in the river ecosystem in a severely dry year.

5.4 ADAPTING TO INCREASING WATER SHORTAGE

The blue-water sharing situation will evidently be most critical in water-scarce basins where agricultural production depends on irrigation to secure the food supply. Below 1,000 $m^3 \cdot p^{-1} \cdot a^{-1}$ a country can be seen as suffering a chronic water shortage, so that further irrigation may be difficult to achieve (SIWI 2007, Falkenmark and Molden 2008).

That region is hosting a total population of 3–5.6 billion (depending on the global change scenario assumed, SIWI 2007), and will be particularly vulnerable by 2050. Geographically, most of these countries are located in the ribbon of latitude lying between 5 and 40 degrees N in Africa and Asia. In that region, the urban water supply will consequently develop into a critical problem.

Most relevant for the theme of this chapter – the urban water supply – is the blue-water situation, while the green cropland situation is of interest when considering potential water sharing tradeoffs with water requirements for irrigation. Irrigation increases the consumptive use of water due to evapotranspiration, and although improved irrigation efficiency (crop-per-drop policies) saves water, it also reduces the blue-water return flows from agriculture. It is a major question whether and how societies will adapt when river basins pass the point of closure. Will there be a hard landing characterized by increasing pollution, damage to ecosystem services, intensified competition and unequal sharing of benefits? Or could the landing be soft, with sustainable, equitable and efficient use of water? Many rivers are already running more or less dry on the way to their outflow, meaning that water is over-allocated and the basin already well beyond closure. Environmental catastrophes like the dramatic shrinking of the Aral Sea may therefore repeat themselves in many other regions unless suitable measures are taken to counteract such developments.

Figure 5.2 offers a graphical overview for judging the degree of blue-water scarcity in a river basin (Falkenmark and Rockström 2004). The horizontal axis shows – instead of per capita availability – the inversed unit water crowding in persons per flow units of 1 million cubic metres per year, the vertical axis the percentage of water availability used (use-to-availability ratio). The diagram also characterizes different degrees of water stress and water shortage (SIWI 2007, Falkenmark and Molden 2008). The per capita water use is shown by the diagonal lines. The diagram has logarithmic scales and presents an overview of the main links between water crowding in person per flow units (horizontal axis), percentage use-to-availability (vertical axis) and per capita level of water use $m^3 \cdot p^{-1} \cdot a^{-1}$ (diagonal lines). The bold vertical line shows the 1,000 persons per flow unit level beyond which a water shortage is characterized as chronic.

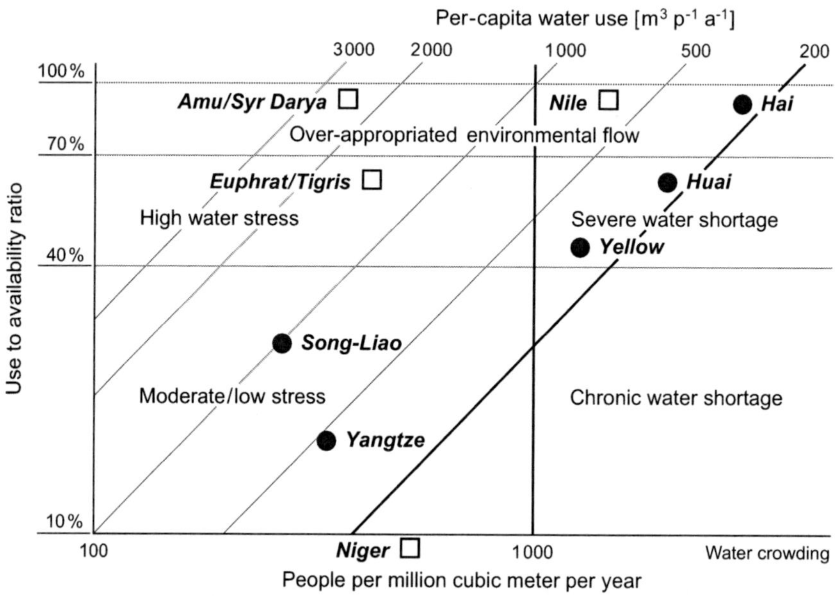

Figure 5.2 Two dimensions of present and imminent water scarcity risks: demand-driven water stress and population-driven water shortage. Where the ratio of use to availability is high (Y-axis), blue-water resources are stressed to the point where no more water is accessible for use and the basin is said to be closed. Beyond 70%, water can be seen as over-appropriated since the last 30% of the blue-water resource should preferably be reserved for aquatic ecosystems as "environmental flow." Where many people depend on a limited amount of water, this is shown as population-driven "water crowding" (X-axis). Beyond the 1,000 persons per flow unit level, the shortage can be seen as chronic. Developed from SIWI (2007).

The diagram distinguishes zones with different types of water stress: one is due to population-driven water crowding (horizontal axis) as opposed to the demand-driven level of use-to-availability, which can also be referred to as the water mobilization level (vertical axis). In the latter dimension, more than 40% is conventionally seen as high water stress. More than 70% is considered unacceptable or over-appropriated, since the remaining amount is needed as reserve for river channel and environmental purposes. By 2075, Oki and Kanae (2006) have estimated the number of people in regions suffering from a chronic water shortage to be between 3 and 7 billion, depending on the scenario. Based on the same scenarios, the population in regions with high water stress will be between 4 and 9 billion.

To illustrate the differences in water scarcity situations, a number of river basins are entered into the graph: they comprise some transnational river basins (□) and a set of Chinese basins (●). It was recently shown (Falkenmark and Molden 2008) that some 1.6 billion people are already living in over-appropriated river basins, including the Amu/Syr Darya basin (upstream of the Aral Sea), the Nile basin, and the Hai river basin in North China.

The right hand side of the diagram includes a diagonal line for 200 $m^3 \cdot p^{-1} \cdot a^{-1}$, indicated above as a conservative level for municipal/industrial water demand (M/I). The area of greatest interest from a water allocation perspective is therefore the triangular zone between the vertical line (1000 p·flow·unit^{-1}) and the 200 $m^3 \cdot p^{-1} \cdot a^{-1}$ diagonal (in bold). This is the area where many of the irrigation-dependent countries are situated. They depend on irrigation for food self-sufficiency, but simultaneously suffer from a chronic water shortage. The large number of countries in this triangular zone motivates taking a closer look at the emerging challenges to water allocation.

5.5 REASONABLE BLUE-WATER ALLOCATION

A generalized blue-water allocation for this triangular zone could follow the following basic relationship:

$$P(M/I + CU) + 0.3B + S = B \tag{5.1}$$

where P is the population, M/I and CU are the per capita requirements for the municipal/industrial water supply and consumptive water use/irrigation respectively and B is the blue-water availability. The term 0.3 B represents the environmental flow reserve and S the remaining surplus flow. The implications of this general allocation are demonstrated in Figure 5.3 when taking an additive approach, that is, seeing urban water supply and irrigation as competitive. Due to the importance of environmental flow for river functions, it has been given priority over irrigation here. Due to the foreseeable planetary constraints for consumptive water use by 2050 (see Rockström *et al.* 2009), water for irrigation is – in view of its consumptive character – to be seen as constrained by the maximum allowable consumptive water use of some 4,000 $km^3 \cdot a^{-1}$. If equally shared by 9 billion people, this works out at an average of 444 $m^3 \cdot p^{-1} \cdot a^{-1}$.

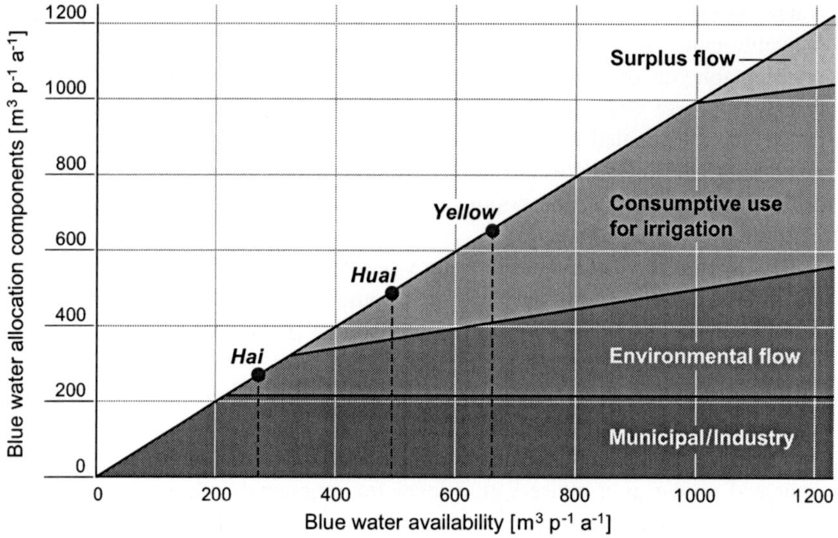

Figure 5.3 Principal allocation at different levels of blue-water availability also showing the predicament of the three H-rivers in China, allowing 200 $m^3 \cdot p^{-1} \cdot a^{-1}$ for municipal/industrial use and taking an additive approach.

Under these assumptions, Figure 5.3 shows that it is only beyond around 1,000 $m^3 \cdot p^{-1} \cdot a^{-1}$ that all the water requirements may be met. This means in other words that water reuse will be increasingly essential in the triangular zone in Figure 5.2. The diagram also illustrates the allocation dilemma below 400 $m^3 \cdot p^{-1} \cdot a^{-1}$ (equivalent to a water crowding level beyond 2,500 p^{-1}·flow $unit^{-1}$), where an M/I supply level would not allow any irrigation at all. Only by reducing the M/I supply can water be made available for irrigation.

Table 5.1 Evaluation of potential water capacity in three severely water-scarce Chinese river basins.

	P/B	B/P	M/I	Env. flow	CU/irrig.	Surplus
	[p·flow $unit^{-1}$]			[$m^3 \cdot p^{-1} \cdot a^{-1}$]		
Yellow river	1,556	643	86	193	206	158
Huai river	2,021	495	89	149	164	93
Hai river	3,711	269	89	81	163	0

This means that other allocation principles would be needed in particularly water-scarce regions. As exemplified in Table 5.1 for the "Three H basins" in China, namely the Hai, Huai and Yellow river basins (Xia *et al.* 2007), national water management limits the M/I allocation to around 90 $m^3 \cdot p^{-1} \cdot a^{-1}$. In this silt-laden region, blue water for environmental flow is seen as essential to meet the requirements for river-based functions and processes, including the flushing of silt. In all these rivers, the M/I supply allocations are in fact limited to less than half the level assumed in this study, that is, considerably lower than the standard levels mentioned in the mid-1990s by both Shuval (Lundquist and Gleick 1997) and Shiklomanov (1996), thereby making water available for other purposes, especially environmental flow and irrigation.

The predicament of these three basins clarifies the dilemma in river basins with less than 1,000 $m^3 \cdot p^{-1} \cdot a^{-1}$ (more than 1,000 p·flow $unit^{-1}$), in other words the triangular zone in Figure 5.2.

5.6 IRRIGATION POTENTIAL

Figure 5.2 and Table 5.1 give an idea of how large a population P can in fact be supported in a river basin by the blue-water availability B. It follows from equation 5.1 that, where $S = 0$,

$$P = 0.7B/(M/I + CU) \tag{5.2}$$

To meet more than just the urban/industrial and environmental flow requirements, that is, $CU > 0$, the basins must have a water-crowding limit of below 3,500 p·flow $units^{-1}$, that is, beyond 285 $m^3 \cdot p^{-1} \cdot a^{-1}$ (the crossing point between the 70% and 200 $m^3 \cdot p^{-1} \cdot a^{-1}$ lines in Figure 5.2).

Figure 5.3 and Table 5.1 clarify the predicament of the three H-rivers in North China and the policy followed to reserve water for irrigation/CU as well. The Hai river has a flow capacity exceeding 3,500 p·flow $units^{-1}$ but is able to provide irrigation by its reduced M/I level. In view of the limited irrigation level chosen, the Huai and Yellow rivers can even produce a certain surplus.

Thus, by limiting urban water supply to no more than 90 $m^3 \cdot p^{-1} \cdot a^{-1}$, this region in China will be able to allocate environmental flow and still afford water for irrigation even in the extremely water-scarce Hai river. In fact, China is currently developing a water policy to this end, known as the "Building Society of Saving Water in China," arguing for a change from water-supply to water-demand management, and from water-quantity management to the integrated management of water quality and quantity. In order to relieve serious water stress, however – besides increasing water-use efficiency and developing scientifically based water-resource management of basins – it is considered necessary to look for additional water resources by means of approaches such as trans-basin water transfer, the reuse of treated wastewater, and desalination.

5.7 ADDITIVE VERSUS COMPETING WATER USES

What emerges from these examples of highly water-stressed situations is a more generalized idea of how this water allocation problem might be growing together with chronic water shortages in a world where both population growth and urbanization are expanding, economic development is raising human water needs, and climate change will increasingly impact a country's water availability. Future water management considerations will evidently also have to incorporate climate change in a scientific evaluation of water system stress and carrying capacity.

The conventional debate sees urban water demands as additive to and competing with agricultural water demands. When cities grow, there is currently an "either/or approach" in the Western world: urban authorities tend to buy water from farmers to obtain enough for their supply, thereby reducing the importance of irrigation water in water allocation. But in countries where food security may be seen as an equally essential water use from a national perspective, attention has to be drawn to what happens to the water after use. In that sense, irrigation is an altogether different kind of use than the urban water supply. Urban water can be reused after treatment. Irrigation on the other hand is a use which is consumptive in the sense that the water evaporates during use and does not return to the river system. This raises the possibility of sequential reuse; urban wastewater can be reused in peri-urban agriculture as already hinted above by Hillel Shuval (Lundquist and Gleick 1997). Additionally, radical innovations in wastewater management technologies as discussed in this book may increase water productivity in cities to a level not possible with conventional technology. Sewers depend on high enough water flow in order to transport particulate (especially faecal) matter and thus present an upper limit to urban water efficiency. Source separation may play a double role, of recovering nutrients so that they can be reused as fertilizer *and* permitting radical water saving and local wastewater reuse in urban areas.

5.8 CONCLUSION

This study has shown that blue-water availability will increasingly prove inadequate for the M/I supply, environmental flow reserve and irrigation in the future. On the basis of previously considered M/I levels – beyond a water crowding level of 3,000 p·flow units^{-1} (i.e. below some 300 $m^3 \cdot p^{-1} \cdot a^{-1}$) – no water will be left for irrigation, a situation already reached in the Hai river basin.

Limitations in terms of acceptable consumptive use/irrigation mean that a region will depend on importing food grown elsewhere and on its ability to pay for such imports (Falkenmark and Lannerstad 2010). Economic development is therefore of fundamental importance. However, such development also opens up the opportunity for the proper treatment of urban wastewater, thus permitting post-treatment water reuse.

The perspectives presented here make it evident that a shift in thinking is urgently called for: it must distinguish between through-flow based water uses, which permit reuse after waste treatment, and consumptive water use needed to secure food production in order to ensure national food supplies.

The post-treatment reuse of municipal water will become increasingly attractive as population-driven water shortages become more acute, wherever feasible combined with source separation – the subject of this book. Nutrient separation for consecutive reuse as fertilizer in peri-urban agriculture may be of particular interest (see also Jönsson and Vinnerås 2013).

Acknowledgement

For both the data from North China and discussions in this paper on scientific concepts from China's three H-basins, we have drawn upon China's Basic National Research Programme (2010CB428406).

REFERENCES

Falkenmark M. and Lundqvist J. (1997). World Freshwater Problems – Call for a New Realism. *Comprehensive Assessment of the Freshwater Resources of the World.* Background Report 1, Stockholm Environment Institute, Stockholm, Sweden.

Falkenmark M. (2003). Water Management and Ecosystems. *Living with Change.* TAC, Report No 9, Global Water Partnership, Stockholm. Sweden.

Falkenmark M. and Lannerstad M. (2010). Food security in water short countries – Coping with carrying capacity overshoot. In: Rethinking Water and Food Security. Fourth Botín Foundation Water Workshop, L. Martinez-Cortina, A. Barrido and E. López-Gunn (eds.), CRC Press, Boca Raton, USA, pp. 3–22.

Falkenmark M. and Molden D. (2008). Wake up to realities of river basin closure. *Water Resources Development,* **24,** 201–215.

Falkenmark M. and Rockström J. (2004). Balancing Water for Humans and Nature. The New Approach in Ecohydrology. Earthscan, London, UK.

Jönsson H. and Vinnerås B. (2013). Closing the loop: Recycling nutrients to agriculture. In: Source Separation and Decentralization for Wastewater Management T. A. Larsen, K. M. Udert and J Lienert. (eds.), IWA Publishing, London, UK, pp. 163–177.

Larsen T. A. and Gujer W. (2013). Implementation of source separation and decentralization in cities. In: Source Separation and Decentralization for Wastewater Management, T. A. Larsen, K. M. Udert and J. Lienert (eds.), IWA Publishing, London, UK, pp. 133–150.

Lundqvist J. and Gleick P. (1997). Sustaining Our Waters into the 21st Century, Stockholm Environment Institute, Stockholm, Sweden.

Molle F., Wester P. and Hirsch P. (2007). River basin development and management. In: Water for Food, Water for Life: Comprehensive Assessment of Water Management in Agriculture, D. Molden (ed.), Earthscan, London, pp. 585–625.

Oki T. and Kanae S. (2006). Global hydrological cycles and world water resources. *Science,* **313**, 1068–1072.

Rittmann B. E. (2013). The energy issue in urban water management. In: Source Separation and Decentralization for Wastewater Management, T. A. Larsen, K. M. Udert and J. Lienert (eds.), IWA Publishing, London, UK, pp. 11–27.

Rockström J. *et al.* (2009). A safe operating space for humanity. *Nature,* **46**, 472–475.

Rockström J, Karlberg L. and Falkenmark M. (2010). Global food production in a water constrained world: Exploring green and blue challenges and solutions. Cambridge University Press, Cambridge, UK.

SIWI (2007). On the verge of a new water scarcity: A call for good governance and human ingenuity. SIWI Policy Briefs. SIWI, Stockholm, Sweden.

Shiklomanov I. A. (ed.) (1996). Assessment of water resources and water availability in the world. Background report to the Comprehensive Freshwater Assessment, State Hydrological Institute, St Petersburg, Russia.

Whittington J. (2002). Working rivers. Watershed. Cooperative Research Centre for Freshwater Ecology, Canberra, Australia.

Xia J., Zhang L., Liu C. and Yu J. (2007). Towards better water security in North China. *Water Resources Management*, **21**, 233–247.

Xia J., Feng H.-L., Zhan C.-S. and Niu C.-W. (2006). Determination of a Reasonable Percentage for Ecological Water-Use in the Haihe River Basin, China. *Pedosphere,* **16**(1), 563–575.

Chapter 6

The issue of micropollutants in urban water management

Klaus Kümmerer

6.1 INTRODUCTION

The products of the chemical and pharmaceutical industries such as medicines, disinfectants, personal care products, detergents, pesticides, dyes, preservatives, food additives, nanoparticles, to name a few, are ubiquitous in everyday life.

Most of these chemicals are used in large amounts in open applications such as personal care, hygiene, plant protection, health and the textile industry. In many cases, their release into the environment after appropriate use is unavoidable. We now know that despite extensive measures taken to prevent pollution of the environment by chemical contaminants, they are still present in the environment.

Improved analytical techniques such as LC-MS/MS have shown the presence of such "micropollutants" in the lower $\mu g \cdot L^{-1}$ levels in the aquatic environment. They permit the detection of pollutants at trace levels and in particular of polar compounds such as pharmaceuticals, including their metabolites and transformation products resulting from incomplete degradation that could not previously be analyzed. Because of their typical concentration range they are often called organic micropollutants.

It can be assumed that the (aquatic) environment is polluted by myriads of still undetected contaminants in the $\mu g \cdot L^{-1}$-range and below. Most of them either failed to be investigated in the past or were undetectable because of the limitations of the analytical equipment. Therefore, they are often called "emerging" contaminants to stress the novelty of their detection. Although this may be correct from an analytical chemist's point of view, most of these compounds have been present in the environment for years or even decades. Sometimes they were even reported in the literature years or decades ago but were largely ignored at the time knowledge is helpful when considering source separation options. The examples may also demonstrate some of the follow-up problems encountered if retention and/or separation at source is not possible. As there are many more micropollutants than the ones described here, and more information is also available about them, the reader is advised to seek further details in the literature (e.g., Kümmerer 2011, Kassinos *et al.* 2009, Schwarzenbach *et al.* 2006 and references cited therein).

6.4.1 Flame retardants

Organobromine compounds such as polybrominated diphenyl ethers (PBDEs) are one of the important groups of organic chemicals used as flame retardants, for example in polymeric materials such as furniture foam, rigid plastics and textiles.

PBDEs are technical, that is, loosely defined, mixtures. The shares of individual isomers in the mixture are not exactly known and research most often focuses on the main constituents. Many of these chemicals are considered harmful, having been linked to liver, thyroid, reproductive/developmental, and neurological diseases. Textiles represent the main source of organobromine compounds found in the aquatic environment: these are “washed out” during the washing cycles, in common with other textile chemicals. Organobromine flame retardants are not biodegradable and partially end up in the sewage sludge. PBDEs are ubiquitous in all compartments including water, sediments and biota. Contamination levels are higher in urbanized regions.

Another important group of widely used flame retardants are organophosphates. Main chemicals belonging to this group are halogenated organic phosphorus compounds such as phosphates and phosphonium salts.

6.4.2 Biocides and pesticides

Pesticides are a subgroup of biocides.[3] Antimicrobials are another. Some of the latter also belong to the group of pharmaceuticals (see below).

Pesticides are present in effluents from manufacturing sites, in effluents and run-offs from farms and private gardens, and from urban run-offs. After being applied, they can reach groundwater by infiltration or surface water as a result of surface run-off, both typical examples of non-point sources. However, residues in pesticide containers may also reach STPs when the vessels are cleaned. Pesticides (or, as they are legally named, biocides) in urban run-offs mainly originate from the surface protection of buildings. Some are used as algaecides for the protection of outdoor surfaces such as facades. Whether they end up as a point source (STP) or a diffuse source depends on the fate of the run-off. Other

[3]Not used here in a legal context (see Footnote 1), but in the sense that biocides are chemicals used to kill organisms.

biocides such as disinfectants are used in medical environments and households.[4] Especially those used for surface disinfection, for instance in households, hospitals and industry, are washed off from the surface and end up in sewage. Moreover, biocides are used in the preservation of textiles. They are also released during laundering and show up in household effluents.

The group of pesticides includes an interesting example of the possible significance of transformation products resulting from the incomplete degradation of the parent compounds: the application and microbial degradation of the fungicide tolylfluanid in soil results in a transformation product, N, N-dimethylsulfamide (DMS). DMS was found in ground and surface waters with typical concentrations in the range of 100–1,000 and 50–90 $ng{\cdot}L^{-1}$, respectively. DMS itself is not of toxicological concern, but it exhibits high mobility in soils and water, enabling it to enter the drinking water treatment process. During ozonation, about 30–50% of it is converted to the carcinogen N-nitrosodimethylamine (NDMA). The discovery of the high yield of carcinogenic NDMA upon ozonation immediately led to the closure of some drinking water treatment facilities. Within six months of the discovery that tolylfluanid was the source of the NDMA, the producer withdrew it from the European market and a legal ban on its use was in place within one year. This is a very good example of source control measures leading to a reduced toxic load for humans as well as on the environment (since ozonation is one of the most efficient means of removing micropollutants in wastewater treatment). However, we do not know what other parent compounds and their transformation products may be present in the aquatic environment. This demonstrates the need for a more extensive discussion of source control and the precautionary principle.

6.4.3 Endocrine disrupting chemicals

Endocrine disrupting chemicals (EDCs) are substances of natural or synthetic origin that may interfere with the body's endocrine system, thus affecting several of its processes, among them sexual differentiation. They may have estrogenic, anti-estrogenic, androgenic, anti-androgenic and other hormone-like effects. Thus feminization of fish has been observed downstream of sewage treatment plants (Sumpter and Jobling 1995). Within the last fifteen years, the discovery of the estrogenic nature of sewage treatment plant (STP) effluents has resulted in great public concern and a huge amount of scientific literature in order to better understand and identify EDCs with a view to determining their occurrence, fate and biological effects on wildlife. An often analyzed chemical of this group is ethinylestradiol (EE2), the synthetic active ingredient in many contraceptive pills. It was identified as the main contributor to estrogenicity of effluents of some STPs. Other hormones discussed as relevant include progesterone and dexamethasone.

---

[4] A versatile database for chemicals used in household products can be found on the internet (Household Products Database, http://hpd.nlm.nih.gov/)

Other synthetic chemicals that mimic hormones originate from various groups of products and chemicals such as pesticides, flame retardants, plasticizers PCBs and dioxins. As EDCs are present in different types of products, effective source control is difficult.

6.4.4 Anti-corrosive additives

Benzotriazole (BT) and tolyltriazole (TT) are widely used as anti-corrosive additives. They are also components of cooling and hydraulic fluids, antifreeze products, aircraft de-icers and anti-icing fluids. The main input of BT into the aquatic environment stems from their use as dishwasher detergent additives for the purpose of silver protection. From there, they are discharged into municipal wastewaters. Their presence in rivers demonstrates their incomplete removal by sewage treatment (Wolschke *et al.* 2011).

6.4.5 Personal care products

Personal care involves products as diverse as deodorants, eyeliners, facial tissues, lipstick, lotions, makeup, mouthwashes, pomades, perfumes, shampoos, shaving creams, skin creams and toothpastes. Some products like shampoos can contain up to 10–20 different ingredients. During or after use, many of these are washed off the skin or hair and end up in the sewage. Typical classes of compounds are fragrances, odorants, disinfectants, preservatives, surfactants and UV filters. This demonstrates the complexity of the modern way of life in terms of chemical products.

6.4.6 Perfluorinated surfactants – PFOS and PFOA

Perfluorooctanesulfonate (PFOS) and perfluorooctanoic acid (PFOA) are perfluorinated surfactants used to produce polymers and telomers. They also play a role in the electroplating industry. They consequently feature in the effluents of such activities. PFOS and PFOA are the end products of many fluorochemical compounds in the natural environment. As they are non-polar and amphiphilic at the same time, some similar compounds of different chain length tend to bioaccumulate in animals and humans whereas others do not. Other fluorinated organic chemicals are used in textiles and are therefore present in wastewater from households and commercial laundries.

6.4.7 Pharmaceuticals

Although the study of environmental pharmaceuticals is still fairly recent, it has already generated a vast literature (Kümmerer 2008a, Ternes and Joss 2006). Pharmaceuticals are classified according to their purpose and biological activity (e.g., antibiotics, analgesics, anti-inflammatory substances, antibiotics, antihistamines). Compared to most other chemicals, pharmaceutically active compounds are often chemically complex molecules with specific properties, such as dependence of their octanol-water partition coefficient (K_{OW}) on pH. Although pharmaceuticals are optimized for biological activity in humans, most of them are incompletely metabolized. In some cases up to 90% of the still active parent compounds are excreted unchanged and are thus contained in wastewater from households and hospitals. Until recently, no emissions from manufacturing sites were assumed. However, very high emissions have now been found from manufacturing sites in developing countries (Larsson *et al.* 2007). Elevated concentrations of active ingredients have also been demonstrated in developed countries such as the US and France (Philips *et al.* 2010, Gilbert 2011).

Pharmaceutical concentrations are higher (up to 100 $\mu g \cdot L^{-1}$ in some cases) in hospital wastewater than in municipal sewage (often in the range of a few up to 20 $\mu g \cdot L^{-1}$). In terms of volume and total load, however, the public (households) is the main source of pharmaceuticals in the aquatic environment. Another, but minor source is effluents from landfills, as date-elapsed medicaments are often disposed of with household waste. Around 200 million people in the world are estimated to have used illicit drugs at least once during the last year. Such compounds have recently been detected in surface water and wastewater (Castiglioni *et al.* 2011), a reason for concern because of their psychoactive properties.

6.4.8 Artificial sweeteners

Artificial low-calorie sweeteners are consumed in considerable quantities with food and beverages. After ingestion, some of them pass through the human metabolism largely unaffected, are quantitatively excreted via urine and faeces, and thus reach the environment together with domestic wastewater. Typical sweeteners are sucralose, acesulfame, cyclamate, and saccharin. The data show that these artificial sweeteners are quite persistent in surface waters (Scheurer *et al.* 2009).

6.4.9 Engineered nanoparticles

Nanomaterials are of increasing technological and economical importance. Their key properties include their size, shape, surface properties, aggregation state, solubility, structure and chemical composition. Thus chemical composition and shape provide the basis for several classes of engineered nanoparticles (ENPs) such as inorganics (e.g., TiO_2, SiO_2, ZnO, Ag), organics such as fullerenes, multi- and single-walled carbon nanotubes and so on. Some of them are chemically modified to carry organic functionalities on their surface. Insoluble material such as fullerenes may become soluble as a result. Because of their widespread use, some of them can also reach the environment via production processes, and even more likely via the routine use of products containing nanoparticles (e.g., sunscreens). Others are the result of wash-off for example, from facades of buildings (Kaegi *et al.* 2010). Some ENPs are also used to remedy environmental pollution, for instance in the treatment of effluents and wastewater (e.g., photocatalysis by TiO_2). In such cases, the introduction of nanomaterials is intentional. Transmission electron microscopy (TEM) analyses have confirmed that nano-scale Ag particles were sorbed to wastewater biosolids in both the sludge and effluent. Their presence in biosolids was mainly in the form of Ag_2S (Kaegi *et al.* 2011), indicating oxidation of the particles.

6.5 MANAGEMENT OPTIONS

As discussed by Kümmerer (2007), there are three main strategies for reducing the input of pharmaceuticals into the environment: technology, education and training, and the "benign-by-design" approach. Source separation, the topic of this book, involves elements from both technology and education/training and will thus be discussed as an additional element.

6.5.1 Technology

In recent decades, much research has been conducted into advanced wastewater effluent treatment (see e.g., Ternes and Joss 2006). Advanced oxidation and adsorption processes have proved to be particularly effective. Despite their apparent success, advanced treatment technologies have important limitations. None of them can remove all compounds, and it is unclear how they will cope with new compounds of the future (Kümmerer 2008b). More serious, however, is the fact that advanced treatment is limited to the small number of countries where tertiary treatment technology is available. Membrane bioreactors (MBR) have been shown to achieve only very slightly increased rates of micropollutant removal, due to the higher sludge age rather than the filtration or the often postulated but unproven change in microbial composition of the sludge. For centralized municipal applications, the additional energy and costs required for MBR compared to conventional treatment are significantly higher than for a post-treatment step (by ozonation or activated carbon). Since micro-pollutant removal is significantly higher in post-treatment steps, it should be made quite apparent that MBRs are not superior to conventional activated sludge treatment in this respect. In decentralized treatment, MBRs are chosen for their process reliability and not for their ability to remove micropollutants. However, a post-treatment step also needs additional energy and may results in toxic transformation products.

6.5.2 Education and training

It is always possible and desirable to avoid the unnecessary input of chemical compounds into wastewater ("pouring down the drain"). If products and individual chemicals are kept separate, this may allow their reuse (after some quality/purity assessment) or at least specific and more effective treatment or destruction. Effective and efficient source collection requires appropriate information and education of all the stakeholders involved as well as a working collection system. Such systems for hazardous chemicals and pharmaceuticals are available for the consumers in several countries (Niquille and Bugnon 2008) and are often implemented in companies. For pharmaceuticals, the relationship between patients and doctors and the involvement of pharmacists is of great importance in order to avoid unnecessary medication (Götz and Deffner 2010). Thus Sweden introduced a classification system for pharmaceuticals in 2005 (Wenmalm and Gunnarson 2010). It allows a doctor to select a more ecological API if there are alternatives. Such approaches may only work after some time, but assessment data indicate already a small shift from hazardous to less hazardous compounds. Furthermore, appropriate package sizes encourage the avoidance of remainders that may be released to wastewater. If waste is incinerated properly in suitable plants, it could also offer an option for collecting date-elapsed chemicals and pharmaceuticals with household waste. However, all chemicals should be handled with care by trained staff only. Obviously this approach would not be suitable for compounds whose designated use implies introduction into water, such as pharmaceuticals, laundry detergents and so on.

Another important measure is proper labelling, but its effectiveness is open to discussion. Its success would probably require a shift in social mindset, as it is impossible to control all possible inputs from individuals. If feasible, a ban or restriction on the production and sale side would be much more effective. In general, we should aim to reduce the diversity and amounts of chemicals used. Thus suitable use of wood for example, as construction material only in dry places in buildings for example would help to reduce the amount of chemicals needed for its protection. We have to think more about the relationships between the materials we use and the conditions in the environment (e.g., wet/dry or shadow/intense light). If materials are regularly exposed to light and moisture, bacteria and algae will unavoidably grow on their surfaces. Although we tend to think that the active molecules in pharmaceuticals should be freely available, this point may be disputed. Thus lifestyle drugs are never necessary, so a restriction on their use could be discussed. These few examples demonstrate that we need a different attitude based on a broader understanding of the potentials of sustainability.

6.5.3 Source separation

Pharmaceuticals are mainly contained in human excreta, which can be collected and treated separately. Such excreta collection is under intensive discussion for hospitals. Thus iodinated contrast media persist after biological treatment and to some extent also after ozonation and powdered activated carbon treatment. They have been detected in drinking water, which is deemed unacceptable. Research has therefore been conducted into the collection of excreta from patients receiving X-ray contrast media (Heinzmann *et al.* 2006). These have the shortest half-life of any compounds administered to humans. A half life of 2h means that patients (in-patients and out-patients) have to stay in the hospital for 2 hours to theoretically collect 50% of the administered dose, 4 hours for 75% and 8 hours for around 90%. For various reasons, such an approach is considered too expensive. Some pharmaceuticals need special attention. Cytotoxics are highly active compounds that may cause cancer. Their presence in excretions poses a risk to the collectors if patients are unable to go to the toilet. There could also be a hazard for others using this toilet. It is therefore not recommended to collect excretions from patients undergoing anti-cancer treatment (Kümmerer and Al-Ahmad 2010).

However, hospitals are only a minor source of pharmaceuticals in terms of the total load, very often below 5% in a specific catchment area, for example, in the case of antibiotics (Kümmerer and Henninger 2003, Heberer and Feldmann 2005, Schuster *et al.* 2008). Radiological practices can use much more x-ray contrast media than a hospital. Source-separating installations would be needed for the collection of excreta from households and public buildings. Lienert *et al.* (2007) showed that on average about 70% of pharmaceuticals are excreted via urine. However, the figure could be as low as 5% for individual compounds. NoMix toilets that only treat urine separately would consequently not remove all pharmaceuticals from wastewater. Experience from conventional biological wastewater treatment suggests that their processes are unlikely to be disturbed by antibiotics and disinfectants. This may, however, be different for decentralized biological treatment of urine and faeces due to the effects of individuals excreting antibiotics or to flushing from cleaning and disinfecting activities.

Industrial wastewater often contains a much narrower spectrum of different chemicals specific to the type of activity involved. These specific wastewaters should therefore be kept and treated separately.

Run-off from roads and other sealed surfaces in urban environments contains yet other chemicals. Their sources are often unknown in view of the diverse activities and infrastructure materials concerned. Accordingly, it contains many different chemicals such as (heavy) metals, particulate minerals – including those resulting from various incineration and abrasion processes. Organic pollutants such as polycyclic aromatic hydrocarbons stem from incineration (of waste, plus wood or oil heating) and traffic. In these cases source control would aim at the prudent application of these chemicals or a change of product or a different concept of mobility. A transition to electric transport and different energy sources for heating would reduce emissions of hydrocarbon-related substances and improve the situation. However, human activities continue to produce emissions. The chemicals involved are currently still emitted in high loads and are found in surface run-off. Its separate treatment in small on-site plants would be advantageous but would require large investments. This is probably not feasible outside urban areas. Also, the run-off would pollute the land and waters close to roads and railways.

6.5.4 Benign by design

It is obvious that source control cannot be applied to the basic pollutants in human excretions (COD, nutrients, and pathogens). An alternative would be source separation, which allows for more resource-efficient wastewater treatment, including recovery of resources within single wastewater streams. The situation is different for micropollutants, where the input of chemicals can be avoided by the prudent use of products, that is, avoiding their misuse or wrong disposal (e.g., date-elapsed medicines, see above).[5]

About 100,000 chemicals are on the market in the European Union. Some of them are only used in non-open applications (i.e., in closed systems contained at the source) and will probably not reach the environment in any significant amounts. However, most of them are used in open applications in households, public buildings or industries and in high tonnages (for all the purposes described above). The introduction of most chemicals into wastewater is thus an unavoidable result of their intended use.

The goal must be to reduce the number and concentration of micropollutants in wastewater. This requires their complete mineralization or at least removal by sorption. Biological treatment seems to offer the best way of reducing their concentration without increasing toxicity, because it generally produces fewer transformation products. Photolysis in particular often produces many different products of this kind, probably due to the reactivity of the intermediate chemical species involved. Biodegradation is mediated by enzymes and therefore does not involve highly energetic and reactive chemical intermediates. In contrast, photolysis involves highly energetic radicals of low selectivity. Reactions that do not involve the formation of radicals, such as ozonation, can be expected to be rather more selective. As a general tendency, the greater the number of diverse products formed, the higher is the probability that they will include toxic ones, but carcinogens can even be formed in metabolic reactions.

This is where the benign-by-design approach comes into play: going back to the real beginning of the pipe brings the products of the chemical industry – namely the molecules themselves – into focus. A new understanding would regard low, that is, slow and/or incomplete, (bio)degradability after use as an unwanted side effect of pharmaceuticals and chemicals. Chemicals should be designed to comply with both the application and the environment along their whole life cycle. The principles of green chemistry imply that the functionality of a chemical should include not only the properties it needs for its application, but also easy and fast degradability after its use, for example, in wastewater treatment. An optimized synthesis and renewable feedstocks are very prominent in the present discussion of green chemistry, whereas the environmental properties of the molecules are

---

[5]Note that by far the most important source of pharmaceuticals reaching the aquatic environment is their use and not their wrong disposal.

somewhat underestimated. This benign-by-design approach (Kümmerer 2007) is quite ambitious, but feasible (Daughton 2003, Kümmerer 2007, Boethling *et al.* 2007) and highly promising for the future in order to overcome the limitations of current effluent treatment and water reuse. Our own biodegradability database of the OECD test series 302 and 301 contains about 2,200 different chemicals, about 500 of which were examined in two or more tests. An analysis of its data shows that 979 chemicals currently on the market are readily or inherently biodegradable (36%), whereas this applies to 29% of the pharmaceuticals from different chemical and therapeutic classes (28 out of 96). Only biodegradability under standardized condition was investigated. These data demonstrate that (bio)degradability and efficacy are not necessarily mutually exclusive. The currently widespread marketing view is that a chemical has to be as stable as possible in order to be suitable for the market. A chemist's view tends to be based on reactivity, which leads to a different understanding: during the life cycle of a chemical, stability is desirable in some phases whereas instability, such as biodegradability, is preferable in others. Many chemical conditions vary along the life cycle of a chemical, and different bacterial diversities and densities are found in different environments. Thus the conditions of a chemical on the shelf (e.g., pure and dry, exclusion of light, room temperature) differ during and after its use (Kümmerer 2007, Kümmerer 2010). If a chemical reaches a sewage treatment plant or surface water, it is exposed to different conditions than those dominating during its application (e.g., moisture, temperature, certain more or less specific bacteria and enzymes, high pH, anaerobic conditions in parts of the sewage treatment plant). They differ from those in surface water (neutral pH, low bacterial density, light access) and sediments (oxygen-rich and anoxic conditions, different sorption matrix). The spectrum and intensity of light and the concentration of ·OH radicals in the ambient air varies in different places and at different times (seasonally, daily, under day and night conditions, at different latitudes).

Bioactive chemicals such as pesticides, disinfectants and pharmaceuticals need a certain reactivity within their range of application. If they were unreactive, they would not be bio-active and would therefore be worthless for their intended use. They exert their reactivity within a particular environment, for example, within the human body, where they are activated in a specific manner. Coatings represent another example of reactive chemicals. A completely stable chemical would rarely be of any use in most cases because it would not undergo any interaction or reaction with its environment, which is often the precondition for its application (Kümmerer 2010). The direct consequence for users would be to allow direct greywater recycling in a single household, which may increase consumer interest in biodegradable products. In some cases, such as several detergents, consumers can already buy them in the supermarket.

A combination of management strategies is likely to be most effective in mitigating the risks presented by chemicals. The benign-by-design ("green" chemicals) approach together with source separation will offer the most promising future strategy to reduce and avoid the presence of micropollutants in wastewater and thus permit increased water reuse. This is urgently needed for a sustainable future, especially in countries suffering from water shortages and/or with less sophisticated wastewater treatment facilities.

REFERENCES

Aherne G. W., English J. and Marks V. (1985). The role of immunoassay in the analysis of microcontaminants in water samples. *Ecotoxicology and Environmental Safety*, **9**, 79–83.

Boethling R. S., Sommer E. and DiFiore D. (2007). Designing small molecules for biodegradability. *Chemical Reviews*, **107**, 2207–2227.

Castiglioni S., Zuccato E. and Fanelli R. (2011). Illicit Drugs in the Environment: Occurrence, Analysis, and Fate Using Mass Spectrometry. John Wiley & Sons, Hoboken, New Jersey, USA.

Daughton C. G. (2003). Cradle-to-cradle stewardship of drugs for minimizing their environmental disposition while promoting human health. I. Rationale for and avenues toward a green pharmacy. *Environmental Health Perspectives*, **111**, 757–774.

Gilbert N. (2011). Drug waste harms fish. *Nature*, **476**, 265.

Götz K. and Deffner J. (2010). Options for a more environmentally friendly handling of pharmaceuticals. In: Green and Sustainable Pharmacy, K. Kümmerer and M. Hempel (eds.), Springer Heidelberg, Dordrecht, London, UK; New York, USA, pp. 149–163.

Heinzmann B., Schuster P., Schwarz R. J., Wiemann M. and Schulz C. (2006). Getrennte Erfassung von jodorganischen Röntgenkontrastmitteln in Krankenhäusern. (Separate Collection of Iodinated X-Ray Contrast Media in Hospitals) final report, May 2006, Berlin, Germany.

Henninger A. and Kümmerer K. (2003). Promoting resistance by the emission of antibiotics from hospitals and households into effluent. *Clinical Microbiology and Infection*, **9**, 1203–1214.

Heberer T. and Feldmann D. (2005). Contribution of effluents from hospitals and private households to the total loads of diclofenac and carbamazepine in municipal sewage effluents – Modeling versus measurements. *Journal of Hazardous Materials*, **122**, 211–218.

Kassinos D., Bester K. and Kümmerer K. (eds) (2009). Xenobiotics in the Urban Water Cycle. Springer, Amsterdam, The Netherlands.

Kaegi R., Sinnet B., Zuleeg S., Hagendorfer H., Mueller E., Vonbank R., Boller M. and Burkhardt M. (2010). Release of silver nanoparticles from outdoor facades. *Environmental Pollution*, **158**, 2900–2905.

Kaegi R., Voegelin A., Sinnet B., Zuleeg S., Hagendorfer H., Burkhardt M. and Siegrist H. (2011). Behaviour of metallic silver nanoparticles in a pilot wastewater treatment plant. *Environmental Science & Technology*, **45**, 3902–3908.

Kümmerer K. (2007). Sustainable from the very beginning: rational design of molecules by life cycle engineering as an important approach for green pharmacy and green chemistry. *Green Chemistry*, **9**, 899–907.

Kümmerer K. (2008a). Pharmaceuticals in the environment. Sources, Fate, Effects and Risks, 3rd ed. Springer, Berlin, Heidelberg, Germany.

Kümmerer K. (2008b). Strategies for reducing the input of pharmaceuticals into the environment. In: Pharmaceuticals in the environment. Sources, Fate, Effects and Risks, K. Kümmerer (ed.), 3rd ed., Springer, Berlin, Heidelberg, Germany, pp. 411–418.

Kümmerer K. (2010). Rational design of molecules. In: K. Kümmerer and M. Hempel (eds.), Green and Sustainable Pharmacy. Springer, Heidelberg, Dordrecht, London New York, pp. 135–146.

Kümmerer K. and Al-Ahmad A. (2010). Estimation of the cancer risk to humans from cyclophoshamide and ifosfamide excreta emitted into surface water via hospital effluents. *Environmental Science and Pollution Research*, **17**, 486–496.

Kümmerer K. (2011). Emerging Contaminants (Pesticides, Pharmaceuticals, Personal Care Products etc.). In: Wilderer P. Treatise on Water Science. Elsevier, Amsterdam, The Netherlands.

Larsson D. G., de Pedro C. and Paxeus N. (2007). Effluent from drug manufactures contains extremely high levels of pharmaceuticals. *Journal of Hazardous Materials*, **148**, 751–755.

Lienert J., Bürki T. and Escher B. I. (2007). Reducing micropollutants with source control: substance flow analysis of 212 pharmaceuticals in faeces and urine. *Water Science and Technology*, **56**, 87–96.

Niquille A. and Bugnon O. (2008). Pharmaceuticals and environment: Role of community pharmacies. In: Pharmaceuticals in the environment. Sources, Fate, Effects and Risks, K. Kümmerer (ed.), 3rd ed., Springer, Berlin, Heidelberg, pp. 467–473.

Phillps P. J., Smith S. G., Kolpin D. W., Zaugg S. D., Buxton H. T., Furlong E. T., Esposito K. and Stinson B. (2010). Pharmaceutical formulation facilities as sources of opioids and other pharmaceuticals to wastewater treatment plant effluents. *Environmental Science & Technology*, **44**, 4910–4916.

Richardson M. L. and Bowron J. M. (1985). The fate of pharmaceutical chemicals in the aquatic environment. *Journal of Pharmacy and Pharmacology*, **37**, 1–12.

Scheurer M., Brauch H. J. and Lange F. T. (2009). Analysis and occurrence of seven artificial sweeteners in German waste water and surface water and in soil aquifer treatment (SAT). *Analytical and Bioanalytical Chemistry*, **394**, 1585–1594.

Schuster A., Hädrich C. and Kümmerer K. (2008). Flows of active pharmaceutical ingredients originating from health care practices on a local, regional, and nationwide level in Germany – Is hospital effluent treatment an effective approach for risk reduction? *Water Air Soil Pollution – Focus*, **8**, 457–471.

Schwarzenbach R. P., Escher B. I., Fenner K., Hofstetter T. B., Johnson C. A., von Gunten U. and Wehrli B. (2006). The challenge of micro-pollutants in aquatic systems. *Science*, **313**, 1072–1077.

Sumpter J. P. and Jobling S. (1995). Vitellogenesis as a biomarker for estrogenic contamination of the aquatic environment. *Environmental Health Perspectives*, **103**, 173–178 suppl.

Ternes T. A. and Joss A. (2006). Human Pharmaceuticals, Hormones and Fragrances. The challenge of micro-pollutants in urban water management. IWA Publishing, London, UK.

Wenmalm A. and Gunnarsson B. (2010). Experiences with the Swedish classification systems. In: Green and Sustainable Pharmacy, K. Kümmerer and M. Hempel (eds.), Springer Heidelberg, Dordrecht, London, New York, pp. 243–249.

Wolschke H., Xie Z., Möller A., Sturm R. and Ebinghaus R. (2011). Occurrence, distribution and fluxes of benzotriazoles along the German large river basins into the North Sea. *Water Research*, **45**, 6259–6266.

Chapter 7

Full costs, (dis-)economies of scale and the price of uncertainty

Max Maurer

7.1 INTRODUCTION

The sewage systems of the 19th century were introduced as an emergency solution to a health crisis. In Britain, an average of 26% of children died before the age of five, in the cities this average was around 48% (Brown 2003). The draining of excrements and excess water in European cities improved the quality of life and health standards dramatically. The powerful sewer system also enabled a freshwater supply network which further improved the functions of the sewers. Since then, this successful solution has been continuously improved, but at its core it remains a conveyance system with some very important characteristics:

- The almost organic growth of the system makes it difficult to assess the full costs of the wastewater infrastructure. The first section of this chapter therefore sheds some light on the costs of conventional wastewater handling. This serves as a kind of benchmark for other approaches.
- The water infrastructure network is very capital intensive and one of the most valuable assets in a community. However, the existing network is often regarded as sunk costs so that there is a strong incentive to build larger, more cost-effective treatment plants that benefit from economies of scale. This leads effectively to increased centralization, emphasizing the conveyance aspect even more. The second section investigates the almost unknown economies or diseconomies of scale in sewer systems.
- Wastewater treatment systems operate as a natural monopoly, that is, there is no competitive market for wastewater services. As a consequence, the universally applied net present value method used to compare the cost effectiveness of different technical treatment options is fundamentally crippled (Maurer 2009). The third section discusses these shortcomings in

detail and introduces an adapted evaluation method that is more adequate to compare fundamentally different system solutions.

- Long life-spans decrease annual capital costs but also mean long planning horizons, which are intrinsically difficult and present large uncertainties. The fourth section takes a quantitative look at the cost of these uncertainties and the price we might pay for a more flexible system.

Due to a lack of large-scale alternatives to conventional sewerage, hardly any research has been done to understand how centralized and decentralized structures can be evaluated in monetary terms. This book makes it very clear that this situation is changing. For these reasons, the first four sections (7.2–7.5) focus mainly on the generic characterization of conventional centralized sewerage. The last section (7.6) uses three examples to illustrate some initial steps to determine criteria to evaluate decentralized and on-site treatment systems.

7.2 CONVEYANCE-BASED WASTEWATER TREATMENT

For the current public wastewater infrastructures in Western Europe and North America, we find typical replacement values of 2600 US$·cap^{-1} for large countries (I, UK, F, D, USA) and 4800 US$·cap^{-1} for small countries (DK, CH, A; Maurer *et al.* 2006). The average annual operational expenditures (opex) are reported to be 3.8% of the replacement values. Table 7.1 compiles typical costs for wastewater treatment based on these data. The term "replacement value" is used here as an approximation for construction costs, either based on historic capital expenditures (capex) or on estimates of present construction costs.

Table 7.1 Typical numbers for replacement and running costs of small and large countries based on literature data (adapted from Maurer *et al.* 2006). WWTP: wastewater treatment plant; cap: capita; opex: operational expenditures. See text for further details.

Item	Unit	Large country		Small country	
		Sewer	WWTP	Sewer	WWTP
Replacement value public	[US$·cap^{-1}]	2,100	500	3,800	1,000
Replacement value private	[US$·cap^{-1}]	630		1,140	
Total replacement value	[US$·cap^{-1}]	3,230		5,940	
Opex	[US$·cap^{-1}·a^{-1}]	15	35	15	35
Depreciation	[US$·cap^{-1}·a^{-1}]	39	17	70	33
Opportunity costs, 3% a^{-1}	[US$·cap^{-1}·a^{-1}]	55	9	100	18
Total annual costs	[US$·cap^{-1}·a^{-1}]	109	61	185	86
Total annual costs	[US$·cap^{-1}·a^{-1}]	170 (5.2%)		271 (4.6%)	

The literature values only concern the public part of the infrastructure, primarily from larger cities and service providers. The costs of private sewers, for example, for connection the house to a public sewer line, can be substantial. In a German study representing 50,875 km of public sewers, about 91,850 km of corresponding private pipes were found (Berger *et al.* 2002). The non-public part of the sewer system may thus represent a considerable value generally not accounted for. For Switzerland we estimated that the replacement value of the private sewer lines is about 30% of the public sewers (Maurer and Herlyn 2006).

Operational expenditures (opex) are directly induced by the daily operation of the system, including energy, chemicals, staff, repairs and minor investments. Typically, the national opex is between 26 and 65 US$$\cdot$cap^{-1}a^{-1} (Maurer *et al.* 2006). The variance is mainly due to differing costs of labour and sludge disposal, and the tricky distinction between maintenance and investments. No detailed information about the costs split between wastewater treatment plants (WWTP) and sewer systems can be found. For Table 7.1 we use a plausible assumption based on Swiss data that approximately 30% of the total opex are used for the sewers and the other 70% for the WWTP (Maurer and Herlyn 2006).

Depreciation depends on the average life span and is almost impossible to estimate, especially for sewers. Besides the pure physical aging or wearing due to environmental stress, functional aging must also be considered. Functional aging occurs when the required performance exceeds the technical specifications, for example, if an increase of population leads to an overloading of the hydraulic capacity of the main sewer or WWTP. Other problematic issues include the definition of the end-of-life of a sewer pipe or a clear distinction between maintenance and replacement. The German guideline (ATV 1996) recommends linear depreciation factors of 1% to 2% for sewers and 4% to 5% for wastewater treatment plants. Applying these guidelines, we assume a life-span of 80 years for sewers, 50 years for private sewer lines, and 30 years for WWTP plants in Table 7.1.

Capital financing costs are normally incurred to access the capital market to finance investments, such as interest payments in the case of loans. These costs can have a substantial impact on the annual cost of a system, considering the large investments and long life spans of the assets. From an economic point of view, it is not required or may even be questionable to charge the customer for capital financing costs based on the replacement values of the entire system. However, the scope of this paper is to provide an estimate of the acceptable costs for alternative technologies. We therefore include opportunity costs. Table 7.1 illustrates some typical cost characteristics of conventional sewerage systems:

- This is an investment-dominated system. Typically, the replacement values are 10–20% of the annual gross domestic product (GDP)[1] in OECD countries, which is a substantial investment for any economy.

[1]Average GDP in these countries: US$ 28,500 (standard deviation: US$ 3,900; data from CIA 2003; conversion based on PPP from OECD 2008)

- Based on assumed life-spans, overall annual depreciation is about 1.7%a^{-1} of the replacement value, indicating an overall lifespan of 58 years.
- Including a 3%a^{-1} interest rate as opportunity costs, the total annual costs are around 5% of the replacement value. This is relatively small and typically substantially less than 1% of GDP.
- The annual costs are sensitive to life span and highly sensitive to interest rates. A decrease of life span by 25% increases the annual costs by approximately 10%, whereas an increase of opportunity costs to 6%a^{-1} increases the annual cost by 48% for large and 55% for small countries.
- The different replacement values for small and large countries in Table 7.1 most probably do not reflect real economies of scale but are an artefact of the data (see also Section 7.3.).

These numbers can serve as a guideline for the willingness of an economy to pay for wastewater services. Any alternative system with ambitions to be a generic solution will be measured against these overall costs. However, these costs can vary substantially on a more local level. The next section discusses economies of scale and shows how the costs of conventional systems scale with size.

7.3 (DIS-)ECONOMIES OF SCALE

Water and wastewater treatment plants show distinct economies of scale (see review in Maurer 2009 and example in Figure 7.1[2]). Unit costs decrease as the capacity increases, reducing marginal costs in larger plants. The consequence of this is that larger plants are usually given preference over several smaller ones.

However, it is less clear how the corresponding sewer system behaves with respect to the size of the catchment area. Shapiro and Rogers (1978) developed a model to determine the area to be served by a central treatment plant, instead of using available on-lot disposal systems. They concluded that the results are most responsive to the sewer cost function, which they found difficult to establish. Compared to treatment plant data, it is much harder to extract comparative cost information from real world sewer data. Conveyance systems have generally grown over decades and are very heterogeneous, even within the same catchment area. Additionally, there are non-linear correlations between population, sewered area and infrastructure costs. In order to overcome these difficulties, we formulated a generic model that is capable of representing the combined sewer infrastructure of a settlement quantitatively (Maurer *et al.* 2010). The urban water infrastructure model (UWIM) used for this analysis contains two main modules: (i) the catchment area module and (ii) the sewer construction cost module. The former calculates the length and size distribution of the required sewer

[2]Note that Figure 7.1 (cost per installed capacity) is not directly comparable with Table 7.1 (cost per inhabitant). In Switzerland the installed capacity in population equivalents (PE) is twice the total population, to accommodate industry and reserve capacity.

pipes on the basis of housing densities and area size, and the latter converts this information into construction costs.

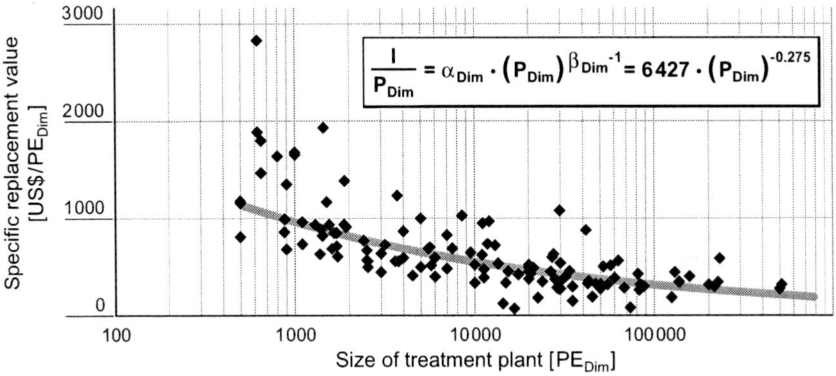

Figure 7.1 Example of economies of scale for the replacement value per installed capacity (PE_{Dim}) of wastewater treatment plants in Switzerland (Maurer and Herlyn 2006). The dataset consists of 128 replacement values for WWTPs of between 500 and 500,000 population equivalents (PE); the values were reported by WWTP operators in the year 2006 and in CHF. Purchasing power parity conversion factor US$:CHF = 0.60, German€:CHF = 0.52 (OECD 2008).

Figure 7.2 (left) shows for an exemplary case that combined sewers do indeed show diseconomies of scale if everything stays constant. As the size of a settlement increases, the sewer system must be able to transport larger amounts (in absolute terms) of water. Greater transport capacity means larger, more expensive pipes, but this is not what we observe in reality when we compare cities of different size. In Swiss settlements, there are apparent economies of scale for combined sewer systems. This is due to the higher population density in larger cities, which decreases the sewer length per capita. However, higher runoff coefficients also increase the amount of storm water runoff, leading to a higher fraction of large and expensive pipes (Maurer *et al.* 2010). The combined effect is seen in Figure 7.2 (right). Lower construction costs in smaller towns only partially compensate the higher costs of longer specific sewer lengths.

The construction costs for sewer systems in larger cities are thus the result of two main opposing cost factors: (i) increased construction costs for larger sewer systems due to larger pipes and increased rain runoff, and (ii) decreasing costs due to higher population and building densities. In Switzerland, the more or less organically grown settlement structure and limited land availability emphasize the second factor so that economies of scale become apparent.

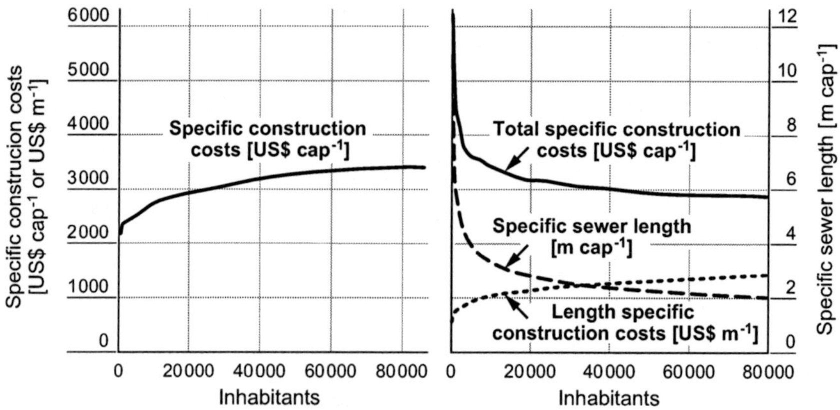

Figure 7.2 Left: Specific combined sewer construction costs for an exemplary case with constant 25 cap·ha^{-1}, 5 buildings·ha^{-1} and a runoff coefficient of 0.25. Right: Specific combined sewer construction costs and lengths for an exemplary case considering typical Swiss settlement characteristics. Adapted from Maurer *et al.* (2010).

The literature on this specific topic is sparse. Clarke (1997) reports that diseconomies of scale are inherent in all pipe networks. However, the extent varies with urban density and is affected by the layout of the piping network, a result which is also supported by other authors.

In conclusion, we do not know whether larger cities have higher or lower total specific sewer construction costs. Systems can show economies or diseconomies of scale, depending on the specific local conditions. Additionally, there is significant heterogeneity within the settlements, which can lead to very diverse local cost behaviour. This means that non-transport based approaches, such as on-site systems, need be assessed within a specific environment. It is also likely that mixed or hybrid systems might be viable from an economic point of view.

7.4 DEFICITS OF THE NET PRESENT VALUE METHOD

The net present value method (NPV) is a standard approach for using the time value of money to evaluate long-term projects (Newnan *et al.* 2004). The NPV sums up all capital expenditures, operating expenses and revenues (NPV = capex + opex + revenues) that occur during the planning horizon. Generally, projects with higher NPV are financially more profitable.

The revenues of monopolistic water businesses are generally equal to their expenditures (revenues = capex + opex). Accordingly, the NPV method used in construction projects for water infrastructures only considers the (present value of) total expenditures. This substantially restricts the usefulness of the NPV because information about the expected demand curve during the planning horizon is missing. For example, the capex of a plant are completely independent of whether it has substantial overcapacities or not. However, the cost per user will differ proportionally to the amount of overcapacity present.

This highlights the fact that the generally applied NPV method cannot identify the most cost-effective option per unit serviced. A more appropriate measure is the *specific net present value* (SNPV) as introduced by Maurer (2009), which expresses the conventionally calculated present value of expenditures divided by the average demand:

$$SNPV = \frac{PV(capex) + PV(opex)}{\frac{1}{T_P} \cdot \int_{t=0}^{T_P} P_t \cdot dt} \tag{7.1}$$

where PV = present value, capex = capital expenditures, opex = operational expenditures, T_P = planning horizon, P_t = demand at time t.

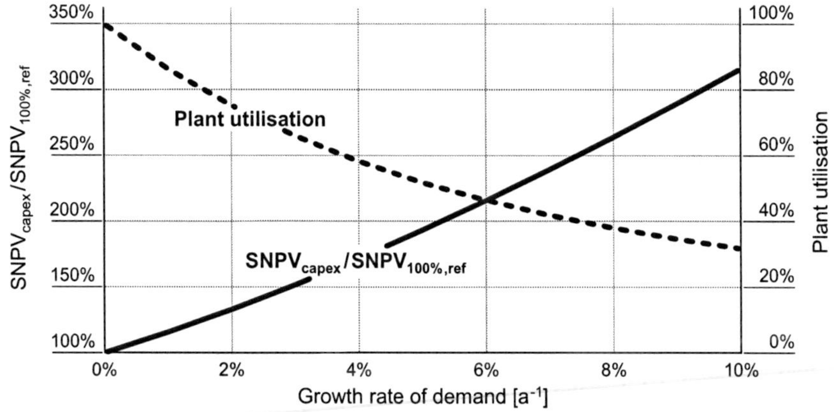

Figure 7.3 The effect of overcapacity on specific investment costs (capex). The solid line shows the increase of specific capital costs ($SNPV_{capex}$) as a function of the (exponential) growth rate of demand over the assumed lifespan of 30 years. The values are normalized with $SNPV_{100\%,ref}$ which is equal to the SNPV with no growth (= present value of capex divided by the maximum capacity). No interest rate assumed. The dotted line shows the average plant utilization as a function of growth rate. Adapted from Maurer (2009).

The SNPV can be calculated for capital expenditures (capex) and operating expenses (opex), the total SNPV being the sum of both. Figure 7.3 shows an

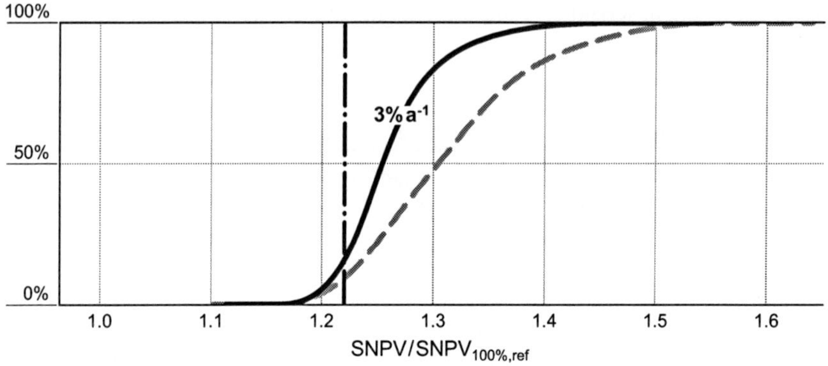

Figure 7.5 Effect of overcapacities (dotted line) on costs shown as cumulative probability density of normalized specific costs (SNPV) for an exponential growth rate of 0.03 a^{-1} and standard deviations of 0.03 a^{-1} with no interest rate. The solid lines are the same as described in Figure 7.4. A safety factor of 1.24 was chosen (see text). Adapted from Hug *et al.* (2010).

The above methodology is a first step in exploring the costs of uncertainty. Many assumptions and simplifications have still not been validated. For example, it is not clear whether a normal distribution describes uncertainties well, or how the construction and lead time for expanding the treatment plant influence the results. Nevertheless, the core methodology is sufficiently robust to be easily adapted to other growth and uncertainty approaches. Hug and Maurer (2010) applied the same methodology to sewer systems with very similar results.

7.6 ON-SITE TREATMENT SYSTEMS

Exploring the costs of uncertainty is equivalent to the quest for identifying the costs of flexibility. If uncertainty increases the probability of higher costs, then this amount might be better spent in investing in more flexible technology, for example, On-Site Treatment Technology (OST). The previous four sections focus on the characterization of a conventional centralized system. This is a necessary precondition for comparing and evaluating alternative or complementary systems. OST systems are neither a specific technological approach nor a mature technology and it is thus impossible to determine their realistic costs and pricing.

This section tries to solve this very common dilemma by indicating at what cost level OST systems could be competitive. This methodology implies that alternative technologies may be adapted when and only when similar performance can be obtained at a similar or lower cost than with the prevailing system. All examples assume that the conveyance system remains unchanged and therefore only compare centralized wastewater treatment with OST concepts. This leads to very conservative results. It is quite obvious that the implementation of OST can have substantial impacts on the conveyance system. Whereas OST reduces micropollutant and nutrient loss over combined sewer overflows and leaks, it may also allow local water reuse, infiltration, open-channel transportation, reduced operation and maintenance of existing sewers, or even the abandonment of a conveyance system altogether. Furthermore, uncertainty is not taken into account. The following examples highlight that OST systems are already competitive in specific niches even with a very conservative approach. More research is needed to assess the financial benefits of entire systems.

Example 1: Urine source separation

The first example explores the allowable costs for on-site urine source separation in areas where nutrient removal is required. By collecting and treating urine separately, it is possible to switch from nutrient removing plants to high-loaded WWTPs. This will cut the capex for treatment plants by half and decrease their opex by 25% (data from Nolting and Dahlem 1997, Bohn 1997). The allowable annual costs for on-site urine treatment can be calculated from the costs for capex (see Figure 7.1) and opex for WWTPs in Switzerland (Maurer and Herlyn 2006) plus a 3% annual interest rate for the opportunity costs. Figure 7.6 shows that annual costs must be below 15 to 50 US$·cap^{-1}·a^{-1} depending on the catchment area. This translates into capex for a urine source-separating installation of between 170 and 600 US$·cap^{-1}, assuming a lifespan for the OST hardware of 15 years and compensation of the running costs by selling the products of value. This accords well with the less differentiated calculations in Maurer *et al.* (2006). It is worthwhile noting that these calculations do not account for additional benefits such as reduced micropollutant and nutrient loss via combined sewer overflows and leaks.

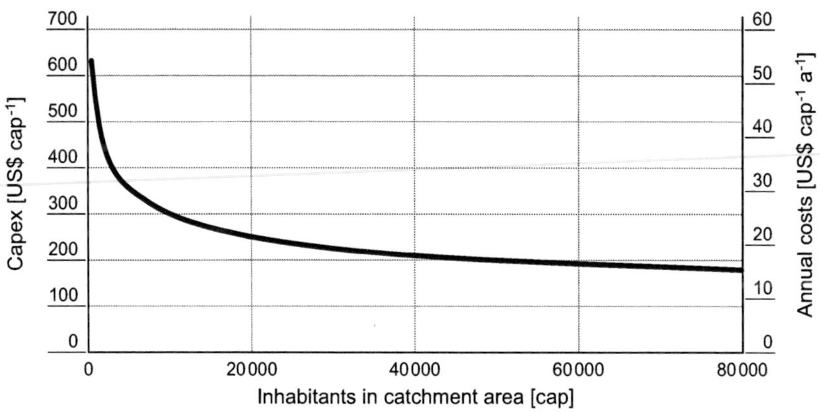

Figure 7.6 The annual costs (right ordinate) that can be used to achieve nutrient removal by urine source separation in a conventional system. This corresponds to the capex on the left ordinate, assuming a lifespan for the OST hardware of 15 years, 0.03 a^{-1} opportunity costs, and no operating costs.

Example 2: On-site treatment with conventional sewer

The second example estimates how much pure OST systems may cost in order to replace a conventional centralized treatment plant, without adapting the existing conveyance system. The results thus illustrate a "worst case" where OST is integrated into an existing system without any benefits from other adaptations.

The numbers are based on Swiss data (Maurer and Herlyn 2006) and assume 3% annual opportunity costs. In order to compensate for the variation of costs in different countries (see Table 7.1), the diagram is normalized with the SNPV (equation 7.1) for a Swiss plant for 10,000 population equivalents (PE). This enables us to draw some general conclusions going beyond Switzerland, as the shape of the economies-of-scale curves depends mainly on generic factors.

The lines in Figure 7.7 are a typical representation of the SNPV variations due to exponential demand growth in the catchment and the absolute size of the plant. The area below a specific line indicates the economic conditions favourable for an OST system. The lowest line ($\lambda = 0\% a^{-1}$) represents the SNPV cost curve in a mature area without growth, showing the economies of scale for treatment plants. The other curves show the effect of growth on the SNPV. Two immediate conclusions are apparent: smaller systems have substantially higher SNPV, and high growth rates have a strong effect on smaller systems.

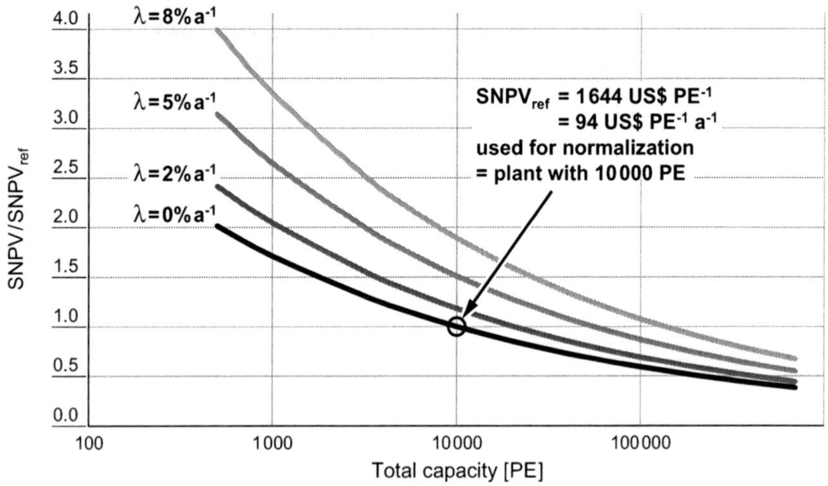

Figure 7.7 The dependency of SNPV (capex + opex) on plant size and exponential growth rates λ (life-span WWTP = 30 a). The numbers are normalized with, $SNPV_{ref}$, the cost of a plant with 10,000 population equivalents (PE), no growth ($\lambda = 0.0\ a^{-1}$) and 0.03 a^{-1} interest rate (opportunity costs). Purchasing power parity conversion factor US$:CHF = 0.60 (OECD 2008). Adapted from Maurer (2009).

High opportunity costs would emphasize the effect of growth even more and increase the spread of the lines in Figure 7.7. In addition, the uncertainty of the growth forecast shown in the figure increases the probability of even higher costs. OST systems can be built as demand develops, with the potential of avoiding large overcapacities. This flexibility to adapt to demand keeps the SNPV of OST systems fairly constant, while the SNPV of centralized systems develop as shown in Figure 7.7. This indicates that small catchment areas with high demand growth rates and correspondingly large forecast uncertainties may be good entry markets for OST systems, especially at high opportunity costs.

Example 3: Membrane-based OST wastewater system

Comparing the current cost estimate for OST plants can yield valuable insight and show the usefulness of the results derived in the last example. Fletcher *et al.* (2007) evaluated the capital and operating costs associated with a small package membrane bioreactor (MBR). The capex of such on-site MBRs are expected to be between US$ 742–1,009 per population equivalent (€645–877; US$:€ = 1.15; OECD 2008) for small units (5 PE) and US$ 296–614 (€ 257–534) for larger plants (200 PE). The average life span was estimated to be 13.8 years.

On the basis of the lower-end figures found in Fletcher *et al.* (2007) and opportunity costs of $3\%a^{-1}$, we obtain annualized capex of 65 $US\${\cdot}PE^{-1}{\cdot}a^{-1}$ and opex of 232 $US\${\cdot}PE^{-1}{\cdot}a^{-1}$. By applying the same normalization as in Figure 7.7, these numbers can be translated into the dimensionless numbers 0.7 and 2.5 (divided by 94 $US\${\cdot}PE^{-1}{\cdot}a^{-1}$; see Figure 7.7), respectively. A comparison of these numbers with Figure 7.7 indicates that even if the monetary benefits from an adapted conveyance system are ignored (see also remarks in the introduction to Section 7.6.), high-end OST systems could already be competitive under specific circumstances.

A comparison of the figures for MBR plants with conventional centralized wastewater treatment also shows clearly that operational expenses are most likely to be the current Achilles' heel of OST systems. While capex of 0.7 are absolutely comparable with centralized systems, the opex are substantially higher. Further research to minimize the opex of OST systems will dramatically increase the chances of MBR systems entering the market in large numbers.

7.7 CONCLUSIONS

The aim of this chapter is to discuss the financial characteristics of wastewater infrastructures and provide some insights into how centralized and decentralized structures can be evaluated and compared in monetary terms. The analysis presented here is by no means a comprehensive cost-benefit analysis or an environmental economic assessment, but reveals some key-strengths of decentralized systems such as flexibility and potentially lower per unit costs.

- The average figures for the national costs for wastewater treatment reveal two main points: Firstly, it is an investment-dominated system. Secondly, the annual costs are fairly moderate if all the assumptions for long life-spans and low interest rates hold true. Large variability occurs within a system. Centralized WWTP show distinct economies of scale and the costs of sewer systems depend strongly on urban geographical parameters. In essence, sewer systems show diseconomies of scale due to the fact that larger catchments require larger pipes.
- Centralized wastewater treatment requires large upfront investments that can satisfy the demand of the subsequent decades. Planning the design capacity is based on forecasting the long-term future of the catchment area. However, forecasting involves large uncertainties and has a substantial impact on system size and ultimately on costs. Combined with demand growth, these uncertainties substantially increase the specific costs of wastewater treatment. The magnitude of this effect is similar or greater than the economies of scale often used to justify merging catchment areas.
- The conventional net present value approach to comparing system alternatives cannot consider the effects of growth and an uncertain future. A specific net present costs (SNPV) method is more appropriate in situations with rapid and uncertain demand growth.

The examples presented here mainly address transitional scenarios in which innovative OST concepts such as urine source separation or on-site wastewater treatment are implemented in existing systems, which is a fairly conservative approach. Nevertheless, we see that in small catchments with high interest rates, high demand growth rates and correspondingly large forecast uncertainties, an OST system with higher overall net present costs but better ability to adapt to growing demand may be favourable. These examples also show clearly that a careful comparison of pure OST and conventional system is still missing. Other open issues are the identification of adequate probabilistic descriptions of uncertainty and the quantification of flexibility to adapt to future events such as innovative technical developments or changing legal requirements. In addition, there is hardly any information about the type of conveyance system needed for pure OST systems, which are currently not seen as competitive alternatives to centralized treatment plants. However, the articles in this book indicate a possible paradigm shift. The results of this chapter will help to establish criteria for the conditions under which OST systems could be a financially viable alternative.

REFERENCES

ATV (1996). Erfassung, Bewertung und Fortschreibung des Vermögens kommunaler Entwässerungseinrichtungen (Valuation of Municipal Wastewater Systems - Systematic Recording, Assessment and Updating, in German). Arbeitsblatt A 133,

Gesellschaft zur Förderung der Abwassertechnik e.V. (GFA), Hennef, Germany. ISBN 3-927729-35-3.

Berger C., Lohaus J., Wittner A. and Schäfer R. (2002) Zustand der Kanalisation in Deutschland. Ergebnisse der ATV-DVWK-Umfrage 2001 (Status of sewer systems in Germany – Findings of the ATV-DVWK survey in 2001, in German). KA – Wasserwirtschaft, Abwasser, Abfall, 49(3), 302–311.

Bohn T. (1997). Aktuelle Betriebskosten von Abwasserbehandlungsanlagen mit weitergehender Reinigung (Current operational costs of wastewater treatment plants with enhanced treatment, in German). In: WAR (1997). Senkung der Betriebskosten von Abwasserbehandlungsanlagen. 52. Darmstädter Seminar am 6. November 1997, p. 17–42, WAR, Darmstadt, Germany.

Brown A. D. (2003) Feed or feedback: agriculture, population dynamics and the state of the planet; International Books Utrecht, the Netherlands. ISBN 90 5727 048X.

CIA (2003). The world fact book. Central Intelligence Agency, Washington, D.C., USA. [www.cia.gov/cia/publications/factbook/index.html, last update: 18.12.2003].

Clark R. (1997). Optimum scale for urban water systems. Report 5 in the water sustainability in urban areas series. Water Resources Group. Dept. of Environment and Natural Resources. SA.

Dominguez D. and Gujer W. (2006). Evolution of a wastewater treatment plant challenges traditional design concepts. *Water Research*, **40**(7), 1389–1396.

Fletcher H., Mackley T. and Judd S. (2007). The cost of a package plant membrane bioreactor. *Water Research*, **41**(12), 2627–2635.

Hotelling H. (1938). The general welfare in relation to problems of taxation and railway and utility rates. *Econometrica*, **6**(3), 242–269.

Hug T., Dominguez D. and Maurer M. (2010). The cost of uncertainty and the value of flexibility in water and wastewater infrastructure planning. Proc. of the WEF conference 'Cities of the Future', 7–10 March 2010, Montreal, Canada, pp. 487–500.

Hug T. and Maurer M. (2010). The economic value of flexibility in wastewater infrastructure planning. Proc. of the IWA Leading Edge Conference, Phoenix, 1–4 June 2010.

Maurer M. (2009). Specific net present value: An improved method for assessing modu-larisation costs in water services with growing demand. *Water Research*, **43**(8), 2121–2130.

Maurer M. and Herlyn A. (2006). Zustand, Kosten und Investitionsbedarf der schweizerischen Abwasserentsorgung (Status, costs and investment needs of wastewater removal in Switzerland, in German). Bafu Final Report, contract-no. StoBoBio/2004. H.15f. Accessed on 2 March 2012: http://www.eawag.ch/forschung/sww/schwerpunkte/infrastrukturen/zustand_kosten/SEI21_Abschlussbericht.pdf.

Maurer M., Rothenberger D. and Larsen T. A. (2006). Decentralised wastewater treatment technologies: At what cost are they competitive? *Water Science and Technology: Water Supply*, **5**(6), 145–154.

Maurer M., Wolfram M. and Herlyn A. (2010). Economy of scale in combined sewer systems. *Water Science and Technology*, **62**(1), 36–41.

Metcalf & Eddy Inc., Tchobanoglous G., Burton F. L., Stensel H. D. (2003). Wastewater engineering: treatment and reuse. McGraw-Hill, Boston, ISBN 978-0071241403.

Newnan D. G., Eschenbach T. G. and Lavelle J. P. (2004). Engineering economic analysis. Oxford University Press, New York, USA.

Nolting B. and Dahlem H. W. (1997). Betriebskostenoptimierung durch sachgerechte Planung von Abwasserbehandlungsanlagen. (Optimization of operational costs by appropriate planning of wastewater treatment plants, in German). In: WAR (1997), Senkung der Betriebskosten von Abwasserbehandlungsanlagen. 52. Darmstädter Seminar am 6. November 1997, WAR, Darmstadt, Germany, pp. 17–42.

Qasim S. R. (1999). Wastewater treatment plants planning, design, and operation. 2nd ed., Technomic Publ., Lancaster, PA, USA, ISBN 1-56676-688-5.

OECD (2008). Purchasing Power Parities (PPPs) for OECD Countries. Obtained from http://www.oecd.org/std/ppp, accessed on 7 June 2008.

Shapiro M. and Rogers P. (1978). Analyzing the impact of land use on sewer service area planning. *Journal of the American Water Resources Association*, **14**(6), 1340–1356.

Vickrey W. (1987). Marginal- and average-cost pricing. In: R. Arnott, K. Arrow, A. Atkinson and J. Drèze (eds.), Public Economics: selected papers by William Vickrey. Cambridge University Press, Cambridge, UK, 1994.

Chapter 8

The rationale for decentralization of wastewater infrastructure

George Tchobanoglous and Harold Leverenz

8.1 TYPES OF WASTEWATER INFRASTRUCTURE

Because there is a continuum of alternatives for wastewater management infrastructure, it is necessary to define the types of systems covered in this review. An overview of alternative wastewater management infrastructure systems and the approximate U.S. population served is presented here; these systems are discussed in more detail in subsequent sections.

Centralized wastewater management systems are used for the collection and drainage of wastewater and sometimes stormwater, from a large, generally urban and peri-urban, area using an extensive network of pumps and piping for transport to a central location for treatment and reclamation, usually near the point of a convenient surface water discharge (65%).

Satellite wastewater management systems are treatment facilities connected upstream in the centralized wastewater collection system and used for water reclamation near the point of water reuse. Satellite treatment plants generally do not have solids processing facilities; solids are returned to the collection system for processing in a centralized treatment system located downstream (4–5%).

Decentralized wastewater management systems are stand alone systems used for treating dispersed small wastewater flows, for example, individual residences, residential clusters, isolated buildings and small communities. Process residuals may be processed on-site or hauled to another facility. The wastewater is collected, treated and dispersed or reused at or near the point of generation (26–28%).

Hybrid wastewater systems incorporate elements of decentralized, satellite and/or centralized facilities to optimize the performance of urban water and wastewater management systems. In the hybrid model, the centralized facility is used for the processing of excess flow, biosolids and source-separated streams; monitoring and management of remote systems; and energy recovery (~1–2%).

8.2 CENTRALIZED TREATMENT SYSTEMS

When the first wastewater collection systems were developed in the United States in the middle and late 1800s, the primary design objective was urban drainage to prevent flooding of cellars and streets. Similarly, agricultural lands were drained to improve farming. Residences were drained for convenience (Waring 1883). Typically, the drainage channels and conduits were directed to the nearest watercourse or a connection to a watercourse for disposal. This drainage practice was carried out with little or no reference to public health. For cities near water bodies, three approaches to disposal were in common use: (1) discharge from individual drainage channels directly into a receiving water body (e.g., onshore discharge to nearby streams, rivers, or sea), (2) interception of the flows from individual drainage channels and conveyance to a nearby body of water for direct discharge and (3) disposal to a water body after some form of treatment, typically primary sedimentation. Cities not near rivers or lakes diverted their wastewater to nearby sewage farms. In large cities situated in flat areas, the city was divided into sectors with the wastewater being directed by gravity or pumped to a nearby sewage farm for disposal. Often, one or more sectors would be connected with an interceptor sewer before being pumped.

With the acceptance of the germ theory and as cities continued to grow, areas for the discharge of untreated wastewater became more limited and the interception of wastewater from individual drainage channels and conduits became more common. In turn, treatment facilities were built at the terminus of wastewater interceptors, typically located in remote areas that could be served by gravity flow and/or near locations where the treated effluent could be disposed of easily. Such treatment facilities were termed "centralized," as the flow from an entire city or a sector of a city was treated in a single location. The practice of building centralized treatment facilities continues today partly because of the prior investment in the collection system infrastructure, but also because of the large expense associated with retrofit of existing buildings and piping systems. While the current investment in centralized infrastructure systems is large (greater than 2 trillion dollars), it is now recognized that centralized wastewater systems have inherent limitations, as discussed in the introductory chapters to this book. Distributed centralized, satellite, decentralized and hybrid wastewater management systems, as discussed in the remainder of this section, have evolved to overcome these limitations.

8.3 DISTRIBUTED CENTRALIZED SYSTEMS

Distributed centralized wastewater systems, as discussed above, were developed to serve large cities that could be divided into discrete sections, typically according to political and or geographic boundaries. New York City, as illustrated in Figure 8.1, is a classic example of the use of distributed centralized facilities. The New York situation evolved from the early methods used for the disposal of stormwater. Of more interest are the reasons for the development of distributed centralized facilities where a single centralized facility was used initially. The following are among the many reasons:

- Future population growth in areas that would be difficult to service
- Unwillingness of citizens to put up with the disruptions resulting from the construction of new large transport systems to accommodate future growth at the existing centralized facility
- Annexation of urban areas which had their own wastewater treatment facilities
- Capacity limitations in the collection and treatment systems to handle anticipated future growth
- Useful life of existing treatment facilities had been reached
- Capacity limitations for the discharge of additional treated effluent to nearby receiving waters
- Increased pumping costs from peripheral urban areas
- More recently, opportunities for the water reuse in the upper reaches of the collection system (satellite systems, discussed in the following section, are used more commonly)

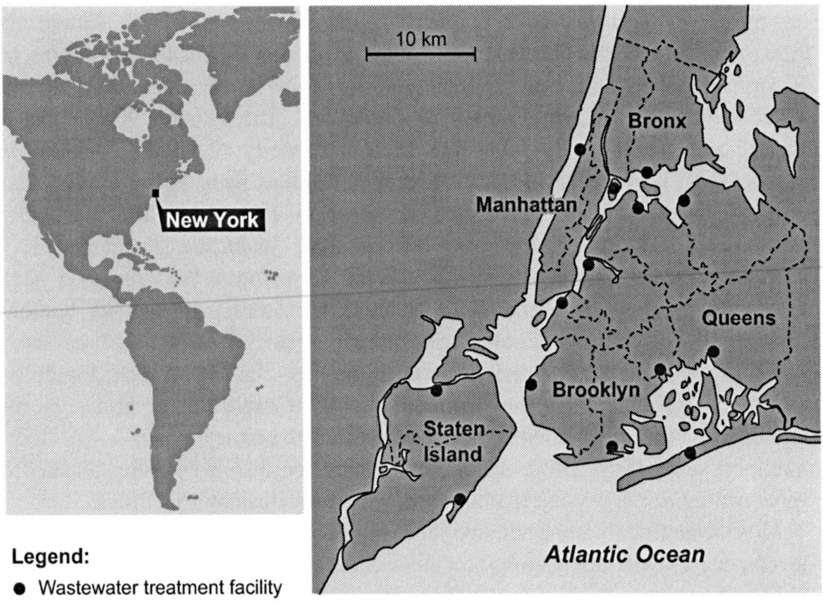

Figure 8.1 Layout of typical wastewater system for New York City employing distributed centralized treatment facilities.

The reasons that distributed centralized systems are implemented are almost always site specific. In the future, as water reuse (see also Falkenmark and Xia 2013) and conservation become more significant, it is anticipated that more distributed centralized and hybrid treatment facilities will be utilized.

8.4 CENTRALIZED SYSTEMS WITH SATELLITES

Satellite facilities are wastewater treatment systems located upstream in, and connected directly to, the centralized wastewater collection system. Process solids and excess flow are discharged to the centralized wastewater collection system for processing at the downstream facility. As a result of the direct connection to the wastewater collection system, satellite systems do not require facilities for flow equalization or solids processing resulting in a reduced footprint and improved odor control. Three types of satellite systems are identified: (1) extraction, (2) interception and (3) individual. The size of satellite treatment systems can range from large systems for flows from upstream communities or cities to small systems for source-separated flows from individual buildings and residences. In some cases, satellite facilities share important characteristics with distributed centralized systems, including the use of similar treatment technologies. The types of satellite systems that have been used are shown on Figure 8.2 and described below.

Extraction Type In the extraction type satellite system (Figure 8.2a), the wastewater to be reclaimed is extracted (mined) from a collection system main, trunk, or interceptor sewer. Historically, the extraction type satellite systems were of large capacity and used to treat and reuse the wastewater from upstream sewersheds as well as from smaller communities. The Tillman WWTP put into operation in 1980 in Los Angeles, with a capacity of about 20,000 $m^3 \cdot d^{-1}$ (80 $Mgal \cdot d^{-1}$), is the largest extraction type satellite plant in the United States. Located upstream in the sewershed, excess flow and solids from the Tillman plant along with the flows from the Glendale plant are discharged to the collection system terminating at the regional Hyperion WWTP, located next to the Pacific Ocean (see Figure 8.3). In another example, the County Sanitation Districts of Los Angeles County operates six upstream extraction type satellite plants to facilitate water reuse projects throughout the county (see Figure 8.3). Treated wastewater from the upstream plants is used for a variety of reuse applications, the most notable being groundwater recharge, which has been in operation since 1962. Large extraction systems are also now being planned and implemented to relieve overloading of downstream treatment facilities.

More recently, a number of smaller extraction type satellite systems have been developed. Typical applications for these smaller satellite systems are for water reuse in landscape, park and greenbelt irrigation; in nearby high rise commercial and residential buildings; and for commercial and industrial cooling tower applications. The quantity of flow to be extracted and reclaimed will depend on the local and seasonal water reuse requirements, especially so for landscape irrigation applications. A view of a satellite plant used in Upland, CA to provide water for golf course irrigation is shown on Figure 8.4. Note that, in this case, (1) the golf course development would not have been feasible without the reclaimed water supply, (2) a dual distribution system from the municipal plant was not required or available and (3) the treatment plant cannot be differentiated from the surrounding homes (Asano *et al.* 2007).

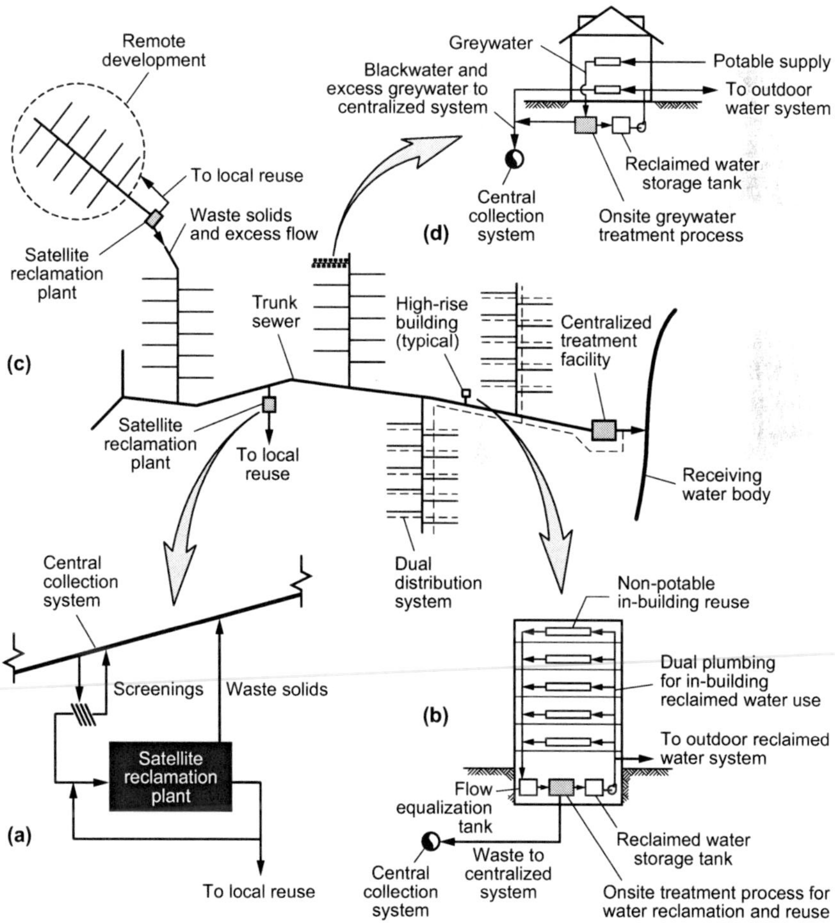

Figure 8.2 Diagram of centralized wastewater collection system with satellite systems: (a) large extraction type, (b) commercial building interception type, (c) upstream interception type and (d) individual home type (adapted from Leverenz and Tchobanoglous 2009).

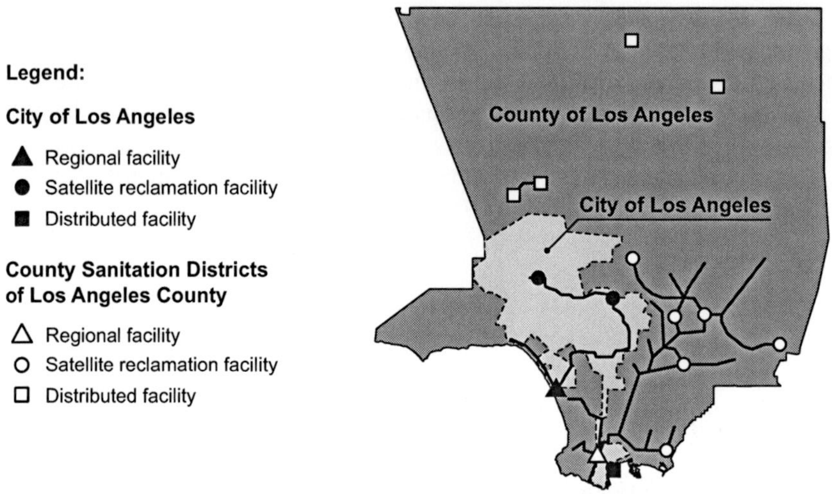

Figure 8.3 Diagram of centralized and satellite treatment systems in the City of Los Angeles and the County Sanitation Districts of Los Angeles County.

Figure 8.4 Extraction type satellite systems used to provide water to nearby golf course. Note: treatment plant (circled) looks like a residence, albeit a bit larger. The system has been in operation for more than 25 years.

Interception Type In the interception type, as illustrated on Figure 8.2b and 8.2c, wastewater to be reclaimed is captured before it reaches the collection system. The most common applications for this type of satellite system are for reuse in high rise commercial and residential buildings and other commercial facilities. A special type of interception system is where all of the flow from a new housing development is intercepted at the extremities of a centralized collection system where opportunities for water reuse (e.g., golf course and median strip irrigation) are available (see Figure 8.2c). In some cases, however, it may be necessary to divert some of the flow directly to the centralized collection system, before or after treatment. The quantity of flow to be intercepted and reclaimed will depend on the local and seasonal water reuse requirements. In some cases, it may be necessary to supplement the intercepted flow with potable water. Should excess flow occur, it is discharged to the collection system along with the biosolids produced during biological treatment. It is important to note that all of the interception type satellite systems are compatible with source separation systems.

Individual Type The individual home satellite system for greywater recovery and reuse, as shown on Figure 8.2d is another example of an intereption type system, which has the advantages of a readily treatable wastewater without the high solids load of combined wastewater and on-site or local reuse. As shown in Figure 8.5, source-separated greywater is treated using a small package treatment system, such as a passive biological filter and then used for landscape irrigation directly or used for indoor non-potable uses after membrane separation. Blackwater and excess greywater are discharged to the centralized wastewater collection system (see Figure 8.2).

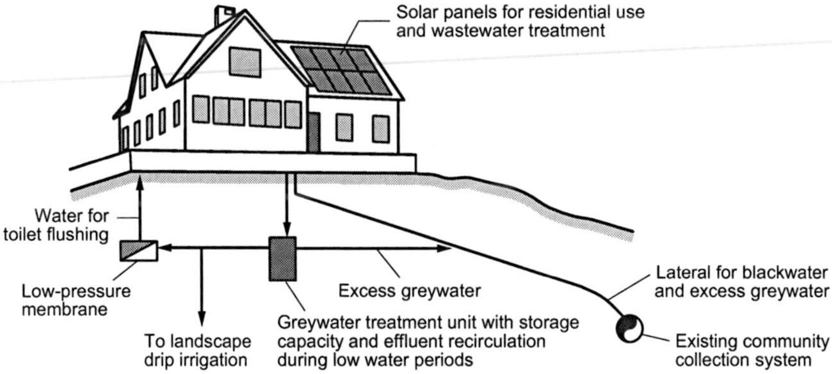

Figure 8.5 Diagram of home satellite system for greywater reuse and connection to collection system for excess greywater and blackwater.

8.4.1 Implementation of satellite systems

Satellite systems have been implemented in response to a number of factors, including:

- Capacity limitations in the collection system to handle current flows and flows from anticipated future growth, leading to localized flooding
- Capacity limitations on the downstream centralized wastewater treatment facility to handle anticipated future growth
- Need to reduce or eliminate an existing surface water discharge to an impaired water body
- The availability of a higher quality wastewater upstream as compared to the quality at the downstream centralized facility (e.g., sea water infiltration in coastal areas, downstream industrial discharges)
- Provision of reclaimed water for local reuse applications in an upstream reach of the collection system

The factors influencing the use of satellite facilities are similar to those for decentralized systems, except that use is made of the existing collection system.

8.5 DECENTRALIZED SYSTEMS

Decentralized systems have been used since the late 1800s for the management of waste from individual residences and isolated facilities. Typically, decentralized wastewater systems rely on centralized facilities for management of process residuals, such as biosolids. It is the purpose here to review briefly the types of decentralized wastewater management systems and the historical and more recent development of these systems. With this background, the future evolution of decentralized systems and their implementation are considered.

8.5.1 Types of decentralized wastewater systems

Decentralized wastewater systems can use an (1) on-site. (2) cluster, or (3) community type model. The difference between most decentralized systems and the satellite wastewater systems, discussed previously, is the nature of the connection to the centralized wastewater system. Because decentralized systems do not have a piped connection to a centralized wastewater facility, they must be designed with alternative management systems for process residuals (for example hauled to a centralized facility by truck). Flow equalization may be needed to manage high peak flows observed in small and community type wastewater systems.

An on-site *decentralized wastewater system* is used typically for the management of wastewater from an individual site or at the immediate site of wastewater generation. On-site systems are designed to accommodate the variability in wastewater generation expected from individual residences or applications.

Because the flowrates are low, effluent may be processed further by soil infiltration, or recycled for a given application.

Cluster type decentralized systems are stand-alone wastewater management system used to treat wastewater from a collection of buildings. Typically, the buildings are located adjacent to each other to reduce wastewater transport distance. Combining treatment into one system may improve the maintainability and performance of a treatment system.

A community-type *decentralized system* is a stand-alone wastewater management system used to treat all wastewater from a small community. As for cluster systems, advantages of scale are possible. A small-diameter, watertight collection system is used for transport of septic tank effluent or untreated wastewater. The collection system is not as extensive as for centralized wastewater systems.

8.5.2 Historical development of decentralized systems

The original decentralized systems were used in rural and urban areas where the implementation of a centralized system was limited by population density, terrain, or distance from the wastewater treatment facility. These systems generally consisted of a simple septic tank and soil-based effluent dispersal system. Because the environmental impact of these systems was not appreciated, they were designed primarily for wastewater disposal. The septic tanks were buried without access ports or other features that would allow for monitoring and/or routine maintenance. In many instances, the septic tank effluent was directed to the nearest water course. Surfacing effluent from overloaded effluent dispersal fields was common along with innumerable other problems. As a result of poor design and installation, septic tank systems were viewed as a necessary evil and not as effective wastewater management systems from a regulatory perspective.

After the implementation of the Clean Water Act of 1972 in the United States, with the exception of remote and rural areas, the use of decentralized wastewater systems were viewed as a temporary measure put in place until a centralized wastewater collection system could be extended. However, in the past 15 years, it has become increasingly clear that many rural, peri-urban and urban locations will never be connected to a centralized collection system due to economic, logistical and political challenges. For these reasons, the U.S. Environmental Protection Agency has now accepted the premise that decentralized systems must be designed for permanent service and has initiated a number of new programs to enhance the reliability and performance of decentralized systems.

8.5.3 Modern development of decentralized systems

Since the mid-1980s, the implementation of decentralized wastewater management infrastructure has evolved due to significant improvements in technology, scientific understanding, materials of construction and design. Perhaps one of the most significant developments of the past 25 years is the septic tank effluent filter. By limiting the discharge of solids, the effluent filter has made it possible to use high-head water pumps, which has eliminated terrain as a controlling variable in the design of effluent treatment and dispersal systems. Improvements in the materials of construction are reflected in the development and use of inert materials (i.e., plastics of various types) not subject to corrosion for collection piping and tankage and other readily maintainable facilities such as the effluent filters, now routinely used in septic tanks.

Scientific understanding has progressed in the characterization of wastewater, the biology of wastewater treatment, the behaviour of soil treatment systems and the environmental impacts resulting from the discharge of partially treated wastewater. Understanding the complex mineral and biological system present in natural soil systems, it is now possible to use filtration through natural soils to achieve essentially complete removal of the residual trace organic constituents. The ability to remove trace constituents makes on-site systems a viable alternative to the conventional biological treatment systems employed at centralized facilities. Other technological breakthroughs include online instrumentation and remote monitoring, real-time alarm systems and programmable controls. Using these and other new technologies, it is now possible to design systems such as shown in Figure 8.6.

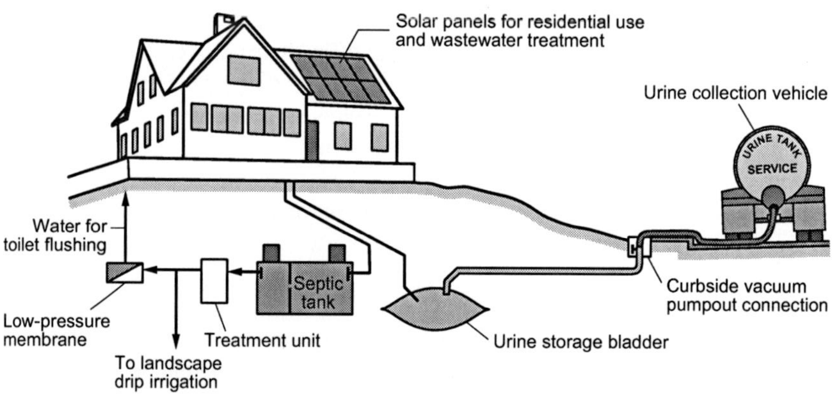

Figure 8.6 Diagram of home reuse system with urine separation and urine storage (Tchobanoglous and Leverenz 2008).

8.5.4 Advantages and disadvantages of decentralization

Modern decentralized systems also have a number of advantages over centralized systems including: (1) the use of shallow, water-tight infrastructure, not subject to corrosion, that can be installed, maintained and repaired easily; (2) the ability to eliminate stormwater and other inflow sources that can overwhelm centralized systems; and (3) the implementation of source separation in decentralized systems is relatively easy compared with the collection and management of separated waste streams with existing centralized collection systems (for an in-depth discussion, see Maurer 2013).

While there have been several important advancements in decentralized wastewater technology, a few inherent limitations that need to be addressed include (1) the need for flow equalization, (2) in many cases, energy input to small systems is also higher on a unit of flow treated basis due to the inefficiencies inherent in the operation of systems that must be overdesigned because they are not maintained or adjusted on an ongoing basis and (3) the physical footprint required for decentralized systems that are operated such that all reactions are carried out at natural rates.

8.5.5 Continued developments in decentralized systems

Much of the development related to decentralized wastewater systems still occurs because it is not cost-effective to extend the centralized collection system, particularly in low density rural areas. However, decentralized wastewater systems have been implemented even when centralized collection was an option. The examples identified below include existing developments on the urban fringe, in natural areas and in urban areas.

Developments on the urban fringe. In some cases, developments around or near the urban fringe have been required to implement a decentralized wastewater management systems instead of connecting to an existing centralized wastewater systems. The development and availability of utility management entities, better management infrastructure and limited capacity in existing centralized systems, have made large-scale developments that utilize decentralized wastewater systems possible. The option to develop a local water reuse system has also been an important driving factor in water short areas.

Developments in natural areas. It was common practice to develop unsewered areas with large lots, where the lot size was determined, in part, by the need for on-site wastewater dispersal. New concepts related to smart growth and sustainable community development call for using decentralized wastewater systems to preserve the character and ecology of natural areas by using clustered wastewater systems that allow for more compact development.

Developments within urban areas. Green building initiatives and rating programs have been used to promote the use of both satellite and decentralized wastewater systems, generally in the form of water recycling to reduce overall water use. In urban areas, satellite systems with a connection to the municipal system have been used, but developing standards have placed greater emphasis on the on-site management and beneficial reuse of all wastes (e.g., zero discharge). Green building standards have been developed for both new construction and retrofits.

chain may vary; in some areas individual household toilets may be the most appropriate while elsewhere shared toilets may have advantages. Very different incentives operate at each point along the value chain. Recent research suggests that households are most highly motivated by aspects of sanitation, which increase utility and convenience and contribute to a sense of prestige. For a comprehensive review of the literature see, for example, Jenkins and Sugden (2006). Local authorities, by contrast, may be more interested in meeting the requirements of public health legislation or in spending available budgets as quickly as possible. This mismatch of incentives was well illustrated by the work of the IWA taskforce on sanitation (Evans and Saywell 2006). The effect is that much public money is spent without increasing access to basic services for poor people.

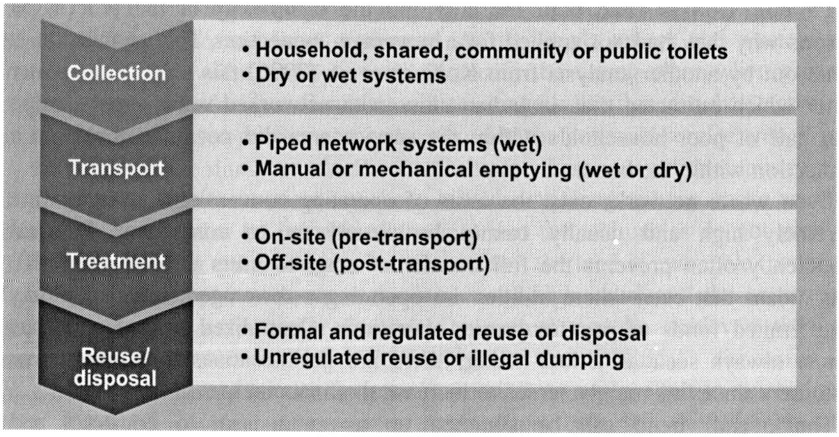

Figure 9.2 The sanitation value chain.

9.3.2 Vertical unbundling

One way of responding to this mismatch in incentives is to disaggregate the system *vertically* along the value chain. Vertical unbundling recognizes that communities may be willing and able to deliver the collection and transport elements of the system up to the community boundaries. They may even have an interest in managing reuse within the community, but they are unlikely to have much interest in the management to wastes generated from other communities.

The best-known example of this approach is the Orangi Pilot Project (OPP) in Karachi, Pakistan. During the early 1990s, the OPP Research and Training Institute supported local communities to plan, finance and deliver sewerage to more than 65,000 people in the large informal settlement of Orangi on the outskirts of Karachi. The approach had several interesting elements, notably the relatively long (two-year) planning period during which the needs and motivations of the community were identified, and the careful organization of communities in small groups to manage local elements of the system. The literature of New Institutional Economics identifies this ability to closely define the group of participants and to exclude others as contributing to the success of many common-property management regimes (Ostrom 1990, Ostrom *et al.* 1993). Specific examples include a farmer-managed irrigation scheme in Nepal and a community-managed ocean fishery in Turkey. It is worth noting that OPP had the advantage of being located on a hillside, allowing for gravity flow, reducing the need to engage with the city sewer network at least in the short to medium term.

Similar long-term success for community management has been demonstrated in the management of community sanitation blocks (providing water and toilets) in Dhaka, Bangladesh (under an approach piloted by a local NGO called DSK known as the Social Intermediation Model), in the construction and management of community toilets in India piloted by the NGO SPARC, and in the management of community toilets in Ghana amongst many other examples.

The key point here is that for institutional sustainability no particular technology is necessarily better than another – what matters is *its relevance to the community and the motivation and ability of the community to manage it.*

9.3.3 Horizontal unbundling

Vertical unbundling has the effect of *decentralizing control of part of the sanitation value chain.* In most of the sanitation examples cited above decentralizing control through community management at the local level was not the major challenge, the real challenge lay in articulating or linking a community-managed system to a wider network of services (main sewers or faecal sludge management).

To reduce this challenge, there is also the potential to unbundle urban sanitation *horizontally* (i.e. using decentralized systems in different areas of the city). Decentralization offers an antidote to some of the challenges of conventional centralized networks and can encourage treatment and reuse or disposal of the waste stream closer to the household. Decentralized networks can be planned and managed more flexibly in response to patterns of urban growth. Decentralized networks may be cheaper to develop, can be constructed in areas where people actually live even if that is far from the city centre and can be properly sized from day one. Horizontal unbundling also allows for the use of least-cost solutions, because it enables the deployment of a range of technologies depending on local conditions including housing density, ground conditions and tenure.

In recent years, there has been some move towards a more strategic deployment of unbundled technologies in urban planning frameworks; Wright (1997) and Tayler *et al.* (2003) demonstrated this in their writing on Strategic Sanitation Planning. Many cities have been supported to develop more flexible planning as a result. Good recent examples have been seen in Ghana, Burkina Faso, Sierra Leone and Uganda.

9.4 DECENTRALIZATION

The result of conceptually dividing a city up into management units depending on both incentives and technical feasibility is that many elements of the system can be developed independently. Community initiatives can become less dependent on city-wide actions, and finance for small elements of the system may be easier to mobilize. It also allows for incremental development (e.g., the addition of treatment to sewer networks developed by the community or the upgrading of shared facilities to household facilities later). Finally, it has the advantage that financial management is simpler – money may be mobilized and spent within the community cutting out the need for centralized fiduciary oversight.

However, there are challenges. While vertical and horizontal unbundling offer increased flexibility they often also require *skilled coordination*. The very different incentives within and outside communities (and, indeed, between different community actors – notably landlords and tenants) can create serious tensions. Some community initiatives simply shift the problem downstream which, while solving the severe public health problems within the community, may place new burdens on cash-strapped public authorities. Furthermore, locally-developed solutions may meet local needs and interests but may be far from "state of the art." A good example of this is in the area of source separation and reuse. Few communities will be aware of the potentials of reuse of treated excreta and, where knowledge does exist, it is often partial and resistance strong. The incentive to simply "get rid of the problem" may be dominant, as has been seen in OPP and its successors, and it is difficult to fault poor communities for feeling this way. External incentives may be needed to counterbalance this effect.

In summary, decentralizing sanitation through vertical and horizontal unbundling has the potential to be a powerful tool for increasing access to basic services particularly for poor people, who were previously excluded. However, strong governance and central planning, along with appropriate incentives are still needed to overcome knowledge gaps and local interests and to meet the public-interest dimensions of sanitation – for example improved treatment standards and increased capture and reuse of nutrients.

9.5 TECHNOLOGIES

Having established in principle that vertical and horizontal unbundling are of interest there still remains the question of what technical options exist for community level services and for their integration into the urban system. For a further more detailed discussion of some options, see Lüthi and Panesar (this volume). In this section, we will briefly review some of the options available to increase access to sanitation services in different types of urban spaces including the use of source separation technologies.

In peripheral growth areas with plenty of space, there is the potential to move towards well-managed on-site sanitation options. It is in these areas that the best opportunities currently lie to explore options for source separation, household level composting latrines, and other options that may appeal to households with space to make use of by-products for urban agriculture.

Where such areas already have high numbers of existing on-site facilities in poor condition, some improvements may be possible through the introduction of urine separation. Scheuen *et al.* (2009) suggest that this remains a rather expensive option in many parts of the global South due to the cost of the technology and the costs of marketing the new approach to users, but costs may fall as these systems become more widely known. Sometimes, the best option may revolve around improved management of the pit waste (either separated at source or, where this is not possible, combined) and its treatment as close to the area as possible (since the costs of transporting wet pit wastes are extremely high).

For urban slums and peripheral growth areas with a reasonable layout and road access community toilets, often combined with water points, bathing blocks or other community facilities, are often a good option. Community toilets, which should not be confused with public toilets, are managed, as the name suggests, by the community and use is usually restricted to community members. The most effective management model has generally proved to be the payment of a regular (monthly or weekly) fee by participating households to pay the salary of a caretaker and to pay for cleaning materials and other running costs (WSUP, 2011).

Where space allows shared or individual household toilets are usually better from a household perspective. The trade-off may lie between household facilities which offer greater utility and provide what people are looking for in terms of convenience and privacy and community facilities which may offer more opportunities for nutrient recovery and reuse because of better access and greater availability of space. In either case, attention to what happens downstream is important. Where space and social conditions allow it, use of technologies which encourage reuse within the community may be attractive as they reduce the amount of coordination required with the city system.

In extremely dense inner city slums and infills space is at a premium (in Dhaka in Bangladesh, for example, landlords not only rent out rooms, but also beds and sometimes tenants must time-share the use of a bed in line with factory shift work). This limits the potential for stand-alone (on-site) sanitation systems (with or without source separation and/or reuse) which require both space for construction, access for emptying and in the case of source separation systems, room for storage and a convenient local market for the product. In many such cases sewerage is the best, and often the only option. The least-cost technologies are generally simplified sewerage or low-cost combined sewerage. Both these options can collect and convey excreta out of the housing area which can then be conveyed to a treatment/ reuse facility – either at the centralized level (often via a connection to an existing conventional sewer network) or at the decentralized level. Sewerage has been very successfully used in conjunction with nutrient and energy capture at the treatment stage.

The main lifetime costs of conventional sewerage are associated with pumping sewage from large depths for treatment or disposal. Simplified sewerage reduces or eliminates these costs by using shallower minimum depths and lower minimum gradients. At high densities, savings can be highly significant (see for example Box 9.2 which compares lifetime costs of on-site sanitation, conventional sewerage and simplified sewerage in Northeast Brazil in 1983).

Box 9.2 Relative costs of sanitation options

Figure 9.3 shows the relative costs of installing and operating conventional sewers, simplified sewers and on-site systems in the low-income peri-urban areas in the city of Natal in northeast Brazil. In this particular case, simplified sewerage became cheaper than on-site systems once population density reached 160 persons per ha (a relatively low level). Note for consistency treatment costs for faecal sludge and/or wastewater for all three options have been excluded. Sewerage costs in both cases are for separate sewers.

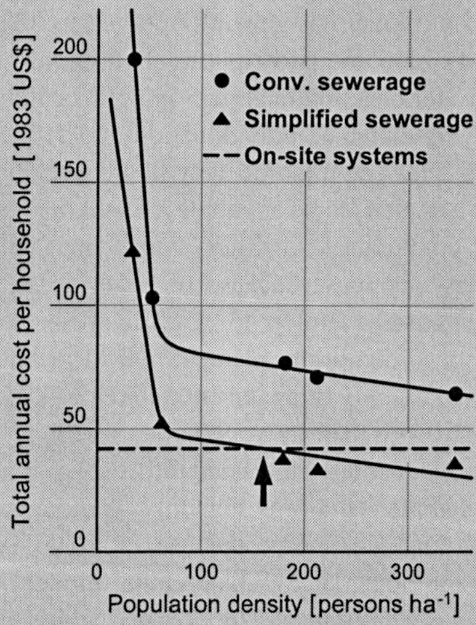

Figure 9.3 Cost comparison.
Source: Mara (1996), WSP (2005)

Finally public toilets, which are open to any passing member of the public, are a useful additional service level in all urban areas, commonly financed by a per-use fee. When properly managed, public toilets have their place in providing sanitation in public areas (markets, bus stations, hospitals etc.) but are not an appropriate solution in slums, where they have generally proved to be too costly and difficult to manage well. Source separation is likely to be quite challenging in public toilets, however, because users may lack familiarity with the system. Other methods to recover nutrients and energy may be possible; biogas reactors for example have been used with some success in community toilets in India, Kenya and Ethiopia.

9.6 CREATING INCENTIVES

Conventional sewerage remains the predominant paradigm for urban sanitation delivery and limited progress has been made to introduce innovations in terms of decentralization, source separation or other non-conventional approaches. Many factors have been put forward to explain this. There is, however, little hard evidence to say, whether this is mostly caused by technical inertia, risk aversion, corruption (and, hence, the preference for high-cost schemes with limited local accountability), political expediency (the need to be seen to be doing something), liking for systems, which are perceived as "modern," or simply from a lack of knowledge. What is clear is that whatever incentives currently exist they tend to encourage local and central authorities and their advisors to stick to conventional top-down planning and conventional centralized sewerage. It is not our purpose to examine how *these* overarching incentives can be changed, but rather to look at what a willing innovator at the *city level* could do to change incentives for communities and local government.

9.6.1 Contractual incentives

One way to create incentives to operators is through the contract under which they operate. This may be a true contract in the case of a private operator (utility or small third party provider) or through a public-service agreement with a public service provider. Evans *et al.* (2002) summarized some of the ways in which contractors could be incentivized as follows:

- Structuring fee payments such that they are linked directly to the numbers of people provided with specified services
- Measuring service delivery rather than specifying types of infrastructure to be built thus encouraging use of cost-efficient technologies and reducing financial waste
- Creating service areas with a requirement for 100% service provision and introducing penalties when coverage falls below this level
- Encouraging services to be delivered through third-parties so that operators can make use of the opportunities for vertical unbundling without losing out on income
- Linking some part of the payment to achievement of public-interest outcomes such as reuse of nutrients and/or energy recovery

Despite much work in this area there have been few documented cases where private operators have been faced with positive contractual incentives to deliver appropriate services to poor people or to be innovative in the use of technologies with public-interest outcomes. This suggests, the first intervention may be to create financial or political incentives for *local governments* to improve access to low income communities, who are currently totally excluded from formal sanitation service provision and pass on these incentives through the contract.

9.6.2 Financial incentives

Financial incentives can be created in a number of ways. Micro-finance and revolving funds have the potential to reduce barriers to household participation in sanitation. Micro-finance and appropriate financial and business services could also encourage more entrepreneurs to enter the market which, in turn, might create incentives for technical innovation and investments in financially sustainable technologies using, for example, source separation. Useful analyses of experiences with sanitation subsidies are available in Evans *et al.* (2009) and Tremolet *et al.* (2010). Business development services for water and sanitation entrepreneurs are discussed in Mehta *et al.* (2006).

Another area is in the use of output-based subsidies. Output-based subsidies pay service providers for the successful delivery of working services. Paid in arrears they have the effect of transferring performance risks to service providers who must successfully construct and operate the elements of a sanitation system before they will be paid. Output-based subsidies can be compared with traditional input-based subsidies which might, for example, fund the construction of a wastewater treatment facility even if no-one is connected to it. In contrast, innovative financing and contracting arrangements can transfer responsibility for providing services to a service provider and encourage the use of cost-efficient decentralized technologies, source separation or, most importantly, the successful delivery of services. For a longer discussion on output-based subsidies and their application to sanitation see Tremolet *et al.* (2010).

9.6.3 Political incentives

Political incentives can best be created through information and empowerment. Decentralization has the potential by itself to be empowering but only if communities are appropriately supported. Participatory planning is one tool, but has limited impact, unless it is backed by the power to change investment decisions.

Decentralized sanitation relies heavily on community engagement as we have seen. We have also noted that communities and their advisors can benefit from access to information on the state of the art in terms of sanitation technologies – see for example Tilley *et al.* (2008). A greater investment in participatory planning could certainly result in better and more flexible urban sanitation plans. For further discussion of this aspect, see Lüthi and Panesar (this volume).

Another area of work could relate to changing national and international targets and definitions of access. Recent work by UN Habitat showed that by tightening up the definitions of acceptable levels of access to water in towns around Lake Victoria in East Africa, reported levels of coverage dropped significantly from those cited by the Joint Monitoring Programme (Graham Alabaster – pers. comm.). Such a shift could lead countries to re-evaluate the investment needs of the urban sanitation sector providing urgently-needed capital funding.

Finally, one of the major constraints to increasing access to sanitation and deploying decentralized community-based solutions is that access to sanitation is often implicitly or explicitly linked to tenure status. Many utilities and local governments claim that they have no responsibility for service provision in areas, where land tenure is uncertain or where communities are illegally settled. However, this objection often turns out to be based on misunderstandings of the legal framework and is more commonly used as an excuse for inaction than as a genuine reason not to act. The simple act of acknowledging rights to services irrespective of tenure status could empower many communities to take action in their own right to invest in sanitation.

9.6.4 Professional incentives

Another area where incentives can be shifted relates to technical norms and standards. In many countries technical norms and standards prevent innovation by describing in minute detail the input-requirements for conventional sewerage. Technicians working in public utilities or consultancy firms then face no incentive to innovate, even if solutions which rely on conventional sewerage are prohibitively expensive or fail to result in households receiving services. The change in the technical standards in Brazil, which resulted in the widespread adoption of simplified sewerage, has had a significant effect on bringing down the costs and therefore increasing access to urban sanitation. Similar changes could potentially shift incentives in other countries to encourage development of solutions that deliver a workable service to all. One effective way of doing this could be to re-write technical norms in terms of outcomes (describing a working *system* rather than an acceptable *infrastructure*).

9.7 SUMMARY

Chaotic rapidly growing poor cities in the global South face different challenges to those faced by the post-industrial cities of the north. Rapid rates of growth, a very low endowment of existing infrastructure and weak institutions represent a significant challenge. Because access to services is so low, and the public health imperative is so urgent, a much stronger focus is needed on sustainably scaling up access to basic services (the household amenity) as compared to developed cities, where the focus is often on optimizing environmental management.

For most poor urban communities achieving sustainable access to sanitation is mostly about finding systems, which are affordable and which can be managed at the local level (assuming that the city is unwilling or unable to manage such services itself). Downstream impacts and wider environmental protection are of little interest. Household and community access can only be achieved, however, if the entire urban sanitation system is viable and it is at this city level that the planner can focus most attention on securing the *ecological* sustainability of the system both by designing the shared elements of the system and by creating incentives for better decision making at the local level.

Urban sanitation lends itself to decentralization through two processes; vertical unbundling, which means that management of different elements of the system can be organized at the lowest appropriate level and which can relieve over-stretched cities of responsibility for operating the entire system right down to the household level; and horizontal unbundling which means different solutions can be applied in different parts of the city depending on the geographical and socio-economic conditions.

Taken together these two processes mean a shift away from centralized conventional sewerage towards offering a flexible range of technology solutions for people living in varied conditions. This offers the opportunity to rapidly scale up access (towards international targets) as well as shift gears towards a more sustainable approach to sanitation which minimizes water and energy use and maximizes productive use of nutrients.

A range of technical solutions already exist which can be deployed in such a decentralized framework including better managed traditional on-site technologies, source separation and resource management at the local level, appropriately managed community toilets, simplified sewerage and low-cost combined sewerage. Where waste resources cannot be managed close to the household, better nutrient and energy recovery from faecal sludge and wastewater treatment in decentralized or even centralized treatment facilities is an option that can be deployed in tandem with decentralized community collection of waste streams. More technological options may remain to be identified and fully tested (sewers carrying urine from source separation, for example), but in the meantime, better deployment of the known options would be a good first step for most cities.

For this to happen, however, cities must first address and then adopt a new set of contractual, financial, political and professional instruments to create the right incentive structure. This would encourage a replacement of the existing paradigm, which limits service delivery to the few and results in an ongoing cycle of non-sustainable investment. Socially sustainable, affordable and workable systems for sanitation service provision to all citizens could be the result.

REFERENCES

Diacon D. (1997). Slum networking – an innovative approach to urban development, building and social housing foundation, Coalville, UK. ISBN 0 9522278 6 X

Economist (2010). Don't Mess with British Bins, 29 July 2010.

Evans B., Kariuki R. and Brocklehurst C. (2002). New designs for water and sanitation transactions: Making private sector participation work for the poor, PPIAF World Bank, Washington DC, USA.

Evans B. and Saywell D. (2006). Sanitation 21: simple approaches to complex sanitation – a draft framework for analysis, International Water Association, London, UK.

Evans B., van der Voorden C. and Peal A. J. (2009). Public funding for sanitation: the many faces of sanitation subsidies, Water Supply and Sanitation Collaborative Council, Geneva, Switzerland.

Komives K., Halpern J., Foster V. and Wodon Q. (2006). The distributional incidence of residential water and electricity subsidies. World Bank Policy Research Working Paper 3878, World Bank, Washington DC, USA.

Jenkins M. and Sugden S. (2006). Rethinking sanitation; lessons and innovation for sustainability and success in the new millennium, Human Development Report Office Occasional Paper, UNDP, New York, USA.

Komives K. (1999). Designing propoor water and sewer concessions: early lessons from Bolivia, Policy Research Working Paper Series 2243, World Bank, Washington DC, USA.

Mara D. (1996). Lowcost Sewerage, John Wiley & Sons, Ltd, Chichester, UK.

Mehta M., Virjee K., Evans B. E. and Wathobio K. (2006). Business development services for community-managed small water enterprises. World Bank Water and Sanitation Program, Nairobi, Kenya.

Morella E., Foster V. and Ghosh Banerjee S. (2009). Climbing the ladder: the state of sanitation in Sub-Saharan Africa, World Bank, Washington DC, USA.

Ostrom E. (1990). Governing the commons: the evolution of institutions for collective action, Cambridge University Press, Cambridge, UK.

Ostrom E., Schroeder L. and Wynne S. (1993). institutional incentives and sustainable development – infrastructure policies in perspective, Westview Press Colarado, USA, ISBN 0-8133-1619-7.

Schuen R., Parkinson J. and Knapp A. (2009). Study for financial and economic analysis of ecological sanitation in Sub-Saharan Africa, World Bank Water and Sanitation Program, Washington DC, USA.

Tilley E. (2013). Conceptualising sanitation systems to account for new complexities in processing and management. In: Source Separation and Decentralization for Wastewater Management, T. A. Larsen, K. M. Udert and J. Lienert (eds.), IWA Publishing, London, UK, pp. 225–239.

Tilley E., Lüthi C., Morel A., Zurbrügg C. and Schertenleib R. (2008). Compendium of sanitation systems and technologies. Swiss Federal Institute of Aquatic Science and Technology (Eawag). Dübendorf, Switzerland. ISBN: 978-3-906484-44-0.

Tayler K, Parkinson J. and Colin J. (2003). Urban sanitation: a guide to strategic planning, Practical Action Publishing, Rugby, UK.

Tremolet S., Evans B. and Schaub-Jones D. (2010). OBA for sustainable sanitation global partnership for output-based aid, Working Paper Series, Paper No. 10, World Bank, Washington DC. USA.

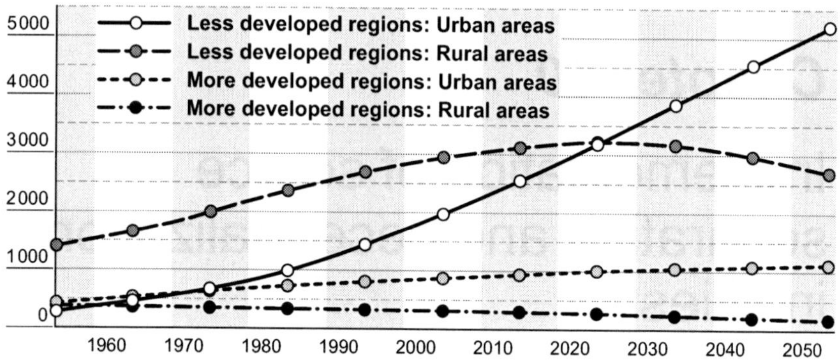

Figure 10.1 Global development of urban and rural populations in millions (adapted from World urbanization prospects, the 2005 Revision).

Source separation is not new; it has long been propagated as an inexpensive and environmentally friendly technology for the poor, especially in rural areas. However, as indicated above, population growth takes place in cities, so the severe consequences of poor sanitation on public health and the environment must also be dealt with in urban contexts. It is no coincidence that source separation and decentralization have historically only been considered in rural areas. A number of technical and organizational problems arise when these technologies are introduced into urban areas, and although it is often claimed that lack of sanitation is not a question of technology, but rather of organization, we believe that this is not true. Setting up resource-efficient *urban* water and waste management on a global scale will require the co-evolution of new technology *and* new organizational structures.

Although the problems of urban water management in industrialized countries are dwarfed by those experienced in developing and rapidly industrializing ones, we also expect the industrialized world to profit from source separation and decentralization. Some of the severe problems normally associated with the global South are also encountered in industrialized countries, especially water scarcity and the difficulties of nutrient management (Cordell 2013; Erisman and Larsen 2013).

When developing decentralized sanitation principles for urban areas, we have to be aware that this is not primarily a problem of making available a new type of toilet or small wastewater treatment plant, but rather of developing an entire cluster of technologies. Factors such as awareness by administrators, legal requirements, rules of trade, technology, organizations for construction and operation, acceptance by engineers, architects and the public, economic competiveness with alternative technologies, must all interact in order to make new sanitation principles possible (see also Truffer *et al.* 2013). The development of such principles is a question of decades. It cannot be generated in a single technology cycle but will require improvements and adaptations over several cycles. Those who work in this area should always remember that the development of today's centralized sanitation systems took over 100 years and they are still not perfect.

Nevertheless, we are convinced that the process of developing decentralized sanitation systems will be much faster than the previous process of developing centralized ones. In fact, we are currently observing a very rapid transition in the community of environmental engineers. We presented our first paper on source separation in 1996 (Larsen and Gujer 1996) at an international conference. At that time, our colleagues did not take us seriously, and only very few friends felt the obligation to follow the presentation. Today, this paper has been cited more than 100 times, which indicates a growing community of scientists interested in different aspects of source separation and decentralized sanitation.

Do not expect to find ideal solutions for decentralized urban sanitation in this chapter, but rather accept that it will take decades to develop such perfection. However, we are convinced that the time is right to start to implement such technologies on a large scale. Our main aim is to point out what is needed for such a development to occur and suggest ways of transitioning from the existing system to a new paradigm of urban water management.

To summarize: environmental engineers, city planners, architects, administrators and others must accept that the sanitary solutions applied in the industrialized northern countries, which is based on centralized sewer and wastewater treatment technology, cannot be the global solution to the sanitation crisis. Decentralized solutions are not downscaled wastewater treatment systems but are rather a new field of technology which requires ingenuity and in many cases the industrial logic of highly engineered mass-produced apparatus. First ideas exist, and this book tries to identify some of them, but many more will have to be developed. In view of the rapid population growth in areas with water scarcity and in many cases still poor sanitation, this sector may well develop into new economic dimensions.

10.2 THE MAIN ADVANTAGES OF SOURCE SEPARATION AND DECENTRALIZATION IN CITIES

The title of this book contains the twin concepts of decentralization and source separation. We believe that their coupling is necessary in order to capture the advantages of both in a new paradigm of urban water management (see also Larsen *et al.* 2009).

On a global scale, water scarcity is the main driver for change (see Falkenmark and Xia 2013). Water saving and reuse will obviously be necessary in water-scarce cities, but the consequences for the systems involved are huge. In centralized settings, problems of acceptance complicate the mixing of purified wastewater with drinking water (as is attempted in Singapore with the concept of "New Water"), and setting up a second piping system for recycled wastewater is expensive. Local recycling of greywater and/or water saving measures often seem more appropriate. The sewer system, however, depends on large amounts of water for transport of feces. With increased water saving and/or local recycling measures, larger amounts of sediments are observed in the sewers, leading to a number of problems such as odour, corrosion and increased "first flush" pollution loads during overflows of combined sewers. Feces make up the single most microbiologically active component in combined sewers and contribute substantial amounts of sediments. Only with their decentralized treatment can transport be made independent of the amount of wastewater.

As shown by Maurer (2013), planning insecurities and the construction of sewers ahead of residential properties may lead to more expensive solutions due to unused capacities. Where cities are expanding too rapidly, planning may even become impossible. Sewers are typically overloaded in rapidly growing cities, resulting in permanently active CSOs. Decentralized solutions follow the construction of new buildings and thus make better use of invested capital. Due to the shorter life cycle of decentralized solutions, new technology and environmental requirements can be implemented more rapidly.

So why will decentralized solutions not be enough? Without source separation, decentralized technology will necessarily represent no more than downscaled advanced treatment plants, and as discussed in Section 10.3.2, these are far too complex for decentralized implementation. Decentralization thus demands source separation in order to reduce the complexity of the treatment process.

However, source separation is also valid on its own terms, especially for resource recovery. Water recovery from greywater occurs spontaneously due to water scarcity, but nutrient recovery from human excreta is also increasingly recognized as an important feature for a sustainable future (Cordell 2013; Erisman and Larsen 2013). As discussed by Jönsson and Vinnerås (2013), nutrients from cities can help boost agriculture wherever nutrient import is too expensive. Only with source separation and adequate on-site treatment will pollutants turn into resources. However, the transport of source-separated fractions in sewers is only partially possible (see Section 10.3.1), and source separation in itself consequently also creates a demand for decentralized solutions.

10.3 CHALLENGES OF SOURCE SEPARATION AND DECENTRALIZATION IN CITIES

Part II of this book is dedicated to the challenges of source separation and decentralization. These challenges are obviously greatest in cities, where the paradigm of sewer-based centralized wastewater management counts as the only efficient way of dealing with sanitation. Most authors of this book are city dwellers and it is difficult to believe that we can live without the usual comfortable water-based "use-and-forget" system for personal and urban hygiene. As discussed by Lienert (2013), however, even city dwellers are ready to accept change in their bathrooms, as long as their personal comfort is not compromised and the costs are acceptable. This means that the user interface must be adapted to local conditions in order to match the expectations of the users.

Furthermore, the technologies must be energy-efficient (Rittman 2013), demand little maintenance (Maurer 2013), fulfill the hygienic requirements (Stenström 2013), and if nutrient recycling is practiced, the logistics of nutrient transport from producers to users is a major topic (Jönsson and Vinnerås 2013). The sheer mass of people in urban areas also demands more effective water pollution control than in rural areas.

Finally, it is obvious that solving global water-related problems constitutes an enormous task for the international community. As pointed out by Wilderer (2005), it is close to impossible to build sewers and treatment plants at the required pace. Neither will decentralized and source-separating technology just pop up from nowhere. The socio-economic pathways which may make such developments possible are discussed by Truffer *et al.* (2013).

This book was clearly produced with the idea of presenting the main advantages as well as the main challenges of a new paradigm based on source separation and decentralization, especially for cities. Although it contains few recipes, these topics are dealt with by competent colleagues, and we will not summarize the various discussions in this chapter. Instead, we will discuss three main engineering aspects of a paradigm change for urban water management ("the three Ts"):

- Transport of water, pollutants and residues
- Treatment process development, operation and monitoring
- Transition from previous system design to new designs

10.3.1 The challenge of transport

Centralized treatment of wastewater requires the transport of large amounts of wastewater over longer distances. Traditional urban water management is one of the largest transport enterprises of Western societies: each year over 100 tons of wastewater per person are typically transported over many kilometers.

Wastewater transport relies on large sewer systems which have been developed over decades and must operate under both dry and wet weather conditions. Design rules require sanitary sewers to be designed for twice the peak daily dry weather flow. In order to keep sediments under control, the peak daily shear stress should reach $2\ N{\cdot}m^{-2}$ and in order to facilitate operation the minimum diameter is normally prescribed as 0.2 m. With water consumption of $0.2\ m^3{\cdot}cap^{-1}{\cdot}d^{-1}$, such a sewer may serve about 1000 inhabitants, and transport via sewers is a well established and efficient technology for the job it does. During periods of severe water scarcity, where only 0.03 $m^3 \cdot cap^{-1} \cdot d^{-1}$ of wastewater are generated (as suggested by Truffer *et al.* 2013), the same sewer would have to transport the wastewater of nearly 10,000 inhabitants. In smaller catchments, only increased maintenance, frequent sewer flushing, or alternatively pressure or vacuum systems could alleviate the sedimentation problems.

Thus, gravity sewer systems suffering from water scarcity will run the risk of sediment accumulation, hydrogen sulfide production and associated corrosion problems. Decentralized water reuse as now often seen in water-scarce areas leads to very low net water consumption, leading to a number of problems (see Tchobanoglous and Leverenz 2013).

Decentralized treatment produces waste streams and residues in different amounts and of different compositions. Some streams may well cause new problems (e.g., urine causes scaling, which must be inhibited with extra effort; see e.g., Udert *et al.* 2003) or must be transported rather inefficiently as concentrated liquids. Reducing volume and weight is thus of prime importance for the development of efficient transport systems for decentralized sanitation (Maurer *et al.* 2006). Local use of *all* residues and treated waste is not possible in modern cities even with large-scale urban agriculture and vertical gardening.

Figure 10.2 compares the mass of waste streams and residues from different waste-handling technologies. Clearly, if decentralized sanitation systems were to produce dry residues, collection together with (but separated from) solid waste would be possible. The transport of fresh urine and diluted feces from flush toilets is feasible only over short distances (as is the transport of thickened sludge from small wastewater treatment plants to larger sludge-handling plants). Suitable collection systems will therefore have to be developed, on the same lines as for green waste, where standard solutions already exist in many European countries, including Switzerland. Small residue volumes are obviously preferable, but even more importantly these must be hygienic and more or less odourless. For instance, transport of fresh feces separated in a dry toilet is not difficult as regards their mass, but rather from the point of view of hygiene and odour.

For implementation in cities, decentralized or even on-site volume reduction and stabilization of the waste fractions thus becomes a primary research goal. Since centralized treatment of combined wastewater has been the main focus of wastewater treatment for many decades, thinking in terms of decentralized treatment of separate waste(water) fractions is rather new within the professional community. From our own experience in the area of urine treatment, we have seen how tedious it was to build up the necessary knowledge and how few ideas we had in the beginning. With time and as some routines were built up in the lab, this sector became just as interesting and productive as any other.

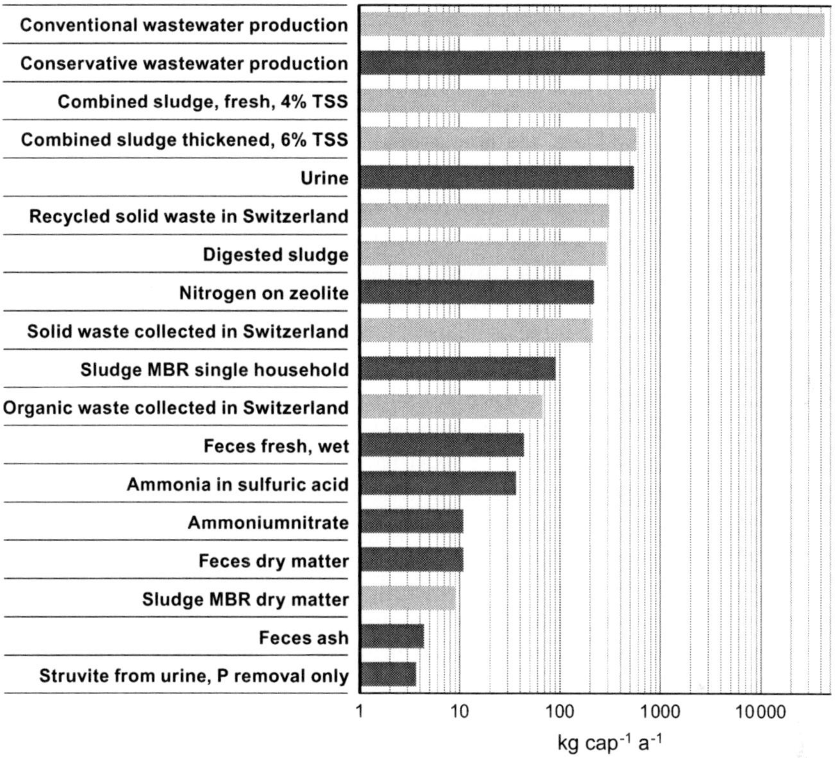

Figure 10.2 Rough comparison of residues from different waste treatment and collection activities (light grey: relating more to conventional wastewater treatment; dark grey: relating to source separation).

10.3.2 The challenge of developing treatment processes

Although we do not normally realize it, centralized wastewater handling infrastructures are based on complex systems. Furthermore, this type of technology depends on the design of prototypes, requires on-site construction, and unfortunately cannot make use of mass production and highly intensive engineering. Modern wastewater treatment relies on sophisticated technologies whose design and day-to-day operation must be left to experts. Sludge handling becomes an ever more complicated task, and research tends towards the integrated modeling and operation of such systems. New challenges such as water scarcity and climate change lead to increasingly complex and resource-intensive management tasks. One example is the dosing of nitrate in sewers threatened by corrosion due to water saving and increased temperatures, calling for complex real-time models in order to avoid overdosing (Mohanakrishnan *et al.* 2009).

Decentralized source-separating systems may appear to be even more complex, but this is not necessarily the case. They will rely on many individual elements, but the goal must be for each element to be easy to operate and adapt to local requirements. Well-engineered small source-separating systems based on mass production require new ideas and new approaches to the problem. A number of these technical ideas are discussed in Part III of this book.

As presented by Larsen and Gujer (2001), one possibility is to combine waste management directly with the apparatus that creates or handles the waste. At that time we used a washing machine as an example, but today a NoMix toilet with phosphorus recovery (e.g., as struvite) and nitrogen elimination (e.g., in an electrochemical cell) seems quite as realistic.

Historically, civil engineers were responsible for the development and design of wastewater treatment processes, whereas today this job tends to be the domain of environmental engineers. These engineers are trained to design large prototype plants, specifically adapted to highly local requirements. In addition, owners of centralized infrastructure investments tend to be conservative and rather risk averse. Under these circumstances, possible investments in the development of technologies are limited, so that progress in this area is quite slow everywhere. Most treatment technologies center around biological processes, sometimes enhanced with chemical precipitation or oxidative disinfection. The recent introduction of membrane technology is no exception; the workhorse of biological treatment still remains in place.

Small-scale decentralized treatment plants are quite different from centralized ones. Rather than single large prototype plants, thousands of small identical units will have to be built. Their construction must be based on industrial logic, optimum engineering combined with efficient industrial production. If environmental engineers want to remain key players in this area, they will have to learn from industrial design and mechanical engineering. A smart phone is not a downsized telephone, TV, photo camera, computer, CD player, and so on but a new device which fulfills its tasks on the basis of entirely new technology and with considerably less material and at less cost than all these gadgets together. Such engineering investment becomes attractive because the result will be produced by the millions. In addition, it was possible to think in generations: networks and technology called 1G, 2G, 3G and soon 4G ("G" for generation) made rapid progress possible and were based on great leaps forward.

If we really want decentralized sanitation to become a success, we must step back and analyze the problem to be solved as a whole. Our task is not to treat a small amount of wastewater on the basis of our experience, it is to provide convenient, efficient, and reliable sanitation with due consideration of local boundary conditions such as water scarcity, environmental requirements and local cultural realities. Once our task has been defined, we can try to devise solutions, which must consider acceptance, function, production, operation and economy.

This way of thinking can best be illustrated in an area in which most "conventional" process engineers (as we consider ourselves) have little experience, that is, in informal settlements. An Eawag team developed a technical and socio-economic concept along these lines (www.diversionsanitation.com) for the "Reinvent The Toilet Challenge" (RTTC) competition of the Bill and Melinda Gates Foundation (BMGF). The concept is based on a urine-diverting dry toilet, but is extended with the possibility of flushing the toilet after diverting dry feces and undiluted urine. Water can also be used for anal cleansing and of course for hand washing. The used water will be treated and recycled on-site. The toilet will be shared between two families and linked to a Resource Recovery Plant (RRP) through a logistics concept for feces and urine. A number of technologies along the lines discussed in Part III of this volume can be implemented at the RRP, depending on the local conditions (see Larsen *et al.* 2010 for a systematic approach to linking technology choice to socio-economic conditions). The business model is adapted to typical conditions in high-density slums, and obviously looks totally different from a possible business model in a western industrialized country. However, the technology choices could be very similar.

There are several reasons why we are convinced that our approach will be successful. First of all, we are cooperating with an industrial designer (www.eoos.com) for the design of the toilet. Designing a toilet is a highly complex task, and just as we would never expect non-professionals to develop modern process engineering, process engineers simply cannot devise an attractive toilet design. Since the toilet will be produced in millions (if successful), the design costs are not significant in the overall process. Secondly, its embedding in the respective socio-economic reality is handled by experts from the Eawag department for Water and Sanitation in Developing Countries (Sandec) (http://www.eawag.ch/forschung/sandec/index_EN). Only with the participation of relevant stakeholders can the business model and technology be adapted to realities, which are so far from our own experience. Finally, we can base our technology choices on more than a decade of experience, both for the water recovery unit and the Resource Recovery Plant (RRP), and the business model is also in the hands of an expert.

Of course, we still do not have any proof that our concept will work, and perhaps it will not. On the basis of Eawag's extensive experience in the area of centralized urban water management, however, we see this project as a good example of the potential of source separation. It *is* possible to come up with just as professional solutions in the area of decentralized source-separating technologies as we are used to from centralized treatment. The approach must be more holistic and involve new and different experts and stakeholders, but the perspectives are extremely attractive.

A rather less successful approach would be to accept today's standard sanitary equipment and water use in households and then to develop a downscaled wastewater treatment plant based on today's technology in centralized systems. Unfortunately this is the principal approach currently visible in industrialized countries, but some actors are gradually realizing that it may not be competitive in international settings (see Truffer *et al.* 2013). In the next section we will discuss opportunities for the development of source separation and decentralization in other parts of the world.

Table 10.1 identifies some characteristic differences between decentralized and centralized treatment processes. It becomes obvious that the differences are large and require alternative approaches to decentralized systems. Today, it is worthwhile to start with a specification of what services the product or process should provide.

Load variation in small waste treatment systems may be extreme. Even single events may affect the performance of very small systems. A longer absence of load (vacation periods), short extreme peaks (at house parties) or dumped medication (antibiotics) may affect the treatment processes. Thus the design of such systems must be robust and their operation must be self-adaptive.

Biological treatment is particularly sensitive to variable loads and may require the extra complexity of load equilibration by storage. It is well possible that the discussion of "biological" versus "physico-chemical" will become topical again. In the 1950s, the answer was "biological," and as a consequence the research and education of the following generations of wastewater professionals emphasized biological process engineering.

Early experience with nitrification demonstrated its tendency to become saturated (to reach a maximum capacity, the maximum growth rate of relevant organisms; see also Gujer 2010). This led to today's design principles for biological treatment, which are typically based on the concept that the maximum required treatment capacity is related to mean loading (which defines the available mass of microorganisms) multiplied by a peak factor for extreme loading. When applied to small units, this will result in rather large and therefore costly reactors. Start-up procedures for such systems are lengthy (they may extend over several solid retention times) and might require costly surveillance even after single extreme events. The fact that microbial processes are autocatalytic (biomass is reproduced in the process itself) is an advantage in centralized plants where the biomass is produced from resources in the wastewater and load equilibration is based on a large number of loading events. In small systems, however, autocatalysis may become counterproductive since load variation is extreme and the time constants for adaptation become too long.

Many physico-chemical treatment processes have a tendency to linear behavior (first order type reaction) or may even be enhanced by increased use of energy, chemicals, flow rates, pressure or membrane surfaces. So this group of processes should be investigated carefully. A disadvantage of physico-chemical processes is the lack of biological stability of the effluent because they typically involve no selective removal of biodegradable organics. Re-growth of microbes may consequently be a severe problem in such systems. However, for nutrient removal there are valid and robust alternatives to microbial processes.

Table 10.1 Characteristic differences between decentralized and centralized wastewater treatment systems (see also Olsson 2013).

Topic	Properties of decentralized systems	Properties of centralized systems
Waste flow and load	Highly variable, subject to individual events	Variable, but individual events not apparent
Rainwater	Hardly an effect	May define hydraulic design load
Waste composition	Rather homogenous conditions between plants Rather concentrated waste	Different for each plant, subject to individual industries. Rather dilute wastewater.
Frequency of attendance	Irregular, long intervals	Daily to permanent
Cost of intervention	Large	Relatively low
Relative cost of sensors	High	Rather low
Calibration of sensors	Very low frequency and relatively very costly	Costly, but rather frequent
Sensor properties	Must be rugged and reliable, accuracy is of secondary importance, very infrequent maintenance	Must be sensitive, accurate and reliable but may require frequent maintenance
Data transmittance and control system	Due to on-going expansion of the number of systems, elements must be based on an adaptive grid	Typically fixed for one technological cycle
Control software	Highly standardized, but due to application in large numbers also highly optimized	May rely on modular design but adaptation to a specific plant typically required
Required process standardization	Very high, only standardized equipment can be produced in large numbers	Individual plants are typically designed as prototypes
Transport of pollutants and residues	Local extraction of concentrated residues and separate transport	Transported in sewers and extracted in the form of concentrated sludge
Handling of residues	May be centralized. An intermediate form may be transported to a central handling station	Typically occurs at the plant. Only small plants connect to larger ones

The future technology for decentralized sanitation may have to combine physical, chemical and biological processes:

- Physical adsorption for load attenuation
- Membrane filtration as a barrier
- Microbial processes for regeneration of adsorption capacity and in order to achieve biological stability and oxidation
- Chemical precipitation, oxidation as well as photo- and electro-chemical transformation of critical compounds could result in interesting and robust options which can deal with substantial load variation

The operation of decentralized technology requires centralized intelligence. Modern cars are complex machines which are used daily by millions of people and provide extremely reliable and convenient service. Their efficient use depends on infrastructure and centralized intelligence (roads, car dealers, service stations, insurance companies etc.). The possibilities currently offered by sensors, data transmission, remote monitoring, and so on permit the development of highly integrated systems which will support the introduction of decentralized sanitation technology (see also Olsson 2013). Some of the problems involved in the start-up of such systems are discussed by Truffer *et al.* (2013).

10.4 TRANSITION

When we started considering source separation in the mid-nineties, our first concern was a possible rapid implementation in urban environments. Our main proposal was to store urine in-house and to discharge it during the night over a short period of time in order to extract a concentrated urine solution at a centralized facility (Larsen and Gujer 1996). The argument was that in some settings, this approach could be more interesting than the expansion of a treatment plant for nutrient removal. We discussed a stepwise introduction of this approach with increasing advantages as more and more users participated in the system.

Our ideas have since been expanded with the concept of peak shaving (Rauch *et al.* 2003) to equilibrate the ammonium load throughout the day and thereby to make better use of the nitrification capacity in existing treatment plants. Initial engineering calculations in the context of an overloaded treatment plant demonstrated that this approach would be cost efficient if it were introduced in a large new development (Borsuk *et al.* 2003). Further developments of urine source separation are discussed for Hong Kong, where urine would be nitrified in decentralized facilities and then denitrified in the sewers (Jiang *et al.* 2011). The intention is to avoid expanding existing wastewater treatment plants for nitrification and denitrification. Since corrosion is an issue in Hong Kong, where seawater is used for toilet flushing, nitrate will also help to conserve the sewers.

Although many ideas for the installation of NoMix toilets have been presented, it is still difficult to buy these toilets. For the sanitary industry, pilot projects are no argument for investing in such a risky business. Initiatives which are not published in academic journals, but developed in practical settings may be more convincing. One of these developments is currently taking place in Durban (South Africa), where urine source separation and urine treatment are investigated on a large, almost city-wide, scale with substantial support from the city and the Bill and Melinda Gates Foundation (BMGF; www.eawag.ch/vuna). See also the example of another BMGF-supported project in Section 10.3.2.

In Switzerland, an on-going initiative to develop and demonstrate sustainable building technology (Next Evolution Sustainable Technology, NEST) has been started, making it possible to experiment with new sanitation systems in a guest house on an academic campus (www.empa.ch/nest). This must be seen in the context of innovative toilets professionally designed by global players.

Centralized sanitation technology has been developed in industrialized societies over more than 100 years and many generations of improvements were necessary to reach today's standard. We accept that it will not be possible to reach equal "perfection" in a single shot, but that decentralized systems will also require evolutionary development. However, given the experience already acquired, this evolution can be much faster.

Today, we see three quite different scenarios for the transition from pre-existing sanitation systems to a potential future system which may be dominated by decentralized sanitation (Table 10.2). In industrialized settings, the growth of cities is slow and renovation of pre-existing sanitation systems and adaptations to new environmental requirements become major tasks. It is possible here to start right away with the introduction of decentralized sanitation. This will decrease the load on existing treatment plants (which can frequently be considered as an expansion); it will reduce problems with CSOs (storm events) and can continuously be expanded as centralized system elements decay. Since sewers already exist, they can be used to transport the treated effluent. In urban areas, sewers expose 2-3 $m^2 \cdot cap^{-1}$ of active biofilm to the flowing wastewater. Compared to RBCs (Rotation Biological Contactors), which are designed for 4-5 $m^2 \cdot cap^{-1}$, this provides substantial in-sewer treatment which will even increase as the load transported in the sewer decreases. Over the lifecycle of the pre-existing system, such an approach will lead towards an entirely new technology based on decentralized sanitation.

In rapidly industrializing settings, cities grow very rapidly, frequently in the form of large new developments. Here, existing sewer systems are mostly overloaded and it will prove beneficial to equip new developments exclusively with decentralized technology. The expected shorter lifecycle of such technology will allow faster adaptation to the highest possible standards and the introduction of more comfortable solutions in the course of development: this is an important goal of such societies.

Table 10.2. Different scenarios for the transition from centralized to decentralized sanitation systems or the introduction of new sanitation systems.

Scenario	Pre-existing infrastructure	Main problem to be solved	Environmental requirements
Industrialized	Fully developed centralized sanitation	Slow growth, upgrading, renovation	High, including nutrient control
Rapidly industrializing	Partially developed centralized system	Rapid growth, overloaded system	Medium, improving present standards
Informal settlement	Marginal, dry system	Basic sanitation	Urban hygiene, local pollution control, cost reduction through resource recovery

In informal settings, manpower is still cheap and may be used to operate decentralized systems: this opens up new possibilities. Here, it will be necessary to develop entire sanitation systems which provide urban hygiene and protect local water resources and the environment (see the example in Section 10.3.2; www.diversionsanitation.com). It will prove interesting to produce such systems on a large industrial scale. This will allow the provision of adequate comfort and reliable service at reasonable cost.

10.5 CONCLUSIONS

We come to the conclusion that the time is right to start introducing decentralized sanitation. We now need vision, courage and last but not least capital.

REFERENCES

Borsuk M. E., Maurer M., Lienert J. and Larsen T. A. (2008). Charting a path for innovative toilet technology using multicriteria decision analysis. *Environ. Sci. Technol.*, **42**(6), 1855–1862.

Cordell D. (2013). Peak phosphorus and the role of P recovery in achieving food security. In: Source Separation and Decentralization for Wastewater Management, T. A. Larsen, K. M. Udert and J. Lienert (eds.), IWA Publishing, London, UK, pp. 29–44.

Erisman J. W. and Larsen T. A. (2013). Nitrogen economy of the 21st Century. In: Source Separation and Decentralization for Wastewater Management, T. A. Larsen, K. M. Udert and J. Lienert (eds.), IWA Publishing, London, UK, pp. 45–58.

Falkenmark M. and Xia J. (2013). Urban water supply under expanding water scarcity. In: Source Separation and Decentralization for Wastewater Management, T. A. Larsen, K. M. Udert and J. Lienert (eds.), IWA Publishing, London, UK, pp. 59–70.

Gujer W. (2010). Nitrification and me – A subjective review. *Water Res.*, **44**(1), 1–19.

Jiang F., Chen Y., Mackey H. R., Chen G. H. and van Loosdrecht M. C. M. (2011). Urine nitrification and sewer discharge to realize in-sewer denitrification to simplify sewage treatment in Hong Kong. *Water Sci. Technol.*, **64**(3), 618–626.

Jönsson H. and Vinnerås B. (2013). Closing the loop: recycling nutrients to agriculture. In: Source Separation and Decentralization for Wastewater Management, T. A. Larsen, K. M. Udert and J. Lienert (eds.), IWA Publishing, London, UK, pp. 163–177.

Larsen T. A., Alder A. C., Eggen R. I. L, Maurer M. and Lienert J. (2009). Source separation: will we see a paradigm shift in wastewater handling? *Environ. Sci. Technol.*, **43**(16), 6121–6125.

Larsen T. A., Eggen R. I. L, Maurer M., Pronk W. and Lienert J. (2010) Decision support in urban water management based on generic scenarios: the example of NoMix technology. *J. Environ. Manage.*, **91**(12), 2676–2687.

Larsen T. A. and Gujer W. (1996). Separate management of anthropogenic nutrient solutions (Human urine). *Water Sci. Technol.*, **34**(3–4), 87–94.

Larsen T. and Gujer W. (2001). Waste design and source control lead to flexibility in wastewater management. *Water Sci. Technol.*, **43**(5), 309–317.

Lienert J. (2013). High acceptance of source-separating technologies, but... In: Source Separation and Decentralization for Wastewater Management, T. A. Larsen, K. M. Udert and J. Lienert (eds.), IWA Publishing, London, UK, pp. 193–207.

Maurer M. (2013). Full costs, (dis-)economies of scale and the price of uncertainty. In: Source Separation and Decentralization for Wastewater Management, T. A. Larsen, K. M. Udert, J. Lienert (eds.), IWA Publishing, London, UK, pp. 85–99.

Maurer M., Pronk W. and Larsen T. A. 2006. Treatment processes for source-separated urine. *Water Res.*, **40**(17), 3151–3166.

Mohanakrishnan J., Gutierrez O., Sharma K. R., Guisasola A., Werner U., Meyer R. L., Keller J. and Yuan Z. (2009). Impact of nitrate addition on biofilm properties and activities in rising main sewers. *Water Res.*, **43**, 4225–4237.

Olsson G. (2013). The potential of control and monitoring. In: Source Separation and Decentralization for Wastewater Management, T. A. Larsen, K. M. Udert and J. Lienert (eds.), IWA Publishing, London, UK, pp. 179–191.

Rauch W., Brockmann D., Peters I., Larsen T. A. and Gujer W. (2003). Combining urine separation with waste design: an analysis using a stochastic model for urine production. *Water Res.*, **37**(3), 681–689.

Rittmann B. E. (2013). The energy issue in urban water management. In: Source Separation and Decentralization for Wastewater Management, T. A. Larsen, K. M. Udert and J. Lienert (eds.), IWA Publishing, London, UK, pp. 11–27.

Stenström T. A. (2013). Hygiene, a major challenge for source separation and decentralization. In: Source Separation and Decentralization for Wastewater Management, T. A. Larsen, K. M. Udert and J. Lienert (eds.), IWA Publishing, London, UK, pp. 151–161.

Tchobanoglous G. and Leverenz H. (2013). The rationale for decentralization of wastewater infrastructure. In: Source Separation and Decentralization for Wastewater Management, T. A. Larsen, K. M. Udert and J. Lienert (eds.), IWA Publishing, London, UK, pp. 101–115.

Truffer B., Binz C., Gebauer H. and Störmer E. (2013). Market success of on-site treatment: a systemic innovation problem. In: Source Separation and Decentralization for Wastewater Management, T. A. Larsen, K. M. Udert and J. Lienert (eds.), IWA Publishing, London, UK, pp. 209–223.

Udert K. M., Larsen T. A. and Gujer W. (2003). Biologically induced precipitation in urine-collecting systems. *Water Sci. Technol.: Water Supply*, **3**(3), 71–78.

Wilderer P. A. (2005). Sustainable water management in rural and peri-urban areas: what technology do we need to meet the UN Millennium Development Goals? *Water Sci. Technol.*, **51**(10), 1–6.

Chapter 11

Hygiene, a major challenge for source separation and decentralization

Thor Axel Stenström

11.1 INTRODUCTION

Urban hygiene and the protection of human health are among the most important goals of sanitation systems. Conventional sewer systems achieve them via transport. Although the microbiological aspects and the risk of disease transmission at the point of discharge to the environment have been recognized, the treatment processes are seldom optimized to reduce pathogens and the dilution of outgoing wastewater in the recipients fails to safeguard human beings from hazardous amounts of these disease-causing organisms.

Any type of sanitation system, whether centralized, decentralized or source-separating, must also aim to reduce or inactivate potentially pathogenic organisms. This lowers the risk of infection for exposed human beings. The relevant risks for all types of sanitation system are assessed on the basis of their reduction efficiency, transmission routes and exposure. Surface waters are the main recipients of effluents from conventional sewer-based end-of-pipe systems, and the main risks are associated with their use for bathing and recreational purposes and the subsequent exposure, as raw-water sources for drinking water production and for any downstream shellfish areas. Soils are the main recipient for many conventional decentralized systems, the final recipient being the saturated groundwater zone after infiltration and natural reduction of pollutants and pathogens in the unsaturated layer. The subsequent use of groundwater as drinking water is the main transmission pathway. This is largely the case in various types of horizontal or vertical infiltration flow units and leach-fields, but the same phenomenon occurs with leakage from conventional sewers placed in the unsaturated layer. The risk profile of such systems depends on their site features, the reduction efficiency of the corresponding soil and the preferential flow characteristics.

Source-separating systems may have the same recipients as above for one or several fractions, but they are often designed to reuse one or several fractions (urine, greywater or excreta). This can be done with or without treatment and/or containment, with the purpose of agricultural recovery of nutrients or irrigation. The main transmission pathways for pathogens will then be intentional (for instance by using treated greywater for personal hygiene) or represent unintentional direct contact for different groups (like the community members or farmers), or through contamination of food products for humans. Naturally, direct contact and food product contamination also apply when wastewater from conventional systems is reused for agricultural irrigation, or sludge is reused. When assessing the health effects of source-separating systems, either centralized or decentralized, the risk is governed by pathogen input and variation together with local transmission routes. Risk reduction is achieved by multiple robust and efficient barriers. In developed countries, decentralized systems are mainly found in sparsely populated areas or to assure more direct use of waste products, such as urine, treated faeces or greywater. A comparison of centralized and decentralized systems, as well as of wastewater systems in general, shows different likelihoods of exposure and risk of disease transmission, but in all cases the risks depend on the level of treatment, handling practices and reuse schemes. This is where the system perspective comes in – all systems need to be evaluated in a systematic way that accounts for local conditions, as further described below.

The health outcome of a sanitation system is linked to both technical and social/behavioural factors. A barrier produces a quantifiable risk reduction. Multiple barriers in a sanitation system are usually needed, the sum of the individual barriers giving the resulting risk reduction of pathogenic microorganisms and hazardous constituents (see Kümmerer 2013), defined at the point of exposure. The reduction capacity via single or multiple treatment barriers must account for the variability, for example in the load, and the resulting ability to withstand variations in reduction efficiency, sometimes defined as "robustness." Moreover, to operate successfully, the sanitation system needs to be socially acceptable (Lienert 2013) at the user interface, be appropriate from a behavioural point of view and account for downstream exposure. Regular operation and maintenance further ensure its proper functionality (see Larsen and Gujer 2013). These factors are naturally of major importance from a hygiene perspective.

Within the WHO Guidelines for the Safe Use of Wastewater, Excreta and Grey Water, Vol. 4 (WHO 2006), the information on the health risks associated with pathogens occurring in human excreta and greywater is presented within a system framework of risk and exposure assessment based on the local situation (in both developed and developing countries). This information must be related to health targets and include health protection measures as part of a risk management strategy. It is applicable to decentralized and source-separating systems as well as to conventional ones. The health objectives are linked to the environmental impact, so that safe containment of excreta, for example, will avoid the contamination of groundwater through infiltration. From a management perspective, this further links to the concept of Sanitation Safety Plans (SSPs; Barrenberg and Stenström 2010) aimed to achieve simpler operation of the guidelines given in the Water Safety Plans. It comprises three parts: (1) a system and exposure assessment, mapping the system and identifying potential risks along the sanitation chain, (2) operational monitoring in order to establish control measures for previously identified and ranked hazards and exposures at critical control points in the chain, (3) a management component, referring to a plan of actions and control measures for normal conditions and incident situations. This broader approach accounts for the recipient effects and thus includes the secondary impact on human health from receiving water bodies via environmental routes to pasture and agricultural land used for food production. It also covers the technical system components, including their robustness and ability to withstand variability in performance.

11.2 HAZARD IDENTIFICATION IN A SYSTEM PERSPECTIVE

A system assessment defines the barriers and their efficiency in reducing pathogen loads qualitatively or quantitatively. It refers to the exposure involved and further helps to define both technical and non-technical interventions designed to reduce the risk.

In the hazard identification process, the microbial agents are identified in the local context and related to the spectrum of human diseases associated with each specific pathogen. This may include the virulence of the microorganisms as well as aspects of acquired immunity and multiple exposures. The range of pathogens includes various bacteria, viruses, protozoa and helminths excreted by humans (and in a more restricted range by animals). The disease incidence of specific pathogens in a population connected to a sanitation system represents its input concentration. Variation occurs between organisms due to their excretion loads, duration of excretion and infective dose. The lower the infective dose for a given organism, the higher the risk of infection. The incidences given in Table 11.1 are representative in a European context. Thus, the higher the frequency of infected people connected to a system, the greater the risk and the more important is the effectiveness of the barriers. A high risk may be typical of some developing countries or represent an outbreak scenario in Europe.

In a larger conventional centralized system with a low prevalence in the connected population, the likelihood of a few individuals being infected is greater than in small systems. Decentralized and source-separating systems are often smaller in size and the number of connected people, normally resulting in a lower likelihood of pathogens. However, when connected individuals are infected, the resulting concentrations may be higher than in large systems due to a lower relative dilution. This time-dependent variability must be accounted for when assessing both the treatment barrier and the reuse practices and risk.

Table 11.1. Epidemiological data for selected pathogens (reduced and adapted on the basis of Westrell 2004, Wheeler *et al.* 1999 and WHO 2006). ID = Infectious Dose.

	Incidence / $100{,}000^{-1}$	Excretion / g_{faeces}^{-1}	Duration [d]	ID_{50}
Campylobacter	78–97	10^{6-9}	1–77	900
EHEC	0.8–1.4	10^{2-3}	5–12	1120
Hepatitis A	0.8–7.8	10^{4-6}	13–30	30
Rotavirus	21	10^{7-11}	1–39	6
Norovirus	1.2	10^{5-9}	5–22	10
Cryptosporidium	0.3–1.6	10^{7-8}	2–30	165
Giardia	15–26	10^{5-8}	28–284	35

The faecal fraction determines the risk. Faeces should always be considered to contain pathogens, although at local level this depends on the prevalence of various diseases in the excreting population. Raw wastewater can be regarded as diluted faeces. Faecal cross-contamination is therefore of major concern in a source-separating system. Few pathogens are directly excreted with the urine, and the risk is initially governed by faecal cross-contamination in the urine fraction. This cross-contamination is normally small in European systems, but depends on the technical design of the components at the user interface and on user behaviours. Schönning *et al.* (2002) estimated faecal cross-contamination in the urine fraction from urine-diverting toilets in Sweden based on the amounts of faecal sterols present as a quantitatively conservative indicator (showing comparative prevalence). The mean faecal contamination amounted to $9 \pm 6\ mg{\cdot}L_{urine}^{-1}$. Accounting for variability between different user groups and systems, a risk reduction of approximately 10^4–10^5 compared to the faecal fraction is achieved. This is not considered entirely risk-free (WHO 2006) but is a substantial improvement, similar to the outlet of treated wastewater from a wastewater treatment plant with a 3 log reduction efficiency. However, the risks also increase with substantially higher faecal cross-contamination. An extended storage period is consequently recommended in the WHO Guidelines, since the pathogen die-off is time-dependent. In greywater systems, the faecal input may come from childcare, laundry handling and shower facilities. Dilution occurs in the water. Ottoson and Stenström (2003) estimated the faecal input in untreated greywater into a source-separating surface-flow wetland system and found that the faecal cross-contamination was in the same range and had a similar risk reduction factor as in the urine example above. The treatment naturally reduced the risk further.

National or international guidelines are most often based on the quantitative assessment of "indicator organisms" such as faecal (thermotolerant) coliforms, *Escherichia coli* (*E coli*) or faecal enterococci (streptococci). *E coli* die off much more rapidly in the urine fraction than some other groups of pathogens. They thus act as an "index" for conventional bacterial pathogens like *Salmonella* and *Shigella*, but underestimate the risk of enteric viruses and parasitic protozoa. The opposite occurs in many greywater systems, where *E coli* have been shown to multiply rapidly on easily degradable organic material. In this case, the use of *E coli* as indicator will over-estimate the risk instead, since viruses, parasitic protozoa and many conventional bacterial pathogens will be unable to grow. Other bacterial indicators such as enteric enterococci (streptococci) have shown to better represent the risk in groundwater impact, recreational impact or greywater systems. Bacterial spores can be better predictors than *E coli* when assessing the barrier function in wastewater treatment systems for parasitic protozoa and bacteriophages (bacterial virus), likewise can act as a predictor for enteric viral reduction. Ascaris is most often used as an "index organism" for helminth risks and acts comparatively well as a conservative risk predictor in this regard. It is also included in the verification monitoring of some guidelines. In the risk assessment approaches taken in the WHO Guidelines (WHO 2006), a few reference pathogens are selected that represent the most environmentally persistent organisms from each group (bacteria, viruses, protozoa or helminths) and should consequently act as a conservative index for these groups. This is combined with using the prevalence of disease within the connected population representative of the local situation to estimate the risk and validate the treatment barrier function.

11.3 HUMAN EXPOSURE ASSESSMENT

As initially stated, human exposure to pathogens and hazardous substances and its reduction, potentially impacting different points of the sanitation system, is central. The exposure assessment follows the flow from households, through the collection and treatment facilities to the point of reuse or disposal and also accounts for the downstream populations (see also Tilley 2013). The risk of exposure to pathogens concerns all the diverse transmission routes and not just the faecal-oral route through contaminated drinking water. Oral exposure also occurs through a lack of hygiene, bathing and recreational activities (WHO 2003). Wastewater effluents indirectly affect human health through pathogens from discharged wastewater accumulated in shellfish, through potential contamination of crops irrigated with wastewater, direct or indirect human contact involving further transmission to the mouth, or through aerosols. In developing countries, we must further consider direct contact through bare skin, e.g., for hookworm infection, or as a breeding ground for biting insects that can act as disease vectors. This is currently of less concern in Europe, due to its low prevalence there, but vector-borne transmission may in the future become locally relevant due to climate change. The various transmission pathways have different magnitudes depending on the points of the sanitation system where exposure occurs, from toilets (or deposition/introduction of faecal material to other waste streams) via the collection, transport and treatment system to the point of reuse or disposal. Thus an exposure assessment is central from a health point of view and involves the overall system. It asks the following questions:

WHO is exposed? This defines the groups potentially at risk. In a source-separating system, this may be individual users, maintenance workers, consumers of crops fertilized with treated excreta, faecal sludge or wastewater, biosolids, greywater or urine, or those indirectly exposed to contaminated soils or surface/groundwater.

HOW MANY people (individuals) are likely to be exposed? In a reuse system, this mainly concerns the end consumers of the crops, whereas the emptying and maintenance of the system most often involves only a few individuals.

HOW frequently does exposure occur? Does it happen daily, weekly or perhaps just once a year? A guesstimate is better than nothing.

WHERE does exposure occur within the sanitation system? For a source-separating and reuse system, this question follows the system from the user to the reuse or disposal steps and also covers any secondary exposures occurring due to environmental pollution.

WHICH routes should we consider?

WHAT is the exposure dose?

The answers to these questions describe the risk and form part of an assessment of the local situation. They obviously depend on prior treatment (the barrier efficiency) as well as on real behaviour patterns.

11.4 TREATMENT BARRIERS AND EXAMPLES OF THEIR REDUCTION EFFICIENCY

In a well-functioning wastewater treatment plant in Europe (without disinfection) the reduction efficiency is in the range of 2–4 logs for different groups of pathogens. This provides a reference point for assessing decentralized alternatives. A main concern is the reduction of viruses and to some extent parasitic protozoa. Ottoson *et al.* (2006) recorded a lower mean removal of noroviruses and enteroviruses of 0.9 and 1.3 log respectively, while *Giardia* cysts were removed with 2.6 logs by conventional secondary wastewater treatment. These figures will show great variability over time due to sewer overflow, emergency outlets and combined wastewater/stormwater systems. The robustness of the system will be a determinant of risk variability.

For well-functioning decentralized package treatment plants, the reduction may reach the same level as for large systems, but is around ten times lower in many cases. This figure may be adjusted upwards or downwards depending on the unit processes involved, the load and interrelated differences in the reduction of microbial groups.

Several other decentralized technical options will achieve similar or better results. Properly designed and operated waste stabilization ponds used in both developed and developing countries show a 2–4 log removal of viral pathogens and 1–2 log of protozoa (Jiménez *et al.* 2010). Similar removal efficiencies are recorded for wastewater treatment and storage reservoirs running in batch mode with retention times of over 20 days.

Pre-treatment in small or medium decentralized filtration systems for combined wastewater or greywater are always needed to avoid clogging in subsequent treatment steps. Since some of the microorganisms are attached or associated to the particulate material, they are reduced in the pre-treatment phase, typically in the range of 50–90% (0.5–1 log). However, this is marginal and requires subsequent treatment of the remaining liquid fraction. Conventional pathogen removal, e.g., via leach-fields or other soil infiltration systems, depends on the load and hydraulic conditions as well as the condition of the infiltrating soil, but is generally high or in a similar range as for conventional wastewater treatment. This normally gives a reduction range of >2 logs for bacteria and viruses and >3 logs for parasitic protozoa (Siegrist *et al.* 2000; WHO 2006). The same or a slightly lower reduction range has been found for vertical-flow sand-filter trenches.

The reduction efficiency in constructed wetlands is variable. The efficiency of surface-flow wetlands depends on the retention time, the flow path, vegetation and other factors. Values between 0.5–3 logs have been recorded for bacteria and somewhat lower for viruses. Subsurface horizontal wetlands normally have higher reduction efficiency. Most information concerns the reduction of indicator bacteria which are in the range of 2–3 log reduction. The effects on viruses were assessed by Heistad *et al.* (2009) in a vertical-flow biofilter and a filter unit with lightweight clay aggregates. A viral reduction efficiency of between 3–4 logs was found for two bacteriophages used as virus surrogates. The risks associated with horizontal or vertical subsurface-soil systems are generally low, but depend on the use of the outlet water and the related exposure. A combination with plant/root re-sorption in horizontal-flow wetlands reduces the human risk to very low levels.

Storage is normally the main way of further reducing the microbial content and the subsequent risks of separated urine. Microbial reduction is a function of storage time, the pH value of the stored urine and the temperature. These relationships have been systematically assessed for different microbial groups (Höglund 2001) and were included in the WHO Guidelines (WHO 2006). The bacterial pathogens and indicators are the most sensitive groups and the risk of bacterial pathogens is almost zero after a month (20°C, pH 8.8–9.0). A risk reduction of at least 1–2 logs also occurred for viruses and parasitic protozoa. Further reductions can be realized in the field and the risk will be far below the WHO target P_{inf} value (10^{-4}) for crops after one month of additional withholding time (Figure 11.1;

Höglund *et al.* 2002, WHO 2006). In single households, the urine can be used more or less directly if only the family members are involved. The risk will increase when others are exposed to fertilized products or these are consumed by third parties. A higher safety level then applies.

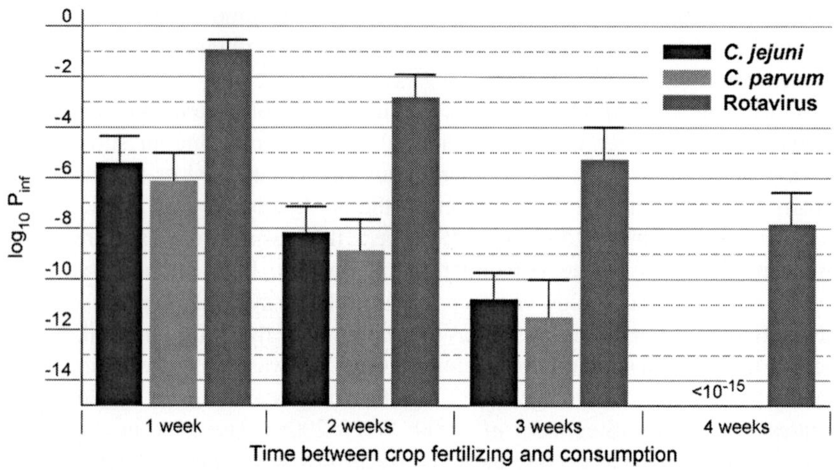

Figure 11.1. Mean probability of infection by pathogens following ingestion of crops fertilized by non-stored urine with varying withholding periods (periods between application and harvest/consumption). Error bars indicate one standard deviation (Höglund 2002; WHO 2006).

For the contained faecal excreta, the risk is again a determinant of storage time, temperature and pH (and reduced moisture content). In most situations, a storage time of 18 months will reduce the content of all pathogen groups by at least 4 logs (WHO 2006). A number of treatment methods may be used to enhance the die-off and shorten any reuse time, including for composting, drying and lime treatment.

11.5 QUANTIFICATIONS OF RISKS AND RISK-BENEFIT STRATEGIES

As described at the beginning of the chapter, the risk after human exposure depends on the concentration in the exposed volume. The risks are described as the probability of infection and relate to WHO target values (WHO 2006). These can either be assessed by an arbitrary relative relationship between indicators and pathogens (Mara and Sleigh 2010), or may be based on the incidence of infection, pathogen organism concentration and barrier reduction efficiency. These alternative approaches are based on Quantitative Microbial Risk Assessment (QMRA). Several studies of decentralized and source-separating systems have

to fertilize them with more phosphorus than is removed by the crop. It is important to ensure that the phosphorus supply is well-synchronized with the plant demand, that is the phosphorus should be supplied in plant-available form just before it is required by the crop, as immobilization increases with time in the soil. Instead of even distribution on the surface, it can be an advantage if the phosphorus is supplied at high concentrations in pockets or lines close to the plants and at a depth where the soil is moist during their vegetative growth. This minimizes the contact between the soil and phosphorus, and consequently the immobilization too. The soil needs to be moist so that the roots can reach the phosphorus and absorb it.

The yield of most crops is limited by the supply of nitrogen. The great quantitative importance of nitrogen is reflected by the fact that world consumption of nitrogen fertilizer is about six times greater than that of phosphorus and five times that of potassium (Table 12.1). The P:N ratios differ between countries for two main reasons: in areas with a long history of over-fertilization with phosphorus, such as Sweden, the P:N ratio is now low. In contrast, this ratio should be high in areas where phosphorus is strongly immobilized, such as Kenya.

Table 12.1 also clearly reflects the fact that crops need larger amounts of potassium than phosphorus, but also that some soils can supply them with potassium. While the proportion of potassium to nitrogen is 18–41% in most countries as well as globally, it is 0% for Pakistan, 5% for Kenya and 13% for India. The different N:P and N:K relationships in various countries are due not only to the different soils, but also to the crops grown and the various price relationships.

Obviously, the choice of fertilizer is greatly influenced by socio-economic factors. Simple fertilizers (N-fertilizer, ammonium phosphate and muriate) are relatively cheap, but carry more mass per unit of N, P, and K. In affluent countries such as Sweden and Austria, there is thus a clear preference for using P and K from complex fertilizers, which may be more expensive to produce but minimize transport costs and the number of necessary spreading operations, as each fertilizer normally needs a separate spreading operation. Combining all required nutrients saves work and is thus preferred by those who can afford it. Furthermore, NPK fertilizers weigh less than less complex fertilizers with the same amount of nutrients, which is important for transport costs. This might explain why Malawi and Zambia, both landlocked countries without fertilizer production, use a high percentage of complex fertilizers (Table 12.1). Faeces also contain significant amounts of organic matter, which is an advantage. If the phosphorus removal by a 3000 $kg{\cdot}ha^{-1}$ wheat grain crop (9 $kgP{\cdot}ha^{-1}$) is replaced by phosphorus from faeces, this would mean the addition of about 1100 $kg{\cdot}ha^{-1}$ of organics if the faeces had just been stored and about half as much if it had been composted or digested. This corresponds to 10–19% of the organics in the root and straw, and as much as 33–67% of the organics in just the root system (Bolinder *et al.* 2007). The organics of the faeces can thus contribute significantly to the soil organic matter.

12.4 ECONOMIC AND GWP VALUE OF NUTRIENTS

Due to the great importance of plant-available nitrogen, it is not surprising that the average farmer spends a large proportion of his fertilizer budget on nitrogen. Based on the price of mineral fertilizer, most of the value of urine and blackwater also comes from nitrogen, and the second largest part from potassium (Table 12.2). So if nitrogen and/or potassium are lost in the treatment of urine, essential parts of its economic and agronomic values are also lost. Fertilizers are neither subsidized nor specially taxed in Sweden. So the prices listed in Table 12.2 should reflect those on the world market fairly closely. Land transport is costly, especially if the infrastructure is poor, and fertilizer prices can consequently be much higher, for example in landlocked countries. On the other hand, fertilizers are subsidized in many countries, so prices can also be lower, especially close to production sites and import harbours. Depending on how the subsidy is calculated and on possible domestic production of some types of fertilizers, the relative prices of nitrogen, phosphorus and potassium also vary. Table 12.2 consequently does not reflect any absolute truth, but gives a reasonable indication of the total and relative values of the different nutrients and fertilizer products.

Table 12.2 The economic fertilizer value of some source-separated sewage products and manure slurries in euro per ton wet weight. Values for urine, faeces and toilet waste are for the product with no flush water. If they are diluted with flush water, their total values per ton wet weight will decrease correspondingly. The table is based on the following fertilizer prizes: 0.77, 1.29 and 0.85 euro per kg N, P and K respectively (Lantmännen Direkt 2010-04-22 for fertilizers at their Norrköping import terminal).

Product	N [% of total value]	P [% of total value]	K [% of total value]	Total value [Euro·$t_{product}^{-1}$]
Urine	68	10	22	7.87
Faeces	31	28	41	11.35
Toilet waste	62	13	25	8.27
Urevit	58	6	36	13.41
Cattle slurry	26	17	57	5.80
Pig slurry	35	32	34	8.24

These source-separated products should replace chemical fertilizers and thus decrease the environmental load from their production. One important impact from the production of chemical fertilizers is the emission of greenhouse gases, which contributes 1.2% to total global GHG emissions. Most of this is due to the production of nitrogen fertilizers. For all fertilizer products shown in Table 12.3, more than 85% of the potential savings in GHG emissions are due to nitrogen, so it is very important that as much plant-available nitrogen as possible is recycled back to agriculture to contribute as much as possible to reducing GHG emissions (Table 12.3).

Table 12.3 The contributions from nitrogen, phosphorus and potassium to the total potential decrease in emissions of greenhouse gases [$kg_{CO2eqv.} \cdot t_{product}^{-1}$] if the nutrients in urine, faeces, toilet waste, urevit, cattle slurry and pig slurry replace chemical fertilizer produced in Europe. Emission data from Jenssen and Kongshaug (2003).

Product	N [% of tot GWP]	P [% of tot GWP]	K [% of tot GWP]	Tot GWP [$kg_{CO2\text{-}eq} \cdot t_{product}^{-1}$]
Urine	98	1	2	38.1
Faeces	86	6	8	28.1
Toilet waste	97	1	2	37.1
Urevit	96	1	4	55.4
Cattle slurry	85	3	12	12.2
Pig slurry	90	5	6	21.8

12.5 URINE IS VERY LOW IN POLLUTANTS

One clear advantage of source-separated products as fertilizers is their low pollution, and this is especially true for urine. This is due to the great care with which we handle food and the strong restrictions we apply to it as well as to the mass balance over the human body. The concentrations of heavy metals (Figure 12.3), hormones and some antibiotics (Winker *et al.* 2009; Kümmerer 2013) are consequently lower in source-separated toilet products from humans than in manure from our livestock, and for metals also lower than in source-separated biowaste from households.

Toilet waste shows approximately the contamination level of what we actually eat, while biowaste consists mainly of foodstuffs that we have discarded, such as peels and other parts more contaminated by dust and soil, and thus by metals. While toilet waste is about half a $\log_{10}$ less contaminated by metals than manure from cattle and pigs, partly because we handle food more carefully than feed, the contamination of faeces is about the same as for the manures. The good job done by our kidneys is shown by the contamination level of urine, which is about 1 to 1.5 $\log_{10}$ less than that of both faeces and manure. This low level ought to make urine very attractive as a fertilizer and even more so as the plant-availability of its nutrients is so high (Kirchmann and Petersson, 1995; Richert Stintzing *et al.* 2011). Both these properties make urine unique among organic fertilizers, and ought to make it attractive for organic farming.

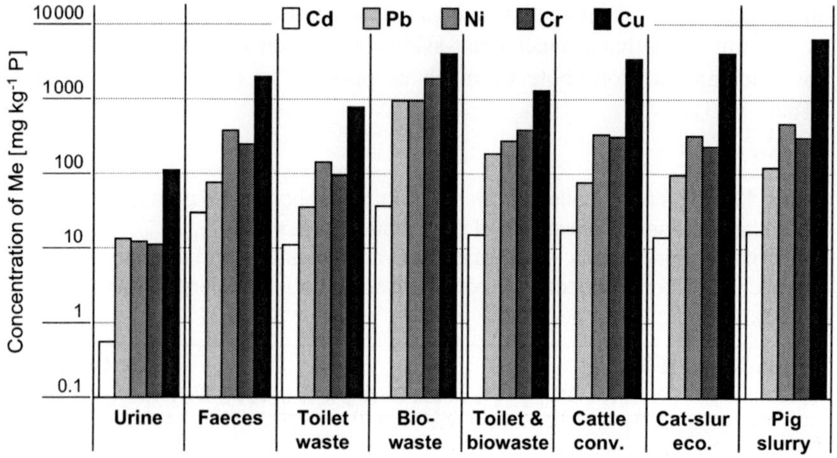

Figure 12.3 Concentrations of heavy metals in urine, faeces and toilet waste compared to their values in cattle slurry from conventional and ecological production and with pig slurry. The concentrations are given in mg·kgP^{-1}. Data from Jönsson *et al.* (2005) and Steineck *et al.* (1999).

Cadmium (Cd) is of special concern, as it is easily taken up by crops and its average exposure in Europe has already reached the tolerable intake limit (EFSA 2009). To stop its current accumulation in European soils, fertilizers should contain less than 46 mg Cd per kg phosphorus (ERM 2001). In sensitive areas such as Sweden, this figure should be less than 20 mg (Kemi 2011). All the products shown in Figure 12.3 are well below 46, and all except for biowaste and faeces are below 20 mg. Urine is exceptionally low, with just 0.6 mg per kg phosphorus. This may be compared with the 83 mg representing the average concentration in 196 samples of chemical phosphorus fertilizers in Europe (Nziguheba and Smolders 2008). Only about 40% of the samples were below 46, while 20% were above 150 mg. Most of the world's rock phosphate reserves are high in cadmium, producing fertilizers with high cadmium levels.

12.6 LOW HYGIENE RISK

Any pathogens contained in the source-separated fractions would represent a risk (Stenström this volume). So to minimize the risk of spreading diseases they need to be sanitized. For urine, this is achieved in a resource efficient way by storage, the sanitizing agent being its intrinsic content of ammonia and its pH of around 9. Most of the nitrogen in urine is excreted as urea but is rapidly degraded to ammonia, which acts as the sanitizing agent that inactivates any pathogens present (Vinnerås *et al.* 2008). Faeces contain similar concentrations of nitrogen, but their pH is lower and they thus contain less free ammonia, which is the main sanitizing agent. Faeces and toilet waste can be sanitized by the addition of urea (the most common nitrogen fertilizer in the world), which degrades to ammonia upon contact with the faeces (Nordin *et al.* 2009 a, b). This is very resource-efficient, as ammonia doubles as both a sanitizing agent and plant-available nitrogen, increasing the fraction of the latter in the product (Figure 12.1). Ammonia is especially efficient in sanitizing bacteria, for example *Salmonella* spp. This is especially important, as *Salmonella* is a zoonos, that is it can be spread from humans to animals and vice versa. At temperatures above 20°C, ammonia is also very efficient in reducing parasitic and viral pathogens.

12.7 SPREADING MACHINERY

It is much easier for formers to accept new fertilizer products if they can be handled with conventional equipment. They can then be introduced without at the same time developing and introducing new spreaders. Three types of fertilizers are common around the world, as is their spreading equipment, that is granular (pelletized or granulated) chemical fertilizers, by far the most common type worldwide, liquid animal slurry and solid farmyard manure. The key data for handling and spreading these fertilizers are given in Table 12.4, which also lists the characteristics of a few source-separated fertilizer products.

Where these products closely resemble those of a common fertilizer, the equipment used for handling and spreading it can also be used for the new fertilizer. The dosage is most often based on the desired amount of plant-available (i.e., mineral) nitrogen, and a reasonable figure for spreading fertilizer onto many crops is around 80 kg of this nitrogen per hectare, on which Table 12.4 is based.

This table clearly shows that source-separated urine, faeces and toilet waste can be handled with existing systems for animal slurry and farmyard manure provided that they are not mixed with too much flush water. For blackwater, excessive volumes would be required to obtain a reasonable nitrogen dosage due to the large dilution used in most current collection systems. However, ongoing developments towards more efficient vacuum systems, which can work well with far less flush water, appear promising. Ammonia sanitization of blackwater has the advantage of also increasing its nitrogen concentration to a level making the use of a slurry spreader suitable.

In regions with little animal production, manure-handling equipment will naturally be scarce, and where the animals are kept outdoors rather than in stables, the manure is often not collected so there will be no equipment to handle it, making it more difficult to introduce manure-like fertilizers. While manure is commonly handled as liquid slurry on large modern farms in temperate regions, this is essentially unknown on small farms in high-evaporation regions. However, if manure and other organic waste is used to produce biogas by anaerobic digestion, the chance of having equipment for handling liquid manure increases, as it is needed to utilize the plant-available nitrogen as efficiently as possible.

Table 12.4 Key data for handling conventional fertilizers (chemical fertilizer, animal slurry and farmyard manure) and a few fertilizers from source-separated sanitation systems.

Product	Form of fertilizer	Plant-available N [$kg \cdot t^{-1}$]	Desirable dosage [$t \cdot ha^{-1}$]	Handling chain
Chemical fertilizer	Solid, granular	100–460	0.2–1.0[a]	Chem. fert.
Animal slurry	Liquid dispersion with solids	2–4	10–40[a]	Slurry
Farmyard manure	Solid	1–3	20–40[a]	Farmyard
Urine, stored	Liquid dispersion with few solids	1.5–2.5–8[b]	10–32–53[b]	Slurry
Faeces	Solid	2–4.5	18–40	Farmyard
Toilet waste	Liquid dispersion with solids	5–8	10–16	Slurry
Blackwater (BW)	Liquid dispersion with few solids	0.4–1	80–200[c]	Irrigation
BW, ammonia- sanitized	Liquid dispersion with few solids	2.5–10	8[d]–32	Slurry
Urevit	Liquid	10	8[d]	Slurry

[a] Easily adjustable dosage limits for spreading equipment. For slurry and farmyard spreaders, the upper limit is set by the restriction that the soil should not become too wet and the manure should be easy to incorporate into the soil.
[b] If the concentration of plant-available N is in the lower range, it is hard to spread 80 kg of it in one go, as this requires a very high dosage.
[c] The nutrient concentration of blackwater is normally too low for spreading with slurry equipment, as this would lead to excessive soil compaction, very wet soil and possibly nutrient run-off.
[d] Dosages below 10 tons are often difficult to spread with a slurry spreader. It would either have to be modified in some way or driven very quickly when spreading.

Urevit, produced from source-separated urine by electrodialysis and ozonation (Boller 2007; Maurer *et al.* 2006), contains about 10 $kg \cdot t^{-1}$ of ammonium and about 12 $kg \cdot t^{-1}$ of total nitrogen, so a suitable dose would be 7–8 $t \cdot ha^{-1}$, which is just about or a little below the level that can be spread with existing equipment without any modification. If a modification is needed, it will probably be fairly simple for most spreaders.

However, no equipment is widely available for spreading liquid fertilizers with concentrations in the range of 15 to 100, or below about 1.5 $kg \cdot t^{-1}$ of available nitrogen. For concentrations above 100 $kg \cdot t^{-1}$, the present field spraying technique, which is available throughout the world, can be used provided that the other properties of the liquid fertilizer do not make it unsuitable for spraying. For all nutrient concentrations, including those below 1.5 $kg \cdot tN^{-1}$, fertigation, that is fertilization via an irrigation system, can be used where such a system is available. However, it is then important to carefully consider the risk of spreading pathogens. The risks of biofilm development, precipitation and clogging in the irrigation system also need to be considered, and this is especially true for a drip-irrigation system. Initially clear solutions, like filtered urine, can also cause precipitation (Udert *et al.* 2003) and thus clogging, when mixed with water. Thus, if urine is to be used in a drip-irrigation system, it is recommended that it be filtered and not mixed with water, but be applied undiluted for short periods with longer intermissions of applying water alone. This minimizes the contact between urine and water and thus also any possible precipitation after mixing.

Equipment for handling and spreading chemical fertilizers relies on these being in the form of fairly spherical granules of certain strength. When new concentrated and solid fertilizer products are developed from source-separated products, it is essential not only that the nutrient concentration is somewhere close to the figures listed in Table 12.4, but also that the product is similar to chemical fertilizers in terms of its diameter, particle strength and friction if spreaders designed for the latter are to be used.

12.8 THE FARMER – BUSINESSMAN, SOIL STEWARD AND ENTREPRENEUR

To recycle nutrients from source-separated fractions to agriculture, farmers using excretion nutrients as fertilizers are just as essential as all other stakeholders of the system. They are the stewards of the arable soil and their acceptance is needed for recycling to take place. Experience shows that it is essential for farmers to participate as early as the initial planning stage of the recycling sanitation system for the recycling to function well (Jönsson *et al.* 2010).

Recycling sanitation systems have many stakeholders, and recognition of the drivers and restrictions of each of them is important for the function of the system as a whole. Farmers are however not only stewards of the soil, but also businessmen. Thus, one important driver for many farmers is for them to increase their business by also acting as entrepreneurs, taking responsibility for the collection and handling of the sanitation-system fertilizer (Jönsson *et al.* 2010). This driver also helps to increase acceptance of the fertilizer product, an important aspect discussed by Lienert 2013. These systems have many stakeholders, and it is important for their long-term performance that a forum for feedback, interaction, improvement and inspiration is organized, as is done by the municipality of Tanum discussed by Vinnerås and Jönsson 2013.

REFERENCES

Bolinder M. A., Janzen H. H., Gregorich E. G., Angers D. A. and VandenBygaart A. J. (2007). An approach for estimating net primary productivity and annual carbon inputs to soil for common agricultural crops in Canada. *Agriculture, Ecosystems & Environment*, **118**, 29–42.

Boller M. (2007). Fertilizer from the library. Eawag News 63e, 17–19. http://www.eawag.ch/services/publikationen/eanews/news_63/en63e_boller.pdf (17.02.2009).

Cordell D. (2013). Peak phosphorus and the role of P recovery in achieving food security. In: Source Separation and Decentralization for Wastewater Management, T. A. Larsen, K. M. Udert, J. Lienert (eds.), IWA publishing, London, UK, pp. 29–44.

EFSA (2009). Scientific opinion of the panel on contaminants in the food chain on a request from the European Commission on cadmium in food. *EFSA Journal*, **980**, 1–139.

Eriksson S., Sanne S. and Thomke S. (1972). Fodermedlen (Feedstuffs), LTs förlag, Sweden. ISBN 91-36-00122-8.

ERM (2001). Analysis and Conclusions from Member States' Assessment of the Risk to Health and the Environment from Cadmium in Fertilizers. For the European Commission – DG Enterprise. Contract No. ETD/00/503201. Final Report.

FAOSTAT (2011). http://faostat.fao.org/site/422/DesktopDefault.aspx?PageID=422#ancor (accessed 11 March 2011).

Hotta S. and Funamizu N. (2007). Biodegradability of fecal nitrogen in composting process. *Bioresource Technology*, **98**, 3412–3414.

Johansson M., Jönsson H., Höglund C., Richert Stintzing A. and Rodhe L. (2000). Urinsortering – en del av kretsloppet (Source separation of urine – a part of the eco-cycle). In Swedish. Report T17:2000, BFR. Sweden. ISBN 91-540-5860-0.

Jönsson H., Vinnerås B., Höglund C., Stenström T. A., Dalhammar G. and Kirchmann H. (2000). Källsorterad humanurin i kretslopp (Recycling source separated human urine). VA-FORSK Report 2000·1. VA-FORSK/VAV. Sweden.

Jenssen T. K. and Kongshaug G. (2003). Energy consumption and greenhouse gas emissions in fertilizer production. Proceedings No. 509, The International Fertilizer Society, United Kingdom.

Jönsson H., Baky A., Jeppsson U., Hellström D. and Kärrman E. (2005). Composition of urine, faeces, greywater and bio-waste - for utilisation in the URWARE model. Report 2005:6, Urban Water, Chalmers. Sweden. Download: http://www.urbanwater.org.

Jönsson H., Richert Stintzing A., Vinnerås B. and Salomon E. (2004). Guidelines on use of urine and faeces in crop production. Report 2004-2, Ecosanres, Stockholm Environment Institute, Stockholm, Sweden. Download: "http://www.ecosanres.org.

Jönsson H., Tidåker P. and Richert Stintzing A. (2010). Role of farmers in improving the sustainability of sanitation systems. In: Social Perspectives on the Sanitation Challenge, B. J. M. van Vliet, G. Spaargaren, P. Oosterveer (eds.), ISBN 978-90-481-3720-6. Springer Science + Business Media, pp. 179–188.

Kemi. (2011). Kadmiumhalten måste minska (The cadmium concentration must decrease). Rapport 1/11, Swedish Chemicals Agency. Download: www.kemi.se.

Kümmerer K. (2013). The issue of micropollutants in urban water management. In: Source Separation and Decentralization for Wastewater Management, T. A. Larsen, K. M. Udert, J. Lienert (eds.), IWA publishing, London, UK, pp. 71–84.

Kirchmann H. and Pettersson S. (1995). Human urine – Chemical composition and fertilizer use efficiency. *Fertilizer Research*, **40**, 149–154.

Lantmännen Direkt (2010). Lantmännen lantbruk.

Larsen T. A. and Gujer W. (2013). Implementation of source separation and decentralization in cities. In: Source Separation and Decentralization for Wastewater Management, T. A. Larsen, K. M. Udert, J. Lienert (eds.), IWA publishing, London, UK, pp. 133–151.

Lienert J. (2013). High acceptance of source-separating technologies, but… . In: Source Separation and Decentralization for Wastewater Management, T. A. Larsen, K. M. Udert, and J. Lienert (eds.), IWA publishing, London, UK, pp. 193–207.

Maurer M., Pronk W. and Larsen T.A. (2006). Treatment processes for source-separated urine. *Water Research*, **40**, 3151–3166.

Nziguheba G. and Smolders E. (2008). Input of trace elements in agricultural soils via phosphate fertilizers in European Countries. *Science of the Total Environment*, **390**, 53–57.

Nordin A., Nyberg K. and Vinnerås B. (2009a). Inactivation of Ascaris eggs in source-separated urine and faeces by ammonia at ambient temperatures, *Applied and Environmental Microbiology*, **75**(3), 662–667.

Nordin A., Ottoson J. and Vinnerås B. (2009b). Sanitation of faeces from source-separating dry toilets using urea. *Journal of Applied Microbiology*, **107**, 1579–1587.

Richert Stintzing A., Gensch R., Jönsson H., Stenström T. A. and Dagerskog L. (2011). Practical guideline on the use of urine in crop production. EcoSanRes Series 2011-2, Stockholm Environment Institute, Sweden. Download: www.ecosanres.org.

Rittmann B. E. (2013). The energy issue in urban water management. In: Source Separation and Decentralization for Wastewater Management, T. A. Larsen, K. M. Udert, J. Lienert (eds.), IWA publishing, London, UK, pp. 11–27.

Simonsson A. (2006). Fodermedel och näringsrekommendationer för gris (Feedstuffs and nutritional recommendations for pig). Rapport 266, Department of Animal Nutrition and Management, Swedish University of Agricultural Sciences, Sweden.

Stenström T. A. (2012). Hygiene, a main challenge for source separation and decentralization. In: Source Separation and Decentralization for Wastewater Management, T. A. Larsen, K. M. Udert and J. Lienert (eds.), IWA publishing, London, UK, pp. 153–161.

Steineck S., Gustafson G., Andersson A., Tersmeden M. and Bergström J. (1999). Animal manure – contents of nutrients and trace elements. Rapport 4974, Naturvårdsverket, Stockholm, Sweden.

Vinnerås B. Jönsson H. (2013). The Swedish experience with source separation. In: Source Separation and Decentralization for Wastewater Management, T. A. Larsen, K. M. Udert, J. Lienert (eds.), IWA publishing, London, UK, pp. 413–422.

Vinnerås B., Nordin A., Niwagaba C. and Nyberg K. (2008). Inactivation of bacteria and viruses in human urine depending on temperature and dilution rate. *Water Research*, **42**, 4067–4074.

Winker M., Vinnerås B., Muskolus A., Arnold U. and Clemens J. (2009). Fertilizer products from new sanitation systems: Their potential values and risks. *Bioresource Technology*, **100**, 4090–4096.

Wohlsager S., Clemens J., Nguyet P. T., Rechenburg A. and Arnold U. (2010). Urine – A valuable fertilizer with low risk after storage in the Tropics. *Water Environment Research*, **82**, 840–847.

Chapter 13

The potential of control and monitoring

Gustaf Olsson

13.1 INTRODUCTION

Recent decades have witnessed extensive discussions on decentralized systems and source separation. Comparisons between conventional centralized and local solutions have been reported and analyzed (Larsen and Gujer 1996, 1997, 2001; Boller 1997; Ødegaard and Skrøvseth 1997; Otterpohl *et al.* 1997; Henze 1997; Harremoës 1997; Wilderer and Schreff 2000). To treat wastewater where it is generated is an attractive option. Orth (2007) presents a detailed discussion and comparison of centralized versus decentralized systems. Detailed accounts of motivations for decentralized systems are given in Part I of this book and a range of international experiences are described in Part IV.

One of the motivations for decentralized systems is that sewer systems connected to centralized treatment plants are inflexible (see also Maurer 2013). The investment costs are prohibitive and future maintenance suffers accordingly. Consequently, there is a great advantage in treating the wastewater close to where it is generated. This also eliminates many of the operational problems caused by the interdependencies of sewers and treatment plants. Decentralized solutions permit significant innovations (Larsen and Gujer 2001; Larsen 2011). Although the wastewater generated in an area served by such systems still needs to be collected, the piping is generally much shorter than that of centralized sewer systems.

The wastewater produced by municipalities and industries of diverse structures is of greatly varying composition, concentration and flow rate. Some of its fractions require advanced treatment, while others can be treated with much less effort. Larsen and Gujer (1996, 1997) have proposed the separation of urine at the earliest stage to facilitate treatment of the remaining wastewater. Since most of the nitrogen and half the phosphorus is contained in urine, separating urine at the source would completely obviate advanced nutrient removal at the treatment plants. This argument for source separation can be extended to other sources. The greywater produced from showers, washing and cleaning contains relatively small concentrations of nitrogen and phosphorous and needs much less treatment. Blackwater and greywater consequently require different treatment efforts and these fractions should be treated separately. Furthermore, since most of the wastewater nutrients are present in urine and faeces, it is quite obvious that the separation of toilet wastewater facilitates nutrient recycling (Larsen *et al.* 2009; Guest *et al.* 2010).

The value of water has been underestimated for too long. Increasing attention is being given to innovative thinking about water supplies in areas of water shortage. It is obvious that different qualities of water can be used for different purposes. Multiple water qualities can be produced and used in the same system. This concept of separate waste streams of highly variable character naturally suggests decentralized solutions: it is in the same spirit as smart water grids. The concept of a smart grid has been embraced in the field of electrical power systems and has inspired its application to other infrastructures. The driving force here is to create a water supply that takes water scarcity into consideration. Not all customers need the same quality of water supply, so we need a much more sophisticated water delivery system than is available today.

The ultimate purpose of a smart water grid is to minimize water consumption, ensure an adequate water quality at all times and operate the system as efficiently and consistently as possible (Olsson 2011). A smart water grid delivers water from suppliers to consumers using a wide range of computer control functions, including sensors, remotely controlled actuators and digital communication. Its measurement and monitoring system keeps track of all the water flowing in the system, that is. it should monitor not only the consumer behaviour and the demand-side requirements but also any bursts and slow leakages. A smart grid should automatically detect, analyze and respond to malfunctions and optimize asset use. It must be able to resist security attacks and empower and incorporate the consumer. It must have high levels of situation awareness and predictive capability.

A smart water supply must also offer smart solutions to wastewater treatment for various specified standards. Reused wastewater is returned via the smart grid to the customers. Huge differences naturally exist between the decentralized solutions found in many developing countries and the advanced treatments available in wealthier parts of the world. The present discussion of control and monitoring techniques is primarily focused on high-tech solutions and will consider alternatives to conventional advanced end-of-pipe wastewater treatment. In order to offer a viable alternative to high performance centralized wastewater treatment, a decentralized system must provide the same water quality for both small and large discharges, utilize resources efficiently, resist disturbances effectively and be competitive in cost. All these factors are a precondition for public acceptance (see Maurer 2013; Tchobanoglous and Leverenz 2013).

Small, decentralized systems must be easy to operate in order to be accepted. The user should only have to handle simple routine tasks. Professional service enterprises should be responsible for keeping the plant operating properly and for handling any failures and faults. At the same time, failures of single units must not cause the whole system to collapse. Remote sensing is a proven technology in many other areas and should also be implemented here, both in household size equipment and larger units.

Many authors in this book point out that the best available scientific advances must be utilized in order to achieve a decentralized solution that can feasibly replace or complement existing advanced wastewater treatment. Adequate performance criteria and regulatory standards must also be adopted.

13.1.1 Instrumentation, control and automation aspects

The ICA (instrumentation, control and automation) aspects can look different depending on the scale of the wastewater treatment unit. The load to a large treatment plant is mostly less variable, since many waste sources are combined. The requirements on measurement accuracy are likely to be greater for a large plant than for a family-sized unit. However, similar parameters have to be tracked. Small systems require some issues to be addressed that differ from those relevant to conventional centralized systems. They include:

- How to handle the more extreme load variations – both concentrations and flow rates – that may appear in a local plant?
- Does the user have sufficient interest to operate the plant?
- How much process knowledge is needed for the operation and who can provide this knowledge?
- Is the system cost-effective?
- Is the decentralized solution more sustainable than the centralized one?
- What kind of infrastructure is needed?

As already noted, a small wastewater treatment plant must supply a similar effluent quality or load to the environment as a centralized one and should be cost and energy efficient as well as easy to operate (Wilderer and Schreff 2000). Small plants showing extreme fluctuations in the inflow state and with no professional engineer on site need an operating method that enables them to react efficiently to these fluctuations while at the same time allowing easy maintenance (Ingildsen and Olsson 2001; Olsson *et al.* 2005). The aim of this chapter is to consider decentralized plants of various sizes and evaluate possible difficulties and opportunities in terms of instrumentation, control and automation. The design should also be adapted to the available control and operation options.

The influent pattern and sewer replacement are first considered. Several technologies are available for small-scale wastewater treatment. Instrumentation used to be considered a major obstacle to better control in wastewater treatment, but newer developments are promising. A large plant with skilled operators requires a different approach to instrumentation than a small package plant. In the first case, more advanced instrumentation can be adequately used and maintained, whereas a small plant requires rugged and low-cost instruments. The accuracy requirements also differ, as do other kinds of requirements, for instance on motors, pumps, compressors and valves.

Seen from a small-plant perspective, automated operation will become increasingly important. The plant must operate autonomously, must be resilient against disturbances and must react to sudden changes in load flow rate and composition. This makes detection and early warning systems important.

Finally we will stress the importance of standardization. Decentralized plants will only be competitive with respect to costs and energy with a high degree of standardization in plant design as well as in control and automation systems.

13.2 THE INFLUENT

In considering decentralized plants it is important to define suitable boundaries, and we start with the influent. In conventional systems, all kinds of wastewater are collected in the sewers, often mixed with run-off water. Leakage into sewers is a major issue and is seldom recognized as an operational problem for the wastewater treatment system. Thus the hydraulic load to the plant is often significantly greater than that of the true waste sources. However, the composition of the wastewater flowing to a large plant becomes less variable due to the many different sources of waste. In contrast, a small plant whose load comes from an individual household or small community will have much greater variations in influent entering the treatment unit (see Larsen and Gujer this volume). In the extreme case where one package plant is used for a single family, the load may be zero for several hours every day. Moreover, if the residents are not at home over an extended period, the plant may simply have to be shut down. Large load variations may mean that the efficiency of the plant is low. If several residents share the same facility, the load variations will be less extreme. In view of the large fluctuations in influent load with no professional engineer on-site, it is crucial that an automatic system enables the plant to react efficiently to these fluctuations (Olsson *et al.* 2005). The most obvious warning is to signal that the load has been below a certain limit during a given period.

From an operational point of view it is useful to know the flow rate. This measurement does not have to be accurate, but significant changes must be monitored and acted upon. A rough estimate of the flow rate can often be obtained, for example by measuring the pump velocity, some liquid level or even the weight of the tank. Blackwater naturally needs more advanced treatment than greywater. In order to achieve the same effluent quality as large centralized treatment plants, the plant must ensure C, N and P removal. This will require more advanced instrumentation and control. The greywater can often be treated successfully with membrane technology, see below, while keeping instrumentation and control to a minimum. However, the equipment should be monitored so that the plant operates reliably. In water-scarce areas, the water should obviously be reused. The delivery of reused water to the consumer certainly offers quite a sustainable solution compared to centralized systems. The thermal energy in the water can also be used locally. From a control perspective, the influent can be automatically monitored to assure adequate treatment. Simple valves allow the wastewater flow between basins to be redistributed. By running one, two or three reactors in series or parallel, more flexible operation may be obtained, depending on the load.

13.3 TREATMENT TECHNOLOGIES

The types of decentralized systems currently used vary from individual on-site wastewater management systems to satellite treatment units integrated with centralized systems (Tchobanoglous and Leverenz 2013). A system may serve one family, one apartment building, one hotel, one industry or a subdivision of a city. Today there is a wide range of available and suitable technologies that can be used to produce the required water quality. Tchobanoglous *et al.* (2004) and Asano *et al.* (2007) present possible collection and treatment systems. As the authors remark, we need a paradigm shift from effluent *disposal* ("wastewater is a problem") to water *reuse* ("wastewater is a resource.") Chapters 18–27 also contain a comprehensive discussion on possible process technologies.

It has to be emphasized that decentralized systems should be modular. As treatment becomes more ambitious, the system must be extended in stages (Wilderer and Schreff 2000). Both sewer and treatment overcapacities have been a common problem in the early life of centralized systems. Wilsenach *et al.* (2003) have emphasized the need for an integrated approach to handle the water system in a really systematic and efficient way. Such an approach has to include source control, water pollution control, nutrient recycling, source separation as well as adaptation to specific urban conditions and existing systems.

The development of membrane technology has been remarkable during the last decade, and its applicability to decentralized systems is evident (Meuler *et al.* 2008; Leslie and Bradford-Hartke 2013). Membrane units are used in package plants suitable for a single household as well as in larger units. It is recognized that membranes can replace conventional settler units. Boehler *et al.* (2007; see also Boller 2013) describe how a membrane bioreactor treats toilet wastewater and reuses it on site in the cable-car station at Zermatt, Switzerland. Grundestam and Hellström (2007) report the results of using an anaerobic membrane bioreactor followed by reverse osmosis for treating domestic wastewater from a new city district in Stockholm, Sweden.

UV disinfection and membrane filtration are important tools for creating microbial barriers. Membranes offer high disinfection efficiency and can retain faecal coliforms and other indicator organisms. An efficient membrane should be able to retain particles larger than 0.02 μm, which implies that even the smallest viruses can be caught. Microbial barriers such as membranes are often checked with indirect methods, such as pressure and bubble tests or by using conventional particle counters or turbidimeters. However, these ways of checking the membrane integrity are too insensitive to check the pore size in a membrane filter. A simple analysis method has been used in deep-sea research to quantify "virus-like particles." The water sample is run through a membrane filter. The virus-like particles are then stained with a fluorescent dye and counted in a fluorescent microscope. They should be larger than 0.02 μm (Rydberg 2011). Automation of this test procedure should be considered. Membrane technology is consequently important both for the hygienisation of wastewater and as part of water reuse systems (see Jefferson and Jeffrey 2013). Membrane cleaning should be considered and biofouling cannot be neglected. Automated cleaning should be developed for small-scale operations. Membrane cleaning can be based on regular time intervals but could also use real measurements and a simple signal (to a remote control room) of the transmembrane pressure (TMP). Cleaning procedures are described by Judd *et al.* (2008).

13.4 INSTRUMENTATION

To measure is to know. This is even more important in an unmanned process. The sensors should provide adequate on-line data. The instrumentation has to be robust, easy to maintain and cost effective. Naturally, only the most relevant sensors should be considered. The performance and reliability of many on-line sensors have improved remarkably in recent years and can be used directly in many control strategies (Jeppsson *et al.* 2002; Vanrolleghem and Lee 2003; Häck and Wiese 2006). Chapter 9 of Olsson *et al.* (2005) presents a survey of the state-of-the-art in wastewater instrumentation. It also provides a list of websites for further information about online sensors and analyzers. It includes a number of manufacturers of sensors/analyzers not only for nutrients but also for organic substances and sludge concentrations. It should not be considered complete, but will certainly help to find information on these sensors/analysers on the manufacturers' websites. The market and supply situation for online sensors/analyzers for wastewater treatment plants are expanding rapidly and are changing all the time due to mergers/acquisitions between manufacturers/ dealers, innovations, price reductions and so on.

One class of devices is used just to confirm that the plant is operating. Simple sensors can keep track of the equipment, and if any failure is noted an automatic alarm can be sent, either to the owner or to the remote operating personnel. Thus simple indicators of electric motor speed can be used to ensure that pumps and compressors are running. Simple level measurements or weight sensors can ensure that tank levels are within acceptable limits. Coarse gas flow measurements can monitor the methane production in a small digester, and conductivity sensors can monitor changes in the influent composition. The whole idea is *not* to obtain accurate information of the plant state, but to ensure normal operation.

It is anticipated that low-cost devices will soon be available to monitor the transmittance, total suspended solids, particle size and particle size distribution, turbidity as well as selected organic and inorganic chemical constituents. Developments in biosensors offer a promising basis for providing information about water quality (Alcock 2004). Thus the European SENSPOL network on "Sensors for Monitoring Water Pollution" (Alcock and Branston 2000) has developed many sensors for water pollution applications.

The development of wireless sensors is gaining considerable interest. As with all electronic devices, these sensors are becoming ever smaller and more versatile. Some of them can now detect everything from the rarest chemical to the most exotic bacteria. It is obvious that sensors should be used in key spots, such as the water outflows of reservoirs. If dangerous contamination is detected, the relevant part of the network can be shut down automatically.

13.5 MONITORING

Monitoring means tracking the operational state of a process via online instrumentation. Since a small treatment unit is usually unmanned, there is a great need to equip the plant with monitoring and early warning systems. This will include sensors providing information on the equipment status. Unlike humans, computers are infinitely attentive and can detect abnormal patterns in plant data. Conventional detection is mostly based on the amplitude or rate-of-change of a single variable.

Qualitative detection via sensors can provide useful information, particularly in a small system. Thus a simple digital signal can indicate if a pump is running or not. Timers added to motors or compressors can signal when normal maintenance would be required. Parameters related to water quality should indicate if the system does not perform normally. For example, if the dissolved oxygen (DO) concentration drops below a threshold value for an extended period or if pH or oxidation reduction potential (ORP) or other quality variables are subject to sudden changes, a warning can be transmitted. In many of these cases, the accuracy of the sensors is not critical. The main thing is that they give an early warning.

Any monitoring system must determine whether the acquired data are meaningful and correct. So any data generated in the plant, for local use or for reporting to a remote computer, has to be screened. By combining information from multiple sensors and mathematical models, other operational parameters that are not directly measureable can be estimated: these are known as soft sensors. This allows powerful monitoring tools to be built up. Examples include the oxygen uptake rate and the nitrification rate. Massive experience has been gained in monitoring, and numerous advanced algorithms are available for interpreting noisy signals or the detecting out-of-normal variations and extreme values (Olsson and Newell 1999; Olsson *et al.* 2005). Monitoring and early warning systems should provide necessary information about the status of the plant operation, and this can also be communicated to a remote location at which professional operators are available.

Given the potential of the Internet and the widespread use of network communications, it is obvious that a large number of small units could contact a control centre staffed by professional operators. This is a proven technology and extensive experience has already been acquired in this kind of remote monitoring (Lee *et al.* 2004).

New regulations are probably needed to verify that a local treatment plant performs adequately for its given task. It is crucial here that the effluent quality can be monitored by online instruments. In view of the number of plants involved, this would be necessary both to ensure the good quality of "conventional" effluents and of the reused water.

13.6 ACTUATORS

Actuators are the devices that translate information into "muscles," such as electrical motors, pneumatic devices and hydraulic valves. The performance of these devices is often neglected in many control systems applications, causing them to fail. Plants can be operated more flexibly by assuring adequate controllability of its pumps, motors and valves. The requirements on the actuators often differ for small units compared with large plants. While it is crucial to control the influent flow rate smoothly in a large plant, lower operating costs may be achieved by using on/off control of small pumps or compressors in a small unit. Intermittent aeration can be adequate in a small package plant. The influent flow can also be split in different ways to the process units, depending on the influent water characteristics or the flow rate. However, remote control of actuators in a small plant from a central service group can also be most useful.

13.7 OPERATING COMPETENCE

If the economic goals are to be met with many small plants, these must be of efficient design and operation. One objection to decentralized systems is their complexity, as the operator may not be sufficiently skilled. However, we should not forget that we are always dealing with highly complex systems in our daily lives. A modern computer is extremely complex. Most car drivers would be quite unable to service their vehicles themselves. And yet these complex systems are routinely used by non-specialists. And as decentralized electrical power generation becomes increasingly widespread, non-professional users frequently utilize small turbines as well as photovoltaic systems for power generation. The challenge is to hide the complexity for the unsophisticated user and provide a user interface that is transparent to him or her. Furthermore, the available communication and control technology allows a large number of small plants to be centrally supervised (Lee *et al.* 2004; Alex *et al.* 2003). This makes it more realistic to hire competent operators and guarantee high quality operation.

The "end user" of decentralized treatment units cannot be expected to be experienced in operating a highly complex bio-technological system or to be motivated to provide any kind of serious supervision. Each local plant should be controlled locally with local digital controllers but should be connected to a professional supervisory service. It should be possible to operate each plant in isolation if the communication line is broken. However, as pointed out by Wölle *et al.* (2007), experts who remotely supervise many small plants are faced with a diversity of predominantly unstructured data which has to be transformed into useable information to characterize and operate a local unit efficiently and safely. Tools have to be developed that enable an expert located at a remote location to supervise multiple systems and analyse critical operating conditions and at the same time provide the on-site user (the homeowner, hotel or apartment building serviceman etc.) with adequately processed information. This can consist of online maintenance guidance or basic process information.

Several large vendors – like Siemens, ABB, Rockwell and Hitachi – offer decision support systems that allow advanced systems such as power plants and process units in the chemical industry to be monitored and controlled remotely – via the Internet or other communication channels. This technology also makes remote asset management possible. For example, the package giant Tetra Pak remotely monitors packaging machines in more than 100 countries from a single location via the Internet. Software modules can readily be updated and the functionality of machine units tested remotely. A similar technology can be adapted to the operation of many small wastewater treatment units. Remote sensing and monitoring systems combined with modern devices for data transfer allow simultaneous supervision of a multitude of small plants in municipal areas. Computer or communication capacity is no limitation, even if thousands of units were to be connected.

13.8 THE NEED FOR STANDARDIZATION

To make these systems economically competitive requires *standardized manufacturing* and modularized units. This demands manufacturing competence that goes beyond plumbing, piping and tank production. It is crucial for the decentralized plant manufacturer to have in-depth process and control knowledge. Carmakers have come a long way towards manufacturing highly complex vehicles from standardized units for the engine parts, electrical system, car body, compartment and so on. This approach can naturally also be applied to small sanitation units, provided that mass production is made possible (see also Truffer *et al.* 2013).

Not only the process units, but the actuators and instruments and their interfaces to the process should also be highly standardized. Standardization must be applied to the sensing systems, the signal levels, the monitoring and control software, the communication protocols and the application interfaces. This means that we have to think much more in terms of "plug and play" for the sensors and instruments. Adding a new sensor should not require a re-design of the treatment unit. Communications with added sensors should be available in the same way as we now add new peripherals to our PCs. This also means that new control algorithms relating to a new sensor should be readily put into operation without having to reprogram the control system software.

Setting the standard too early might hamper innovation. On the other hand, the lack of a standard may inhibit further investment. Manufacturers may be worried that they might apply the wrong technology. It is probably a wise strategy to implement advanced decentralized systems step by step and not introduce all features at once. Otherwise, introducing a new technology too quickly can create a backlash, as experienced by some power utilities. Smart meters and new pricing schemes were installed at the same time in some places: the consumers were overwhelmed and blamed the new technology for higher prices, since the reasons for their higher power bills were obscured. Because public acceptance is central for any wastewater-related technology that affects people directly, for example, in their own household, only economically competitive source-separating technologies will have a chance in practice. Prototypes applied today are often too expensive to compete with existing end-of-pipe technology, and only mass production will be able to reduce prices (Truffer *et al.* 2013). Maurer (2013) estimated the investment levels at which source-separating technologies would achieve a break-even with conventional technology.

Standardization of instrumentation specifications makes it possible to specify, compare and select the optimum instrumentation, not only in technical terms but also in economic terms by calculating their costs (Olsson 2008). Increased confidence in instrumentation is now driven by the fact that clear definitions of performance characteristics and standardized tests for instrumentation have become available (ISO 15839:2003). The investment costs for the device itself are often a minor part of the costs incurred during the lifetime of the instrumentation. Measurements derived from the instrumentation must be all around the clock every day of the week. Information needs to be properly extracted from the measured data. Thus, as already emphasized, the instrumentation must always include adequate data screening, measurement processing and extraction of features from the measurements.

Software for monitoring and control purposes must be standardized to be economically feasible. Many large vendors have proposed modular packages for data acquisition, data screening, signal processing including filtering, data reconciliation, fault detection and diagnosis as well as control algorithms. The databases for all online data must also be standardized. The problem with many large real-time control systems is simply that they are too large. Smaller and low-cost modules should be made available in order to make the software systems competitive for small package systems. The availability of this type of software will also simplify the calculation of key reaction indicators (Irizar *et al.* 2008; Olsson *et al.* 2005). The control algorithms can often be kept quite simple. It has been recognized that advanced controllers seldom offer better performance than conventional PI (proportional-integral) controllers in wastewater treatment control (Olsson *et al.* 2005).

13.9 CONCLUSIONS

Decentralized systems offer new flexibility when compared with traditional centralized systems. Small systems allow the separation of waters of different qualities as well as the recovery of nutrients, energy and water. Instrumentation, control and automation should be an integrated part of small system design. Since these units are designed to be unmanned, rugged solutions are necessary. The instruments and sensors to be used in such systems will typically differ from those found in large centralized systems, and the same applies to sensor accuracy and actuator flexibility. Monitoring and early warning systems are key components in an automated plant. We also have emphasized the importance of standardized plant design, the manufacturing and sensor requirements, the software structure for monitoring and control and computer communications. These factors are crucial if decentralized plants are to be competitive in cost and energy.

A new water treatment system goes beyond decentralized technology traditionally used in rural areas. It involves many customers who will not always behave as rationally or logically as an engineer would like. Water professionals may have to learn more about social engineering in order to persuade their customers to be more committed to their systems (see also Lienert 2013). If customer behaviour is not taken into consideration, an otherwise well-designed automatic system may fail completely. This lesson has been learnt in the process industry and in the context of centralized wastewater systems. Many failures have been reported when the system operators were not involved in the design process. As Larsen (2011) points out, "soft" issues of acceptance and compliance have to be considered at the same time as the more "rational" estimates of cost and performance. The ICA can play an important role in transferring responsibility to remotely located professional operators who can support individual system owners.

REFERENCES

Alcock S. J. and Branston L. (2000). SENSPOL: Sensors for Monitoring Water Pollution from Contaminated Land, Landfills and Sediment. http://www.cranfield.ac.uk/biotech/senspol.htm

Alcock S. J. (2004). New developments in sensor technology for water quality surveillance and early warning, *Water Sci. Technol.*, **50**(11), 1–6.

Alex J., Jumar U. and Schütze M. (2003). On-line measurement data of wastewater system via WAP mobile phones. *Water Sci. Technol.*, **47**(2), 205–211.

Asano T., Burton F. L., Leverenz H., Tsuchihashi R. and Tchobanoglous G. (2007). *Water reuse: Issues, technologies and applications*. McGraw–Hill, New York.

Boehler M., Joss A., Buetzer S., Holzapfel M., Mooser H. and Siegrist H. (2007). Treatment of toilet wastewater for reuse in a membrane bioreactor. *Water Sci. Technol.*, **56**(5), 63–70.

Boller M. (1997). Small wastewater treatment plants – a challenge to wastewater engineers. *Water Sci. Technol.*, **35**(6), 1–12.

Boller M. (2013) Source control and source separation: the Swiss experience. In: *Source Separation and Decentralization for Wastewater Management*, T. A. Larsen, K. M. Udert and J. Lienert (eds.), IWA Publishing, London, UK, pp. 439–446.

Grundestam J. and Hellström D. (2007). Wastewater treatment with anaerobic membrane bioreactor and reverse osmosis. *Water Sci. Technol.*, **56**(5), 211–217.

Guest J. S., Skerlos S. J., Daigger G. T., Corbett J. R. E. and Love N. G. (2010). The use of qualitative system dynamics to identify sustainability characteristics of decentralized wastewater management alternatives. *Water Sci. Technol.*, **61**(6), 1637–1644.

Häck M. and Wiese J. (2006). Trends in instrumentation, control and automation and the consequences on urban water systems. *Water Sci. Technol.*, **54**(11–12), 265–272.

Harremoës P. (1997). Integrated water and waste management. *Water Sci. Technol.*, **35**(9), 11–20.

Henze M. (1997). Waste design for households with respect to water, organics and nutrients. *Water Sci. Technol.*, **35**(9), 113–120.

Ingildsen P. and Olsson G. (2001). *Get more our of your wastewater treatment plant – complexity made simple*. Danfoss, Denmark (www.danfoss.com/analytical) ISBN no 87-87411-01-6 (http://212.242.79.105/danfossAN/shop/aboutthebook.asp)

Irizar I., Alferes J., Larrea L. and Ayesa E. (2008). Standard signal processing using enriched sensor information for WWTP monitoring and control. *Water Sci. Technol.*, **57**(7), 1053–1060.

Jefferson B. and Jeffrey P. (2013) Aerobic elimination of organics and pathogens: greywater treatment. In: *Source Separation and Decentralization for Wastewater Management*, T. A. Larsen, K. M. Udert and J. Lienert (eds.), IWA Publishing, London, UK, pp. 275–290.

Jeppsson U., Alex J., Pons M. N., Spanjers H. and Vanrolleghem P. A. (2002). Status and future trends of ICA in wastewater treatment – A European perspective. *Water Sci. Technol.*, **45**(4–5), 485–494.

Judd S., Kim B.-G. and Amy G. (2008). Membrane Bio-reactors. Chapter 13 in *Biological Wastewater Treatment – Principles, Modelling and Design* (M. Henze, M. van Loosdrecht, G. Ekama and D. Brdjanovic, Editors) UNESCO, IWA Publishing, London.

Larsen T. A. and Gujer W. (1996). Separate management of anthropogenic nutrient solutions. *Water Sci. Technol.*, **34**(3–4), 87–94.

Larsen T. A. and Gujer W. (1997). The concept of sustainable urban water management. *Water Sci. Technol.*, **35**(9), 3–10.

Larsen T. A. and Gujer W. (2001). Waste design and source control lead to flexibility in wastewater management. *Water Sci. Technol.*, **43**(5), 309–318.

Larsen T. A., Alder A. C., Eggen R. I. L., Maurer M. and Lienert J. (2009). Source separation: will we see a paradigm shift in wastewater handling? *Environ. Sci. Technol.*, **43**(16), 6121–6125.

Larsen T. A. (2011). Redesigning wastewater infrastructure to improve resource efficiency. *Water Sci. Technol.*, **63**(11), 2535–2541.

Lee H., Min Y. M., Park C. H. and Park Y. H. (2004). Automatic control and remote monitoring system for biological nutrient removal on small wastewater treatment plants in Korea. *Water Sci. Technol.*, **50**(6), 199–206.

Leslie G. and Bradford-Hartke Z. (2013). Membrane processes. In: *Source Separation and Decentralization for Wastewater Management*, T. A. Larsen, K. M. Udert and J. Lienert (eds.), IWA Publishing, London, UK, pp. 367–382.

Lienert J. (2013). High acceptance of source-separating technologies – but…. In: *Source Separation and Decentralization for Wastewater Management*, T. A. Larsen, K. M. Udert and J. Lienert (eds.), IWA Publishing, London, UK, pp. 193–208.

Maurer M. (2013). Full costs, (dis-)economies of scale and the price of uncertainty. In: *Source Separation and Decentralization for Wastewater Management*, T. A. Larsen, K. M. Udert and J. Lienert (eds.), IWA Publishing, London, UK, pp. 85–100.

Meuler S., Paris S. and Hackner T. (2008). Membrane bio-reactors for decentralized wastewater treatment and reuse. *Water Sci. Technol.*, **58**(2), 285–294.

Ødegaard H. and Skrøvseth A. F. (1997). An evaluation of performance and process stability of different processes for small wastewater treatment plants. *Water Sci. Technol.*, **35**(6), 119–127.

Olsson G. and Newell B. (1999). Wastewater Treatment Systems. Modelling, Diagnosis and Control. IWA Publishing, London.

Olsson G., Nielsen M., Yuan Z., Lynggaard-Jensen A. and Steyer J. P. (2005). *Instrumentation, Control and Automation in Wastewater Systems*, IWA Publishing, London.

Olsson G. (2008). Process Control. Chapter 15 in *Biological Wastewater Treatment – Principles, Modelling and Design*, M. Henze, M. van Loosdrecht, G. Ekama and D. Brdjanovic (eds.), UNESCO, IWA Publishing, London.

Olsson G. (2011). Smart Water & Power Grids – drivers, opportunities and challenges. Invited presentation, the 1st *Int. Workshop on Smart Water Grids*, Seoul, Rep. of Korea, March 23–25, 2011.

Orth H. (2007). Centralised versus decentralised wastewater systems? *Water Sci. Technol.*, **56**(5), 259–266.

Otterpohl R., Grottker M. and Lange J. (1997). Sustainable Water and Waste Management in Urban Areas. *Water Sci. Technol.*, **35**(9), 121–133.

Rydberg H. (2011). Checking the microbial barrier efficiency (in Swedish). *Svenskt Vatten* (Swedish Water), **1**, 12.

Tchobanoglous G., Ruppe L., Leverenz H. and Darby J. (2004). Decentralized wastewater management: challenges and opportunities for the twenty-first century. *Water Sci. Technol.: Water Supply*, **4**(1), 95–102.

Tchobanoglous G. and Leverenz H. (2013). The rationale for decentralization of wastewater infrastructure. In: *Source Separation and Decentralization for Wastewater Management*, T. A. Larsen, K. M. Udert and J. Lienert (eds.), IWA Publishing, London, UK, pp. 101–116.

Truffer B., Binz C., Gebauer H. and Störmer E. (2013). Market success of on-site treatment: a systemic innovation problem. In: *Source Separation and Decentralization for Wastewater Management*, T. A. Larsen, K. M. Udert and J. Lienert (eds.), IWA Publishing, London, UK, pp. 209–224.

Vanrolleghem P. A. and Lee D. S. (2003). On-line monitoring equipment for wastewater treatment processes: state of the art. *Water Sci. Technol.*, **47**(2), 1–34.

Wilderer P. A. and Schreff D. (2000). Decentralized and centralized wastewater management: a challenge for technology developers. *Water Sci. Technol.*, **41**(1) 1–8.

Wilsenach J. A., Maurer M., Larsen T. A. and van Loosdrecht M. C. M. (2003). From waste treatment to integrated resource management. *Water Sci. Technol.*, **48**(1), 1–9.

Wölle J., Steinmetz H., Hansen J., Einsfeld K. and Ebert A. (2007). An intelligent visualisation and decision support system for decentralised wastewater treatment plants. *Water Sci. Technol.*, **56**(5), 183–191.

Chapter 14

High acceptance of source-separating technologies – but …

Judit Lienert

14.1 INTRODUCTION

For a long time, the international community dealing with source-separated wastewater seemed to agree: the idea is great, but a major obstacle is acceptance. Likewise, acceptance of urine-based fertilizers by consumers and farmers appeared to be highly critical. Source-separating wastewater technologies challenge users because they break with the "flush-and-gone" paradigm of central wastewater management by actually entering peoples' bathrooms. This is especially true for the toilet. The example of urine separation (NoMix) is used to discuss the acceptance of source separation technologies. This chapter is based on social science studies carried out in Switzerland as part of the cross-cutting Eawag project Novaquatis (www.novaquatis.eawag.ch; 2000–2006) and additional European studies (review: Lienert and Larsen 2010).

Compared with other source-separating technologies such as greywater re-use (see Jefferson and Jeffrey 2013), the toilet is presumed to be the most sensitive area. Unlike the end-of-pipe approach, where human excreta are considered as waste, source-separating approaches in most cases regard excreta as a resource worth recycling. This ancient idea has been practiced for millennia. The use of human urine is known among many peoples, including the Celts, the ancient Romans and the Chinese (historical overview: Bracken *et al.* 2007). However, human excreta is a delicate issue that evokes strong emotions such as disgust (Dellström Rosenquist 2005). The water closet proved to be far superior to lower-tech sanitation solutions because it allowed this issue to be physically and mentally avoided. To overcome such psychological and cultural obstacles, either a mentality shift or a technology shift is required. The first approach is taken by some authors, such as Dellström Rosenquist (2005), although she acknowledges that experience and research indicate how hard or even impossible it is to change "the issue of comfort, cleanliness, feelings of disgust, mental and physical avoidance and perception of safety (which) are to a certain extent products of social ideas and perception." The Novaquatis project team decided that rather than trying to adapt humans to the requirements of source-separating technologies, efforts would be put into designing solutions that are equivalent to today's flush-and-gone technology.

At the end of the 1990s when the Novaquatis project started, the assumption that user acceptance of NoMix toilets would be critically low was actually based on invalid comparisons. The early "modern" urine-diverting toilets developed in Sweden and other Nordic countries were often compost units that required manual handling of faeces, with a urine-diversion component, or pioneering technology lacking a well-adapted design. These early systems were fully appropriate for the setting for which they were developed, namely remote summer houses on islands near the Baltic Sea with no connection to a centralized sewage system, as well as eco-villages (Kvarnström *et al.* 2006). However, they have some drawbacks (see below) for mainstream households in urban areas. Sweden has a long history of urine diversion: around 20,000 urine-diverting "Marino closets" were already in use in Stockholm at the end of the 19th century. They were introduced to reduce smell and the filling speed of latrine buckets (Kvarnström *et al.* 2006). So it's not surprising that Sweden is a pioneer in modern urine-diverting toilets. However, the transfer of low-tech solutions, also known from poor rural environments in developing countries, into a modern, urban setting was quite rightly seen as critical. Another striking factor at the end of the 1990s is that nobody had designed NoMix implementation projects that aimed to receive scientifically sound feedback from a broader public. I suggest that this is because most research was driven by engineers without the expertise to carry out social science research. This has recently changed to some degree.

One aim of this chapter is to introduce engineers to social science methods that have been applied in the context of NoMix technology. I will first give an overview of quantitative and qualitative approaches. A second aim is to summarize the feedback by users of NoMix technology. This includes acceptance by farmers and consumers of urine re-use. The chapter concludes by pointing out the main technological drawbacks of the existing technology, based on the feedback, and the need for future research and technology development.

14.2 SOCIAL SCIENCE METHODS

Feedback from users is needed in order to improve technological innovations. Typical research questions for NoMix toilets first address their idea and acceptance, such as: "Are NoMix toilets equivalent to conventional ones?" Product developers need detailed assessments of technological features, for instance how well the flush works, or if the design is appealing. Sociologists and engineers may want to know whether users would change their habits and, for instance, sit on NoMix toilets to urinate. If not, the technology has to be changed or social scientists must ask follow-up questions to understand the underlying reasons. Psychological intervention strategies can be designed to increase behavioural change. Such experiments again require feedback from trial participants. Generally, it may be interesting to find patterns, for example, systematic differences in acceptance based on gender or environmentally friendly attitudes.

Ideally, a strategy is designed to address research questions in different stages. When little is known, qualitative methods (e.g., focus groups, interviews) are useful. The intensive exchange between users and researchers can then help to set up quantitative surveys that collect representative but less-detailed population data. A subsequent experiment can explicitly test a research hypothesis.

14.2.1 Quantitative questionnaire surveys

A common way of obtaining feedback is to carry out surveys. To ensure that unbiased and representative data is collected, some points which are extensively discussed in the sociological and psychological literature must be considered. The classic textbook by Dillman *et al.* (2009) on how to conduct internet, mail and mixed-mode surveys is strongly recommended. This outline follows the recommendations given in that book. Survey methods are strongly influenced by technological change. It used to be standard to ask a (representative) sample of people via interviews or written questionnaires about their opinions and behaviours (Dillman *et al.* 2009, Chapter 1). Recently, telephone surveys became standard, but are now being replaced by computer-assisted and web-based surveys. Examples are *computer-assisted telephone interviewing* (CATI), *computer-assisted personal interviewing* (CAPI) and *computer-assisted self-interviewing* (CASI). The internet has induced the biggest change. The advantages of *computer-assisted web interviewing* (CAWI) are obvious: many respondents can be easily reached. Costs are low since interviewer wages, printing and postage are obviated; and results can be delivered quickly. A drawback is that some user groups such as elderly people may not have internet access. Professional survey firms account for this by methods such as avoiding the "self-selection" of respondents. Often, *panel surveys* are used, that is, a representative set of people agrees to respond periodically to surveys. Moreover, numbers are changing fast: at the end of 2008, 85% of 15–74 year old Swiss residents had internet access (2006: 78%), and 91% of the working population (survey by LINK-institute, www.link.ch). Recent technologies also allow different survey modes to be mixed and interactive design concepts to be used.

To design good surveys, the reasons why they can fail should be understood (Dillman *et al.* 2009, Chapter 2). The four basic errors are: coverage, sampling, nonresponse and measurement error (Groves 1989). *Coverage error* means that only a part of the population is surveyed (e.g., excluding those without internet access or not in the telephone directory; Dillman *et al.* 2009, Chapter 3). *Sampling error* (not everyone is sampled) is overcome by random sampling; but nevertheless error remains. Dillman *et al.* (2009, Chapter 3) discuss probability sampling as well as formulas to measure this error and to determine sample sizes. *Nonresponse error* occurs because not everyone responds. A check should be made for systematic differences between those that do and do not answer; and study differences (minimization of non-response error: Dillman *et al.* 2009, Chapter 7). Finally, if respondents answer imprecisely, this produces a *measurement error*. This is often due to poor questionnaire design or construction such as unclear wording (Dillman *et al.* 2009, Chapters 4–6).

Dillman *et al.* (2009, Chapter 2) propose the *"Tailored Design Method"* that uses multiple motivational features to encourage a high quantity and quality of responses. Important features are: (A) *increase the benefits of participation* (e.g., provide information and incentives, make questionnaire interesting, say thank you); (B) *decrease the costs of participation* (e.g., easy to respond, use language that subordinates the interviewer by asking for assistance, make questionnaire short, minimize requests for personal information); and (C) *establish trust* (e.g., support by authority, token of appreciation in advance, confidentiality).

A survey is fundamentally about how to formulate *good questions* (Dillman *et al.* 2009, Chapters 4, 5). This depends on factors such as the survey mode (telephone or mail) or the type of information (e.g., it is easier to give your age than offer an opinion about a complex problem). *Open-ended* question formats offer a blank for the answer. This is appropriate if the respondent is to be influenced as little as possible. However, these options are more likely to be skipped or provide inadequate responses. Open-ended questions also require coding. *Closed-ended* formats or *scalar questions* provide a list from which the preferred answer is chosen. The presentation of these options strongly influences their interpretation by respondents. Various other problems should be considered such as using vague quantifiers ("very often" is anything from daily to monthly). Such problems can be overcome by using different formats for the same question and checking for inconsistencies. Further guidelines include asking only one question at a time, using simple words or choosing an appropriate number of categories. The latter also depends on the statistical analysis. I will not discuss statistics here, as there are many good textbooks. For instance, Bortz and Döring (2009) focus on social psychology and cover both quantitative and qualitative approaches, including statistics.

It is important to create a *respondent-friendly questionnaire*. Dillman *et al.* (2009, Chapter 6) emphasize that the first question determines whether people will respond to the survey at all. Sensitive questions should be placed near the end. A body of research indicates that answers can differ significantly depending on the order of questions. Sources of order effects may be cognitive.

For example, earlier questions can bring certain issues to the mind more readily later on (priming effect); or people may answer questions that are perceived as related on the basis of similar considerations (carryover effect). Order effects can also have a normative basis; for example, respondents want to appear neutral or they try to answer questions because they wish to present themselves in a good light.

In Chapter 6, Dillman *et al.* (2009) give very useful advice on *designing a visually attractive questionnaire*. People usually process information in three stages: they (1) understand the basic layout of a page, (2) organize the information and (3) complete the task. A good questionnaire design supports all three steps, for example, with consistent visual properties across pages such as colour and size so that respondents quickly grasp the basic layout. Other aspects may include avoiding visual clutter or minimizing the complexity of matrices. It is mandatory to *pre-test* the questionnaire by having a varied selection of people who report any problems. Specific procedures for pretesting have been developed (see Presser *et al.* 2004).

Finally, *implementation procedures* are discussed in Dillman *et al.* (2009, Chapter 7). These researchers have developed methods that easily ensure high response rates of 50–70%. They emphasize that implementation must be designed just as carefully as the questionnaire itself and in such a way to appeal to many different respondents. Three basic guidelines should be followed and strategically timed: (1) *personalize* all contacts, (2) send a *token of appreciation* with the survey request and (3) use *multiple contacts*, each with a different appeal (e.g., pre-notice letter, questionnaire mailing, thank-you postcard, replacement questionnaire and final contact with a different mode of delivery). I strongly encourage following these and further recommendations.

Finally, the study question, the *hypothesis*, and its analysis must be defined from the start. The *explanatory variables* to explain the data and the appropriate *statistical methods* to test the hypothesis must be clear. I emphasize this because, unfortunately, many surveys on the acceptance of NoMix technology were carried out without appreciating these basic principles; many of them are therefore non-representative (Lienert and Larsen 2010). Nevertheless, we do believe that these data are very valuable, which is why we made them accessible in our review. The main findings are summarized in Section 14.3.2.

14.2.2 Qualitative methods

Qualitative research can evaluate specific topics in detail. Its drawback is a loss of representativity and a difficulty to generalize results. Ideally, qualitative research should be complementary to quantitative surveys.

As an example, when entering a new field it might be important to first identify the people involved and their interests. A *stakeholder analysis* helps us to determine who is important for decision-making, or who is principally affected (Grimble and Wellard 1997). Once the main actors have been identified, first approaches to sketch out the study topic could include: *semi-structured expert interviews* (for an example on urine source separation see Krantz 2005), *moderated focus-group discussions* (e.g., Jäger *et al.* 1999) or *diary surveys* (for an example on WC usage behaviour see Friedler *et al.* 1996; for urine source separation see Krantz 2005). In the event of large uncertainties, a *scenario analysis* might be useful to discuss possible pro-active measures based on different pictures of the future (e. g., Schoemaker 1995). Such qualitative approaches help to structure the research area, to frame a more-quantitative follow-up study and to develop specific study questions.

Qualitative methods can also be applied *after* a first quantitative, representative survey in order to better understand the results. Other good methods could include *in-depth interviews*, desk-top analyses based on theoretical considerations such as *Diffusion Theory* (Rogers 1983), *historical analyses* (e.g., Bracken *et al.* 2007) or *gender research* (for a feminist approach to sanitary engineering see Rydhagen 2002). If specific alternative measures have been identified and something is already known about the attitude of the stakeholders, *Multi-Criteria Decision Analysis (MCDA)* can be applied (e.g., Keeney 1982, 1992). MCDA assesses how well each alternative performs with respect to "objective criteria" (e.g., low costs or high comfort) as well as "subjective stakeholder preferences." If a high potential has been identified for some alternatives, *round table discussions* are a possible next step to bring actors together to further develop or implement a specific technology (see e.g., Truffer *et al.* 2013).

It is obviously not my intention to review qualitative social science research methods. Rather, I aim to illustrate their potential. For further reading, I would recommend certain textbooks (e.g., Scholz and Tietje 2002; Bortz and Döring 2009). Below, I present some qualitative approaches that were applied in the context of NoMix technology in Switzerland. This is followed by user feedback on this technology obtained from quantitative surveys.

14.3 ACCEPTANCE OF NOMIX TECHNOLOGY

14.3.1 Some results from qualitative approaches

In Switzerland, consumer attitudes towards NoMix technology were first addressed by focus groups (Pahl-Wostl *et al.* 2003). *Focus groups for integrated assessment* use a participatory method drawing on elements of *marketing* and *public opinion research* (Jäger *et al.* 1999). Moderated discussions were performed in ten gender-specific groups; the participants were a representative sample of the Swiss population. They received relevant information by visiting a NoMix toilet and using the NoMix Tool (www.novaquatis.eawag.ch/tool), an *interactive computer-based information system*. They then discussed the advantages and disadvantages of the technology. As in later studies, acceptance of this technology and a urine-based fertilizer was high under the condition that NoMix toilets offer the same comfort as conventional toilets at existing prices and that urine-based fertilizers are safe. Some interesting points were addressed; for instance, main arguments against NoMix technology put forward by women were the additional effort required for maintenance, urine collection and transport. For men, technical problems were of major concern.

Larsen and Lienert (2003) used a theoretical approach to analyze strategies that could lead to a rapid diffusion of NoMix technology by combining a *stakeholder analysis* with *diffusion theory* (Rogers 1983). Main stakeholder groups are households, farmers, sanitary firms and wastewater professionals. The attitude of households was based on the focus groups (Pahl-Wostl *et al.* 2003) and that of farmers on a questionnaire survey (Lienert *et al.* 2003). It was concluded that the attitude of these two groups is not critical, a conclusion backed by later research summarized below (Section 14.3.2). Consumers in particular have a surprisingly positive attitude provided that safety and high technological standards are ensured. Informal contacts with large sanitary firms indicated that these producers *do* believe that the technology can be developed to reach a high benchmark. However, they doubted whether more decentralized NoMix approaches would be able to compete with the existing centralized system. To reduce the uncertainty for these firms, a much stronger commitment from wastewater professionals would be required. *Diffusion Theory* (Rogers 1983) helps to explain why this is not (yet) the case and indicates that early diffusion amongst wastewater professionals is the most critical factor. To overcome this barrier, Larsen and Lienert (2003) proposed to use transition scenarios as a stepping stone. These reduce the uncertainty and complexity of the NoMix technology, but increase its capacity for trials and observation.

Once different technological or management alternatives can be sufficiently well specified, *Multi-Criteria Decision Analysis (MCDA)* can support the choice between options. MCDA is able to compare and rank many alternatives for different stakeholder groups, based on a set of goals (criteria). At the core of *value-focused thinking* (Keeney 1992) lies the combination of *objective predictions* of how well each alternative achieves each goal (e.g., low costs, high public acceptance, etc.) with the *subjective preferences* of the decision maker (e.g., how important are costs compared to public acceptance). The MCDA approach was applied to a case study near Zürich (Borsuk *et al.* 2008). The idea was to postpone expensive upgrades of a wastewater treatment plant by using NoMix technology for peak shaving of nutrients, among other options. The analysis found considerable technological "lock-in" caused by an unbalanced distribution of short-term benefits; that is, low payoffs for some stakeholders. Given current priorities, it was considered unlikely that a fixed course of action (including the status quo) would be desirable to all stakeholders. Borsuk *et al.* (2008) proposed a way forward that is open to future options and is not significantly disadvantageous to any stakeholder at the present time, namely peak shaving with conservative CSO control.

I will not go into further detail. I hope to have demonstrated with this short summary that qualitative approaches provide deepened insight far more than quantitative surveys alone. It cannot be emphasized enough that the research question should drive the choice of analytic method and not vice versa.

14.3.2 Results from quantitative approaches

The good news from the questionnaire surveys in pilot projects is that NoMix toilets are generally very well accepted by the large majority of users (Lienert and Larsen 2010). Our review included feedback from 2720 questionnaires from 33 surveys in the following seven European countries: Austria, Denmark, Germany, Luxembourg, Sweden, Switzerland and The Netherlands. These surveys are extremely heterogeneous (e.g., sample sizes, questions) so that the sample size for many questions is considerably lower than 2720. On average, 84% (standard deviation ±13%) of 1212 questioned users found the idea of urine source separation convincing, 79% (±4%) found NoMix toilets in public buildings or at their workplace a good idea, and 79% (±13%) would move into an apartment with a NoMix toilet or would consider installing it at home.

Unfortunately, many surveys are not representative because they fail to meet the methodological requirements (Section 14.2.1). Thus poor data prevented us from carrying out a statistical meta-analysis to identify the reasons for lower acceptance of NoMix toilets by certain users. However, we do regard the feedback obtained as very important. We are confident that it permits a good understanding of the results to be expected from representative national surveys, since they all point in the same direction despite coming from very different kinds of projects (e.g., universities, eco-villages, exhibitions or urban apartments). Moreover, we obtained very similar numbers in our own representative Swiss surveys, for instance the one carried out in a public library by a survey firm.

The review indicated some differences between countries (Figure 14.1). However, the differences between institutional projects and private homes were much more striking. This was especially evident for the practical aspects of NoMix toilets. For instance, 85% (±12%) of users in public buildings were generally satisfied with NoMix toilets, but only 71% (±8%) of those in private homes. Similarly, 80% (±7%) of users in public buildings found the design of NoMix toilets the same or better than that of conventional toilets, but only 70% (±31%) in homes. On the other hand, some requirements of NoMix toilets were easier to comply with in a private setting: 78% (±20%) usually sit on NoMix toilets to urinate at home, but only 68% (±6%) in public places.

As discussed in Lienert and Larsen (2010), people support the idea of urine source separation irrespective of the setting. However, having a NoMix toilet in ones' home is more demanding than using it occasionally elsewhere. This fact is amplified because toilets in public buildings are maintained by professionals so that some drawbacks are less evident. The data also support the hypothesis that people with longer, more-frequent use of NoMix toilets are more critical than visitors to Swiss public libraries or vacationers in Swedish holiday homes, for instance. The drawbacks of NoMix toilets are presented below (Section 14.5.1).

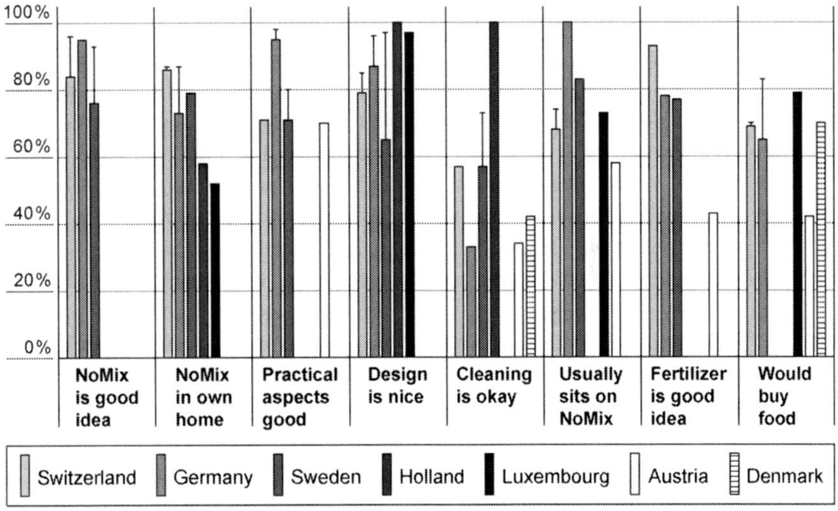

Figure 14.1 Results of surveys on users of NoMix toilets in seven European countries; for details (e.g., sample sizes, references to original data) see Lienert and Larsen (2010). We show mean responses (%) to: idea of NoMix is convincing / would accept NoMix toilet at home / is satisfied with practical aspects of NoMix toilets / design / cleaning of NoMix toilet is same or better than conventional toilet / usually sits on NoMix toilet to urinate / urine as fertilizer is (very) good idea / is willing to purchase food fertilized with urine. The answers for sitting behaviour came only from men in Austria and Sweden; all other samples are from both sexes. Error bars: standard deviation in case of several surveys.

In our own surveys, we had sufficient data to test for explanatory variables with logistic regression. Our results come from NoMix toilet users at our institute, Eawag (Lienert and Larsen 2006), a vocational college (Lienert and Larsen 2006; Lienert *et al.* 2006) and a public library (Lienert and Larsen 2010). At Eawag and the college, apart from the requirement of odour-free, hygienic toilets, the most promising measures to further increase the already high acceptance of NoMix toilets (e.g., 86% would move into an apartment with NoMix toilets) were information and discussions with peers (Lienert and Larsen 2006). Negative discussions about NoMix toilets with colleagues were most strongly correlated with lower acceptance. In social psychology, the "*Theory of Planned Behaviour*" (*TPB*, Ajzen 1991) is often used to explain environmentally friendly behaviour. TPB postulates that behavioural choices are mainly influenced by the *attitude* towards the behaviour, the *perceived behavioural control* (i.e. ability to perform the behaviour) and the *subjective norm*. The latter means the perception of social pressure, and many studies reported that this influenced behaviour (or acceptance as in our case). We proposed that future NoMix pilot projects could organize information events offering an opportunity for peer discussions.

As in other studies, being well-informed about NoMix toilets was positively correlated with acceptance in all three Swiss pilot projects. At Eawag and the school, the instructions in the toilet cabin containing a very short rationale for urine separation already sufficed to significantly increase acceptance, peer discussions, perception, behaviour and knowledge about NoMix toilets (Lienert and Larsen 2006). It also makes sense to offer several arguments since different user groups had different reasons for finding NoMix technology convincing.

In the library, softer personality factors such as openness to innovations and environmentally friendly behaviour were also correlated with acceptance (Lienert and Larsen 2010). Of course, such traits are beyond the control of project managers. However, our review showed that people must be convinced about the ecological reasons to accept living with NoMix toilets. We found indications for low acceptance in projects without re-use of urine that were carried out primarily for research purposes. Moreover, some Swedish communities actively follow a nutrient recycling policy, subsidizing alternative sanitation. Active strategies by the authorities and the recognition of their central role in mediating between stakeholders, for example, in Tanum in Sweden, increase adoption rates of NoMix technology and probably also its acceptance (Kvarnström *et al.* 2006).

Logistic regression shows correlations but not *cause-effect relationships*. I recommend that future research projects should design experiments to *test hypotheses* about the acceptance of household innovations such as the NoMix toilet and to *change (environmental) attitudes* (e.g., Mosler and Martens 2008).

14.4 ACCEPTANCE OF URINE-BASED FERTILIZERS

Not much sociological research has been carried out on urine recycling. I know of four questionnaire surveys addressed to the general public and four to farmers that elicited the acceptance of re-using human urine in agriculture (Lienert and Larsen 2010). However, the few results that were compiled for the general public are very encouraging. In seven surveys of 908 respondents (545 from Switzerland), 68% ($\pm$11%) answered that they would not object to purchasing food fertilized with human urine. Generally, 85% ($\pm$13%) of 765 respondents found re-use of urine in agriculture a (very) good idea. Presumably, concerns with respect to fertilizer quality would have been mentioned, but they were only reported for the Swiss library and a survey at a German agricultural exhibition.

Acceptance of urine fertilizers was markedly lower in an Austrian pilot project than in Sweden, Germany and Switzerland (Figure 14.1). A likely explanation is the attitude of the authorities. In Switzerland, re-use of human urine as fertilizer is not allowed; the questionnaire explicitly stated that the urine would be processed to remove pathogens and micropollutants (pharmaceuticals). So presumably, the 93% positive respondents assumed that the fertilizer is safe. In Sweden, urine has to be stored to remove pathogens, but not processed for micropollutant removal. Here, urine recycling is promoted by the authorities and is a well-known practice, which explains the high acceptance of 77%. The legal situation in Germany is unclear; not surprisingly, 61% of 175 respondents mentioned concerns with respect to micropollutants (only 8% in Switzerland), but acceptance was still 78%. However, it was initially planned to re-use urine in the Austrian pilot "SolarCity" project. The authorities later intervened because of safety concerns. This might well explain the low acceptance of only 43%.

The feedback from farmers was somewhat less enthusiastic (Lienert and Larsen 2010). Only 50% of 150 farmers questioned found urine as fertilizer a (very) good idea, only 34% of 216 farmers in three surveys would use or purchase a urine-based fertilizer, but only if it were cheap or free, and 34% *might* use it. However, here one must definitely consider the low response rates of all these studies and the possibility of biased returns. We tested this for 467 Swiss farmers and found the response rates to show highly significant differences between the various farmer groups, being highest for organic farmers (Lienert *et al.* 2003). In all surveys, the farmers were worried about fertilizer quality, many expected reduced sales and low consumer acceptance of food products. Of 137 farmers, 65% mentioned a fear of liability claims and that regulations should be changed first. From the Swiss survey we also know that farmers have specific fertilizer needs and that the best strategy would be to tailor different nutrient products from urine to meet their requirements (Lienert *et al.* 2003).

From these first surveys, I conclude that the idea of bringing human urine to agriculture is not perceived as critically as many had expected. However, people are anxious about fertilizer safety and want to avoid any risk to human health or the environment coming from pathogens or pharmaceuticals. Farmers are also worried about liability claims. To overcome such barriers, safety issues must be discussed and regulated by the authorities of each country so that the managers of NoMix pilot projects can clearly communicate their strategy.

14.5 TECHNOLOGY REQUIREMENTS AND OUTLOOK

14.5.1 Drawbacks of NoMix toilets for users

One of the most serious problems of NoMix toilets are blockages of urine-conducting pipes, reported to occur after 6–12 months of usage by around 60% of the respondents in the questionnaire surveys (Lienert and Larsen 2010). This problem is well known to anyone conducting a NoMix pilot project. It is caused by urea hydrolysis, which leads to phosphate precipitation and build-up of scaling. It is fairly easy to overcome scaling problems either mechanically (e.g., removal with a steel brush) or chemically, also as a preventive measure (e.g., regular flushing of the urine drain with 10% citric acid). Practical aspects such as these are discussed in Lienert and Larsen (2007), including building requirements, suggestions for technical improvement of NoMix toilets as well as references to data sheets and legal documents. Scaling increases the need for maintenance of NoMix toilets, which is a nuisance in private homes and raises costs in public buildings. The loss of nutrients via precipitates is equally problematic. Nutrients are also lost in ventilation pipes. A loss of up to 40% of the ammonia was estimated in some systems. This becomes detectable to users by a bad smell emanating from NoMix toilets. However, this is presumably not a major problem, as 77% (±18%) of 1903 respondents in 19 surveys judged the smell of the NoMix toilet to be at least as good as that of a conventional toilet (Lienert and Larsen 2010).

A fairly disagreeable phenomenon is that faeces can land in the urine compartment or stick to the divider between the two bowls. Consequently, only 52% (±17%) of 366 users in 14 pilot projects found cleaning NoMix toilets to be as easy as conventional toilets (Lienert and Larsen 2010; Figure 14.1). This probably depends on the type of NoMix toilet: faeces are more likely to end up in a large front bowl, but we have no data to support this assumption. The problem is more pronounced for smaller children, and only 41% (±14%) of the respondents found NoMix toilets easy to use for children (Lienert and Larsen 2010).

A popular argument against NoMix toilets is the need to sit on them. The data suggests that this is not a major obstacle: 68% (±7%) of 1732 users in nine pilot projects answered that they usually sit on NoMix toilets to urinate (Figure 14.1). As already mentioned, sitting is easier at home than in public places, where many dislike it for hygienic reasons. However, in public toilets this mainly affects women, since men can usually use urinals. If the user is not correctly positioned on the NoMix toilet – or does not fully sit so that the closing valve is open (Roediger NoMix toilet; www.roevac.com) – urine and nutrients are lost.

NoMix toilets have several other drawbacks, mostly design-related, such as poor hydraulics resulting in multiple flushes or the need to dispose of toilet paper in a separate bin after urinating to save water (for details see Lienert and Larsen 2007, 2010). Motivational measures and socio-psychological interventions can enhance user awareness and the acceptance of adapting their behaviour to the requirements of NoMix toilets. However, I am convinced that in the long run it will be necessary for sanitary manufacturers to improve NoMix toilets to overcome these drawbacks. This is a challenge – but also an opportunity, because advanced technology will allow new features and devices to be integrated into the toilet (e.g., direct measurement of medical parameters in urine).

14.6 CONCLUSIONS

Source separation and decentralization of wastewater treatment to a point where people are affected in their own homes require new approaches, not only technical ones, but also in the area of social science. It is important to realize that technology development as well as social science research must be conducted professionally and based on sound principles.

Dellström Rosenquist (2005) has pinpointed some of the psychological and cultural reasons that explain why the flush-and-gone technology of water closets is superior to lower-tech solutions that require people to think about excreta. Although I do not agree that a mentality shift will be needed to overcome psychological barriers to excreta, I fully concur that there is ample room for exciting psychological and sociological research in this area. Such research should help to determine which technologies might equal the centralized approach from a user perspective and which could be even more successful. For instance, various ecological sanitation projects indicate that sanitation must not only satisfy physical needs, but also social comfort, safety and status (Cordova and Knuth 2005; see also Dellström Rosenquist 2005).

Moreover, there are ways of increasing acceptance developed by marketing research and behavioural psychology. A growing body of experience from environmental psychology is available to support the design of information or intervention strategies, which can be adapted to decentralized wastewater technologies (e.g., De Young *et al.* 1993; Mosler and Martens 2008; Tamas *et al.* 2009). Such practically oriented research should be complemented by historical research, for instance. As Bracken *et al.* (2007) argue, to effectively address the global sanitation crisis and induce a paradigm change, it is important to understand the history of sanitation and the reasons that have led to the unchallenged dominance of the end-of-pipe sewer system.

The sociological research on NoMix technology indicates that the public and users are surprisingly open. Given good arguments, people are likely to support this innovation, even in an early implementation phase. Research should focus on understanding the reasons for the success of the current system and analyse user preferences to support engineers in developing better technologies.

REFERENCES

Ajzen I. (1991). The theory of planned behavior. *Organ. Behav. Hum. Decis. Process.*, **50**(2), 179–211.

Borsuk M. E, Maurer M., Lienert J. and Larsen T. A. (2008). Charting a path for innovative toilet technology using multicriteria decision analysis. *Environ. Sci. Technol.*, **42**(6), 1855–1862.

Bortz J. and Döring N. (2009). 4th edn., Forschungsmethoden und Evaluation für Human- und Sozialwissenschaftler (Research methods and evaluation for human and social scientists). Springer Medizin Verlag, Heidelberg, Germany.

Bracken P., Wachtler A., Panesar A. R. and Lange J. (2007). The road not taken: how traditional excreta and greywater management may point the way to a sustainable future. *Water Sci. Technol.: Water Supply*, **7**(1), 219–227.

Cordova A. and Knuth B. A. (2005). User satisfaction in large-scale, urban dry sanitation programs in Mexico. *Urban Water J.*, **2**(4), 227–243.

Dellström Rosenquist L. E. (2005). A psychosocial analysis of the human-sanitation nexus. *J. Environ. Psychol.*, **25**, 335–346.

De Young R., Duncan A., Frank J., Gill N., Rothman S., Shenot J., Shotkin A. and Zweizig M. (1993). Promoting source reduction behavior – The role of motivational information. *Environ. Behav.*, **25**(1), 70–85.

Dillman D. A., Smyth J. D. and Melani Christian L. (2009). 3rd edn., Internet, Mail, and Mixed-Mode Surveys. The Tailored Design Method. John Wiley & Sons, Hoboken, New Jersey, USA.

Friedler E., Butler D. and Brown D. M. (1996). Domestic WC usage patterns. *Build. Environ.*, **31**(4), 385–392.

Grimble R. and Wellard K. (1997). Stakeholder methodologies in natural resource management: A review of principles, contexts, experiences and opportunities. *Agric. Syst.*, **55**(2), 173–193.

Groves R. M. (1989). Survey Errors and Survey Costs. Wiley, New York, USA.

Jaeger C., Schuele R. and Kasemir B. (1999). Focus groups in integrated assessment: A microcosmos for reflexive modernization. *Europ. J. Social Sci.*, **3**, 195–219.

Jefferson B. and Jeffrey P. (2013). Aerobic elimination of organics and pathogens: greywater treatment. In: *Source Separation and Decentralization for Wastewater Management*, T. A. Larsen, K. M. Udert and J. Lienert (eds.), IWA publishing, London, UK, pp. 275–290.

Keeney R. L. (1982). Decision-analysis – An overview. *Oper. Res.*, **30**(5), 803–838.

Keeney R. L. (1992). Value-Focused Thinking. Harvard University Press, Cambridge, MA, USA.

Krantz H. (2005). Matter that matters: A study of household routines in a process of changing water and sanitation arrangements. PhD thesis, Dept. of Water and Environmental Studies, Linköping Univ., Sweden. ISBN: 91-85297-65-8. http://liu.diva-portal.org/smash/record.jsf?pid=diva2:20799 (accessed: 13.03.2012).

Kvarnström E., Emilsson K., Richert Stintzing A., Johansson M., Jönsson H., af Petersens E., Schönning C., Christensen J., Hellström D., Qvarnström L., Ridderstolpe P. and Drangert J.-O. (2006). Urine diversion: One step towards sustainable sanitation. EcoSanRes Publications Series, Report 2006–1. http://www.ecosanres.org/news-publications.htm (accessed 13.03.2012).

Larsen T. A. and Lienert J. (2003). Societal implications of re-engineering the toilet. *Water Intelligence Online*. March 2003. UNIQUE ID: 200303006. www.iwaponline.com/wio/2003/03/default001.htm (accessed 13.03.2012).

Lienert J. and Larsen T. A. (2006). Considering user attitude in early development of environmentally friendly technology: A case study of NoMix toilets. *Environ. Sci. Technol.*, **40**(16), 4838–4844.

Lienert J. and Larsen T. A. (2007). Pilot projects in bathrooms: A new challenge for wastewater professionals. *Water Practice & Technol.*, **2**(3), doi10.2166/wpt.2007.0057. http://www.iwaponline.com/wpt/002/03/default.htm (accessed 13.03.2012).

Lienert J. and Larsen T. A. (2010). High acceptance of urine source separation in seven European countries: A review. *Environ. Sci. Technol.*, **44**(2), 556–566.

Lienert J., Haller M., Berner A., Stauffacher M. and Larsen T. A. (2003). How farmers in Switzerland perceive fertilizers from recycled anthropogenic nutrients (urine). *Water Sci. Technol.*, **48**(1), 47–56.

Lienert J., Thiemann K., Kaufmann-Hayoz R. and Larsen T. A. (2006). Young users accept NoMix toilets – A questionnaire survey on urine source separating toilets in a college in Switzerland. *Water Sci. Technol.*, **54**(11–12), 403–412.

Mosler H.-J. and Martens T. (2008). Designing environmental campaigns by using agent-based simulations: Strategies for changing environmental attitudes. *J. Environ. Manage.*, **88**, 805–816.

Pahl-Wostl C., Schönborn A., Willi N., Muncke J. and Larsen T. A. (2003). Investigating consumer attitudes towards the new technology of urine separation. *Water Sci. Technol.*, **48**(1), 57–65.

Presser S., Rothgeb J. M., Couper M. P., Lessler J. T., Martin E., Martin J. and Singer E. (eds.) (2004). Methods for Testing and Evaluation of Survey Questionnaires. Wiley–Interscience, New York, USA.

Rogers E. M. (1983). Diffusion of Innovation. Collier Macmillan Publishers, London, UK.

Rydhagen B. (2002). Feminist sanitary engineering as a participatory alternative in South Africa and Sweden. PhD thesis, School of Technoculture, Humanities and Planning, Blekinge Inst. of Technol., Karlskrona, Sweden. ISBN: 91-7295-017-X. http://www.bth.se/fou/forskinfo.nsf/textpages/diss-and-lic (accessed 13.03.2012).

Schoemaker P. J. H. (1995). Scenario planning: a tool for strategic thinking. *Sloan Manage. Rev.*, **36**(2), 25–40.

Scholz R. W. and Tietje O. (2002). Embedded Case Study Methods. *Integrating Quantitative and Qualitative Knowledge*. Sage Publications, Thousand Oaks, CA, USA.

Tamas A., Tobias R. and Mosler H.-J. (2009). Promotion of solar water disinfection: Comparing the effectiveness of different strategies in a longitudinal field study in Bolivia. *Health Commun.*, **24**(8), 711–722.

Truffer B., Binz C., Gebauer H. and Störmer E. (2013). Market success of on-site treatment: a systemic innovation problem. In: *Source Separation and Decentralization for Wastewater Management*, T. A. Larsen, K. M. Udert and J. Lienert (eds.), IWA publishing, London, UK, pp. 209–223.

Chapter 15

Market success of on-site treatment: a systemic innovation problem

Bernhard Truffer, Christian Binz, Heiko Gebauer and Eckhard Störmer

15.1 INTRODUCTION

On-site treatment could offer major advantages for future sustainable urban water-management systems (Larsen *et al.* 2013). Recent technological advances in fields such as membranes, chemical sensors and remote control open up a broad and promising range of new system configurations (Olsson 2013). So far however, the response to these opportunities has mostly been lukewarm among industry, water utilities and regulators (Nelson 2008). These technologies are obviously not yet considered to be a fully fledged alternative to centralized urban water management systems. This is because of the deficiencies of currently available configurations as well as great uncertainty about their functionality, reliability and cost-effectiveness in the near future. Besides technological uncertainties, on-site systems represent a major challenge to the current competencies of utilities for providing urban water management services and organizing their value chain. They therefore represent a potentially disruptive innovation for the established industry (Christensen 1997). Finally, on-site based urban water management structures are likely to show a mismatch with many institutional conditions (regulations, professional codes or user expectations) developed for supporting the established urban water management systems. Accordingly, on-site technologies also call for innovations in the societal context in which this new technology would have to operate. A future large-scale application of on-site systems to urban water management thus depends on the successful organization of innovation processes in three domains: (i) technological components and system integration, (ii) value chain formation and the development of new business models, and (iii) institutional innovations to create appropriate conditions under which these systems can reliably operate. So what is at stake is the management of a systemic innovation process (Markard and Truffer 2008).

Infrastructure sectors such as urban water management are particularly hard cases for systemic innovations to occur and succeed (Dyner and Larsen 2001; Markard and Truffer 2006). Historical examples include the emergence of centralized piped freshwater systems (Geels 2005) and the transition from cesspools to centralized wastewater treatment (Geels 2006) in the early 20th century. These studies show (along with many other cases from industrial history) that systemic innovations were far from smooth diffusion processes from an existing to an obviously superior technological solution. Rather, they needed an interrelated set of innovation processes in technologies, competence structures and the institutional environment – often coordinated by strategic actions of companies, business associations or governments – before they could be implemented on a broader scale. In the present chapter, therefore, we analyze the success conditions for the widespread implementation of on-site based alternatives from a systemic innovation perspective.

The chapter is structured as follows: In the next section, we will work out in more detail the kinds of innovation processes that on-site technologies would have to undergo in order to reach widespread adoption. In Section 3, results from an empirical study will be presented to exemplify the innovation challenges faced by the German on-site industry. In the fourth section, we will report the results of a road-mapping workshop run with 20 representatives from German companies and research institutes where dominant future designs, business models and potential development trajectories were elaborated. Finally, we will identify the major institutional challenges facing industry and policy makers in developing on-site systems as a contribution to a more sustainable urban water management sector.

15.2 THE SYSTEMIC INNOVATION PROBLEM

On-site systems show a large number of potential advantages for future more sustainable urban water management concepts, as reported in this book (Larsen *et al.* and Maurer 2013). The reasons for their slow take-up are partly due to the specificities of infrastructure sectors (Flyvbjerg 2007; Dyner and Larsen 2001). These are characterized by high upfront investments, life expectancy of several decades for core technological components, and a strong focus on an uninterrupted and homogenous quality of service provision (Hegger *et al.* 2007). As a consequence, decisions to implement a specific system configuration depend crucially on adequate assumptions about the longer term context conditions and the reliability of the implemented technologies (Dominguez *et al.* 2011). Infrastructure sectors therefore show a strong preference for incremental innovations in well-tested components over radical system transformations (Markard and Truffer 2006).

If we look back at the successful introduction of advanced centralized wastewater treatment in Europe in the 1970s, uncertainties could only be kept low by relying on strong state regulation and substantial technological guidance from central governments (Dominguez and Gujer 2006). The management of specific systems is often delegated to local communities, whose strategic planning and technology implementation relies strongly on local civil engineering companies (Dominguez *et al.* 2009). This focus on limiting uncertainties has resulted in strategic planning processes with a strong tendency to ignore radical alternatives and consider sustainability aspects in a highly restricted way (Truffer *et al.* 2010; Milly *et al.* 2008). As a consequence, on-site systems will always have a low chance of success when presented to specific local decision makers, because their uncertainties are too great in too many respects.

The following innovation challenges have to be tackled simultaneously before decision makers would be prepared to consider these alternatives: (i) the development of technological components as well as their integration into reliable system configurations, (ii) setting up new business networks able to organize the mass production of specific components as well as system integration, installation, operation and servicing models, and (iii) adapting regulatory frameworks. In the following, we will briefly elaborate the core problems to be solved in these three domains.

First, reliable on-site systems still depend heavily on research and development efforts for individual components (e.g., improved water treatment reactors, water quality sensors, membrane modules, water saving toilets, etc.). However, the success of on-site systems will not only depend on reliable and effective components but has to overcome thorny problems of integration of the components into working systems. System integration has to be considered at different levels: the level of specific appliances (e.g., water-efficient urine-separating toilets), the room/apartment/house level (e.g., integration of different water uses and water qualities) and the neighborhood/city scale (e.g., greater or lesser reliance on the sewer infrastructure).

Second, besides the reliability of the technological systems, costs are a crucial factor for widespread implementation. Future costs of on-site systems will depend on the successful management of three cost components: (1) costs for manufacturing integrated on-site systems, (2) project costs for installing on-site systems in specific local contexts and (3) costs for operating, maintaining and controlling on-site systems. All three cost components decrease strongly with the size of the market: Project costs depend on "economies of repetition," achieved by installing a large number of appliances in similar building types or within a specific neighbourhood type, for example. Costs for manufactured appliances are subject to "economies of scale," that is, specific costs decrease substantially with high production volumes. Maintenance costs depend on "economies of scale related to market size": an increasing number of on-site systems within a geographical region would increase the service density. As a consequence, we suffer from a typical development trap: costs will only decrease substantially when implementation is carried out on a large scale. But implementation can only be extensive if costs are low.

There are a couple of requirements for an industry structure able to manage these interrelated economies of scale problems (Gebauer *et al.* 2012). The centralized system paradigm with its core component of a centralized wastewater treatment plant has a strong civil engineering logic. The industry is dominated by construction and engineering companies, technology providers and public utilities. Technology implementation is strongly project-oriented, customized and decided by local monopoly organizations only once in a couple of decades. A large-scale introduction of on-site technologies would radically depart from this logic. The core components are produced by large-scale manufacturers, components need high standardization and markets will have to absorb high volumes. Project implementation is oriented to construction companies, architects and sanitary installers, and the life cycles of the system components are likely to be much shorter. Due to economies of scale, cost competitiveness depends crucially on the availability of sufficiently large markets, which could develop both in industrialized and emerging economies with high water scarcity and rapid urbanization problems such as northeastern China or urban regions in Australia (Larsen *et al.* submitted; Binz *et al.* 2012). Market formation, however, depends on the capability of business networks to offer reliable service, operation and control models. Appropriate competencies for delivering, servicing and replacing appliances have to be built up. New contracting models, pricing schemes and appropriate accounting systems have to be developed that allow a large number of on-site systems to be operated in a profitable way. However, these kinds of capabilities are rare in existing water utilities.

Third, additional developments have to take place to improve institutional context conditions, for example, by adapting regulatory frameworks in a way that do not *a priori* exclude on-site systems. Examples include loosening currently mandatory connection regulations, tightening discharge conditions for effluents and the like. Furthermore, both acceptance by end consumers and the general public image of the emerging technology must be considered (Lienert 2013).

Summarizing, we state that on-site system alternatives need to be improved in a wide range of aspects. To develop on-site systems into a wholesale alternative would need an interconnected set of radical and potentially disruptive innovation processes and would most likely have to draw on competencies from outside the group of established actors in the urban water management sector. Furthermore, the currently perceived superiority of centralized urban water management is largely due to the fact that this paradigm benefits from a historical head start and that high sunk costs prevent competitors from entering the market. When anticipating the sustainability challenges of the sector, however, path dependency should not be the decisive argument for continuation of the established paradigm (Larsen *et al.* submitted; Larsen and Gujer 1997). The problem of adequately assessing on-site systems has thus to be understood as a dynamic and systemic innovation problem rather than a static problem of optimal technology choice.

15.3 THE GERMAN ON-SITE INDUSTRY

In order to identify preconditions for potentially successful systemic innovation processes, we build on recent insights from the literature on emerging environmental industries (Markard and Truffer 2008). Here the emphasis is placed on the actor configurations, networks and institutions as well as on the interplay among these elements to support systemic innovation processes. This analysis allows potential future dominant designs as well as likely development trajectories to be identified (Markard *et al.* 2009), which will provide lessons for innovation management and technology policy in this field (Bergek *et al.* 2008).

As regards the actors, an analysis of the innovation activities in related economic fields offers an appropriate start for identifying an emerging industry. The industry dealing with current on-site water systems mostly comprises European, US and Japanese suppliers. In Europe, the strongest competence base is found in Germany (Sartorius 2008). We have restricted our analysis to the emergent German industry, where we conducted a systematic company survey via an extensive internet search, snowballing expert interviews and analyzing the companies listed by the national certification body for on-site wastewater appliances. This way, about 60 companies were identified for a detailed industry analysis. These companies can be divided into three groups according to the technological complexity of their main products and the geographic outreach of their market activities.

First, there are small and medium-sized enterprises selling small sewage treatment plants or septic tanks for single-family houses in rural areas. Their market is strongly supported by current EU law requiring a refurbishment of existing on-site wastewater systems up to 2015. As connection rates in some German federal states (especially in eastern Germany) are as low as 70%, this regulation has created a sizable market. The more innovative companies in this group are also reaching out into international markets. As a consequence, German companies in this group reach a global market share of about 40% (Henzelmann *et al.* 2007). Many of these firms focus on producing concrete or plastic tanks and have only recently entered the wastewater market. The majority are small to medium-sized enterprises ranging from 10 to 100 employees. Over 70% of them develop specific treatment systems covering a broad set of technologies ranging from constructed wetlands to trickling filters, sequential batch reactors (SBR), fixed bed systems and membrane bioreactors (MBR). The market leader sells about 7000 SBR systems per year, compared to a total German market volume of 15,000 to 30,000 units per year. Nearly all these companies buy components such as pumps, sensors and electronic parts from specialized suppliers. Their core competence is thus the integration of these components into reliable system configurations and the design of the treatment process. Service and maintenance is offered either directly by on-site system manufacturers or specialized service companies. So far, these services are mainly provided by small regional companies which have contracts with individual customers. One major global water company, namely Veolia, entered this market in a single German region for the first time as recently as 2008.

A second group of firms focuses on recycling either combined wastewater or greywater. This group typically produces small MBR-based wastewater treatment plants and in-house greywater recycling systems for non-potable water use. These systems are mostly installed in newly constructed buildings. Consequently, these companies focus on the planning phase and maintain close interactions with the building equipment sector. Their major markets are hotels and residential buildings in water scarce regions. Their markets are thus much more international than those of the first group.

The third group comprises large transnational water companies with strong in-house capacity for research, development and marketing. They are primarily engaged in selling technology and offering advice for centralized systems on a global scale. Their activities with on-site systems focus largely on industrial wastewater treatment, including processwater recycling. In general they also have in-house competence in remote monitoring and control of multiple plants and are experienced in organizing appropriate maintenance and service structures (Gebauer *et al.* 2012). However, as the markets for on-site systems are still immature, the activities of these large companies in the municipal on-site market have been restricted to a few pilot and demonstration plants.

As regards networks and institutions, the German on-site sector and current market conditions are relatively mature. For instance, technical reliability is guaranteed by a certification procedure run by the German "approval body for construction products and types of construction." Certified plants benefit from a simplified authorization procedure by local government bodies. Additionally, two national associations provide some basic coordination among this rather disparate set of companies: DWA (Deutsche Vereinigung für Wasserwirtschaft, Abwasser und Abfall e.V.) is the general agency for German water professionals: it maintains a specific commission for innovative sanitary systems (see also Londong 2013). In addition, the association for the promotion of decentralized wastewater technologies (BDZ) develops performance criteria for on-site systems, defines the standards for mandatory maintenance services, offers education courses to wastewater professionals and certifies service and maintenance companies operating in this field. Further important institutional structures are provided by the water agencies of the German federal states, which *inter alia* define the standards for effluent water quality for on-site wastewater treatment systems.

In summary, a somewhat diverse landscape of actors exists in Germany, engaged in various aspects of on-site system development. There is a sizeable market and institutional structures are in place to manage potential coordination problems. In order to assess the capabilities of this sector for organizing the three main innovation processes identified in the introduction, we have to analyze its innovation management in more detail.

15.4 MAJOR INNOVATION CHALLENGES

In order to identify the innovation capabilities of the German on-site industry, we conducted an interview campaign with 45 experts, half of them representatives of on-site technology companies, the other half representatives of water utilities and academia. These experts were asked about urgent development needs, most promising future designs, most likely growth markets and challenges for the German industry emerging in this sector.

As regards the development of new technological components and their system integration, most experts claim that the basic technologies for reliable and efficient on-site systems are already available today. However, they also point out some important fields that need further development. For instance, some core technological components in today's on-site systems are standard products from suppliers to the centralized wastewater industry. As a consequence, these components often lack sufficient adaptation to the particular needs of small wastewater systems. This leads to inefficient and unstable configurations. Sensors for the continuous measurement of effluent water quality offer one example. The currently available sensors were developed for large wastewater treatment plants. As a consequence, their prices are too high for most on-site applications and they would have to operate under sub-optimal conditions. Experts identify a strong need for miniaturization and better adaptation of this and other components to on-site systems.

Companies from the first group in particular see system integration as a major challenge. In order to achieve high water saving efficiency, the entire in-house water system has to be coherently designed, including water saving taps, multiple piping and outputs for the treated wastewater. Experts have argued that companies in the greywater recycling market are better prepared for such system integration because they are used to configuring their on-site systems with a more holistic approach. In addition, the interviewees also stressed the importance of further system integration, that is, the coherent combination of separated on-site blackwater and greywater treatment systems. Related innovations such as energy recovery via heat exchange, the production of biogas and nutrient recovery were also mentioned as visionary components of future on-site systems. However, especially technology providers from group one are critical towards such highly integrated source-separating solutions. They claim that these do not make sense

favouring the more ambitious configurations (OST2 and OST3) emphasized a number of shortcomings in the available competence portfolios such as insufficient company size, limited investment in research and development or deficiencies in system thinking. They therefore identified the need for bringing new partners with complementary technological know-how into the field (more extensively, see Gebauer *et al.* 2012).

OST2 and OST3 promoters were also particularly concerned about the lack of regulatory support for on-site concepts. This is likely to lead to difficulties in running advanced water recycling pilot projects. In all three cases, policymaking was identified as a key field of innovation with the need to develop water recycling standards, integrate on-site concepts into urban planning procedures and revise the current guidelines for the service and maintenance of on-site systems.

A comparison of the innovation preconditions for the three potential dominant designs appears to show a logical sequence of development stages starting with incremental improvements of currently existing decentralized wastewater treatment plants (OST1) and moving on to increasingly ambitious configurations (OST2 and OST3). However, further scrutiny of the capability deficits, innovation problems and context conditions necessary in each case suggests that the synergies are much more limited. The shared problems comprise improved water disinfection, new sensors as well as new monitoring and maintenance concepts. OST2 and OST3 would require the concerted involvement of additional actors with complementary capabilities, including heating and cooling-system experts, sanitary technology specialists and house automation experts.

Both trajectories seem feasible for German companies for the time being. However, when judged from the point of view of sustainable system transformation, OST1 mostly focuses on the incremental improvement of existing decentralized wastewater concepts and would only contribute marginally to a sustainable transformation of urban water management. The other two designs promise much higher efficiency gains but also make much higher demands on managing and coordinating the necessary innovation processes.

15.6 CONCLUSIONS

The core argument expounded in this chapter was that an adequate assessment of the contribution of on-site systems to a future, more sustainable urban water management sector has to be understood as a systemic innovation problem and not as a static decision problem about choosing a single "right" technology. If the focus is on innovation, the critical question is which actors would be able and willing to undertake the necessary development processes and under what conditions success seems more likely. We saw that despite the potential benefits of on-site technologies, a corresponding industry has not yet fully developed. Our industry analysis shows a fragmented landscape of actors dealing with on-site systems development in Germany comprising (i) small and medium-sized enterprises focusing on conventional on-site treatment of combined wastewater, (ii) sanitary appliance manufacturers developing concepts for source separation with an emphasis on greywater treatment, and (iii) transnational companies focusing mainly on centralized wastewater systems and industrial on-site treatment. To date, these three groups of actors have shown very little interaction and synergy. They are active in different market segments and largely differ in their expectations on future dominant designs and the necessary support conditions for on-site systems. Nevertheless, the core difficulty – namely the concurrent management of a complex set of innovation processes – represents a shared challenge for all these actors. In particular, it seems that the problem of developing new capabilities in manufacturing and servicing as well as institutional embedding represents a big challenge for German companies (see Gebauer *et al.* 2012). The same seems to be true for small and medium-sized enterprises in countries like Japan and the US (see Nelson 2008; Scheele 2008).

Looking at the three dominant potential designs that were elaborated in the roadmapping workshop, it seems that German actors are confronted with a strategic choice of either following an incremental path of technology development or leapfrogging into more ambitious configurations. When judged by its contribution to a more sustainable future urban water sector, the second strategy shows much greater promise. It implies that German companies will have to access markets far outside their home countries. For instance, new developments in countries such as China, where new recycling water standards have recently been introduced, could create promising market prospects (see Funamizu *et al.* 2008, for instance). However, accessing spatially and culturally remote market areas would conflict with the need to go through stepwise learning processes and coordinate a wide variety of innovation processes. Workshop participants and interviewees stressed time and again that reference projects in their home countries would be a key success condition for introducing on-site concepts to other parts of the world.

Our analysis therefore shows clearly that on-site industry development could build on existing industrial infrastructures in Germany. This is also underlined by the recently increased interest by the German government in choosing the water technology sector as a potential target for national technology policy (Sartorius 2008). However, in order to develop these promising technologies into mature alternatives, highly concerted action would be needed, probably of the same order of magnitude that was necessary to build up the centralized systems in the first place. Such a program would require a broad political discussion about the future challenges of urban water management on a global scale. It would involve engineers as well as social scientists, industry representatives and policymakers. Such a program has to truly build on interdisciplinary competencies as it must simultaneously anticipate social, technical and economic aspects of a systemic innovation process, that is, a veritable sustainability transition in urban water management.

REFERENCES

Bergek A., Jacobsson S., Carlsson B., Lindmark S. and Rickne A. (2008). Analyzing the functional dynamics of technological innovation systems: A scheme of analysis. *Res. Policy*, **37**(3), 407–429.

Binz C., Truffer B., Li L., Shi Y. and Lu Y. (2012). Conceptualizing leapfrogging with spatially coupled innovation systems: The case of onsite wastewater treatment in China. *Technol. Forecast. Soc. Chang.*, **79**(1), 155–171.

Christensen C. M. (1997). The innovator's dilemma: when new technologies cause great firms to fail, Harvard Business School Press, Boston, Massachusetts.

Dominguez D. and Gujer W. (2006). Evolution of a wastewater treatment plant challenges traditional design concepts. *Water Res.*, **40**, 1389–1396.

Dominguez D., Worch H., Markard J., Truffer B. and Gujer W. (2009). Closing the capability gap: Strategic planning for the infrastructure sector. *Calif. Manage. Rev.*, **51**, 30–50.

Dominguez D., Truffer B. and Gujer W. (2011). Tackling uncertainties in infrastructure sectors through strategic planning – The contribution of discursive approaches in the urban water sector. *Water Policy*, **13**, 299–316.

Dyner I. and Larsen E. R. (2001). From planning to strategy in the electricity industry. *Energ. Policy*, **29**, 1145–1154.

Flyvbjerg B. (2007). Policy and planning for large-infrastructure projects: problems, causes, cures. *Environ. Plan. B-Plan. Des.*, **34**, 578–597.

Funamizu N., Huang X., Chen G., Hu J. and Chettiyapan V. (2008). Water reuse in Asia. In Water Reuse – An International Survey of current practice issues, needs, B. Jimenez and T. Asano) (ed.), 1st edn, IWA Publishing, London, pp. 142–160.

Gebauer H., Truffer B., Binz C. and Störmer E. (2012). Capability perspective on business network formation: Empirical evidence from the wastewater treatment industry. *European Business Review*, **24**(2), 169–190.

Geels F. W. (2005). Co-evolution of technology and society: The transition in water supply and personal hygiene in the Netherlands (1850–1930) – a case study in multi-level perspective. *Technol. Soc.*, **27**, 363–397.

Geels F. W. (2006). The hygienic transition from cesspools to sewer systems (1840–1930): The dynamics of regime transformation. *Res. Policy*, **35**, 1069–1082.

Hegger D. L., Van Vliet J. and Van Vliet B. J. (2007). Niche management and its contribution to regime change: The case of innovation in sanitation. *Technol. Anal. Strateg.*, **19**, 729–746.

Henzelmann T., Mehner S. and Zelt T. (2007). Innovative environmental growth markets from a company perspective. The Federal Environment Agency, Berlin, Germany. http://www.umweltdaten.de/publikationen/fpdf-k/k3337.pdf (accessed 12 March 2012).

Hiessl H., Stemplewski J., Toussaint D., Hetschel M., Geisler S. and Kersting M. (2007). Abwasserentsorgung einer Streusiedlung mit MBR-Kleinkläranlagen im Rahmen eines Contracting-Modells (Wastewater disposal in a dispersed settlement with small scale MBR plants, based on a contracting service model), *KA – Abwasser Abfall*, **54**(12), 1238–1244.

Larsen T. A., Binz C., Maurer M., Stoermer E. and Truffer B. (submitted). Assessing the future role of on-site technologies in urban water management. *Environ. Sci. Technol.*

Larsen T. A. and Gujer W. (1997). The concept of sustainable urban water management. *Water Sci. Technol.*, **35**(9), 3–10.

Larsen T. A., Lienert J. and Udert K. M. (2013). Source separation and decentralization. Editorial. In: *Source Separation and Decentralization for Wastewater Management*, T. A. Larsen, K. M. Udert and J. Lienert (eds.), IWA Publishing, London, UK, pp. 1–10.

Lienert J. (2013). High acceptance of source-separating technologies, but… In: *Source Separation and Decentralization for Wastewater Management*, T. A. Larsen, K. M. Udert and J. Lienert (eds.), IWA Publishing, London, UK, pp. 193–207.

Londong J. (2013). Practical experience with source separation in Germany. In: *Source Separation and Decentralization for Wastewater Management*, T. A. Larsen, K. M. Udert and J. Lienert (eds.), IWA Publishing, London, UK, pp. 423–430.

Markard J., Stadelmann M. and Truffer B. (2009). Prospective analysis of technological innovation systems: Identifying technological and organizational development options for biogas in Switzerland. *Res. Policy*, **38**(4), 655–667.

Markard J. and Truffer B. (2006). Innovation processes in large technical systems: Market liberalization as a driver for radical change? *Res. Policy*, **35**, 609–625.

Markard J. and Truffer B. (2008). Technological innovation systems and the multi-level perspective: towards an integrated framework. *Res. Policy*, **37**, 596–615.

Maurer M. (2013). Full costs, (dis-)economies of scale and the price of uncertainty. In: *Source Separation and Decentralization for Wastewater Management*, T. A. Larsen, K. M. Udert and J. Lienert (eds.), IWA Publishing, London, UK, pp. 85–100.

Milly P. C., Betancourt J., Falkenmark M., Hirsch R. M., Kundzewicz Z. W., Lettenmaier D. J. and Stouffer R. J. (2008). Stationarity is dead: Whither water management? *Science*, **319**(5863), 573–574.

Nelson V. I. (2008). New approaches in decentralized water infrastructure. Coalition for Alternative Wastewater Treatment, Gloucester MA. http://sustainablewaterforum.org/new/newappreport.pdf (accessed 12 March 2012).

Olsson G. (2013). The potential of control and monitoring. In: *Source Separation and Decentralization for Wastewater Management*, T. A. Larsen, K. M. Udert and J. Lienert (eds.), IWA Publishing, London, UK, pp. 179–191.

Phaal R. and Müller G. (2009). An architectural framework for roadmapping: Towards visual strategy. *Technol. Forecast. Soc. Chang.*, **76** (1), 39–49.

Sartorius C. (2008). Comparing the competitiveness of technology providing countries. Water 2050. The Fraunhofer Institute for Systems and Innovation Research ISI, Karlsruhe, Germany. http://www.wasser2050.de/ftp/ap2.pdf (in German, accessed 12 March 2012).

Scheele U. (2008). Transformation of urban water infrastructure: international experiences. netWORKS-Paper, Nr. 25, Forschungsverbund netWORKS, Berlin. http://edoc.difu.de/edoc.php?id=FU08RK6B (in German, accessed 12 March 2012).

Truffer B., Störmer E., Maurer M. and Ruef A. (2010). Local strategic planning processes and sustainability transitions in infrastructure sectors. *Environmental Policy and Governance*, **20**, 258–269.

Managing Strategic Innovation and Change: A Collection of Readings, Oxford University Press, New York.

Larsen T.A. and Gujer W. (1997). The concept of sustainable urban water management. [illegible], 35(9), 3–10.

[illegible] Source separation and decentralization [illegible] IWA Publishing, London [illegible]

[illegible]

Part III

Potential technologies for source separation

Part III

Chapter 16

Conceptualizing sanitation systems to account for new complexities in processing and management

Elizabeth Tilley

16.1 INTRODUCTION

The technological advancements in decentralized technologies have been rapid and impressive. However, the tools needed to describe the elements and interdependencies between different technologies in order to build holistic systems have not developed to the same degree. As a result, it has become increasingly difficult to communicate and plan the complex systems which decentralized and separation technologies require. Whereas classical sanitation planning focused on the transport and treatment of a single waste stream, future sanitation systems will need to address several parallel processes, each carried out at different scales and managed by different stakeholders.

The goal of this chapter is to outline a method of conceptualizing the components of, and interdependencies between technologies which comprise a system. Within a system boundary, technical and operational responsibilities can be identified to improve the overall system efficiency to remove gaps and redundancies.

The chapter begins with a short description of the system components, elaborates the concept of linking components into a system, and finally discusses the importance of system boundaries and their implications for system management.

16.2 EMERGING PRODUCTS

Describing the interdependencies between different technologies requires a way of describing the materials, or wastes that they were designed to treat. For example,

16.3.1 User interface

The *user interface* is the functional group which includes the technologies that allow the user to access the system; toilets and urinals as well as water-based appliances and faucets, are *user interface* technologies. The *user interface* determines the volume, concentration and composition of the input products. The *user interface* is responsible for partitioning, that is, separating, the input products which will then dictate which, and how many technologies and functional groups are required to complete the system. For example, a classic high-volume flush toilet produces a more dilute blackwater than a micro-flush toilet. A urine-diverting flush toilet produces both undiluted urine and brownwater – both of which require subsequent, but parallel storage, transport, treatment and use or disposal technologies. Water-saving fixtures in the house (e.g., shower, washing machine, etc.) can be efficient to the point that subsequent technologies suffer as a result. For example, if the amount of greywater is insufficient or separated completely, gravity sewers may suffer blockages more frequently; if greywater is used for irrigation, but the concentration of chemicals and/or nutrients increases, additional treatment steps may be required before it can be safely used. The choice of the *user interface*, and indeed the user, exercise a significant amount of control over the remainder of the system and what is expected of it. Yet, because that same user is usually disconnected (physically, legally, and financially) from the remainder of the system, s/he has little incentive to modify her/his behaviour. This is compounded by the fact that household infrastructure is costly and difficult to replace, and therefore the user, unless offered significant incentives, remains locked-in to the technologies that s/he inherited. Since source separation most easily occurs at the household level, but most engineers work at the semi-centralized or decentralized level, it is clear that greater attention must be paid to the important role that the user and the *user interface* play in the overall system design.

16.3.2 Collection and storage

Collection and storage technologies act as reservoirs until the intermediate products can be transported or used. Through the function of time, they may also provide some treatment. A urine storage tank is a simple example of such a technology, in that it holds the urine until it is needed for use in agriculture. Additionally, if urine is stored for a sufficiently long time it becomes hygienized and becomes safe for use on a wide range of crops (World Health Organization 2006). The problem with storage technologies is that they require space and space is often expensive or unavailable, especially in urban and/or multi-storey living environments.

The static nature and fixed volume of storage technologies means that emptying and loading must be carefully planned to prevent failure (i.e., overflowing or overloading, respectively). Storage is often required at several points (e.g., the household and a depot), therefore transport between the two must be coordinated even more so.

The design of the technology and duration of the storage will affect the characteristic of the product. Urine will increase in pH and become safe for use, though depending on the tank design, a large portion of the ammonia may be lost due to volatilization (Siegrist *et al.* 2013). Rainwater, collected from roofs (and diverted from combined sewers) may be stored in tanks or cisterns so that it can be collected during the rainy season and used during the dry one. Unfortunately, without disinfection or rapid use, bacterial growth is inevitable, and small apertures may clog. Though simple in concept, *collection and storage* technologies often require more vigilance than is initially foreseen, making a storage-based system, a maintenance-heavy one.

16.3.3 Conveyance

Conveyance technologies are those which allow products to be moved between different functional groups. Aquaducts (above ground) and gravity sewers (below ground) were the original technologies used for transporting liquids to and from a city, respectively. Now, more flexible transportation technologies are needed in order to address the specific needs of the greater range of products requiring transport.

Faecal sludge (from decentralized reactors, e.g., anaerobic baffled reactors) is viscous and heavy and may not flow well. The transport of sludge is one of the greatest problems facing urban sanitation in most of the world (Ingallinella *et al.* 2002). Exhauster trucks must transport fixed quantities over land though the energy, time, and investment required makes it far from an ideal technology, despite the fact that it is entirely necessary.

Urine, on the other hand, flows well and lends itself well to transportation via pipes. However, the propensity for calcium and magnesium minerals to precipitate spontaneously means that unless the pipes maintain a significant slope, clogging becomes a serious issue (Udert *et al.* 2003).

Cities are nutrient hotspots: concentrated sources of the nutrients that are lacking in rural agriculture. Given the huge and growing demand for low-cost sustainable nutrients around the world, more affordable and efficient *conveyance technologies* are needed to support the decentralized transfer of products between producers and consumers.

16.3.4 Treatment

Treatment technologies are specifically designed to change the composition of input or intermediate products. They may extract chemicals, inactivate pathogens or remove solids. Product-specific *treatment technologies* have evolved quickly over the last decades and there now exists an impressive array of treatment options for nearly every type of solid or liquid generated. This book is a testament to the efficiency of treatment that can be achieved when technologies are tailored to the specific properties of the products they were designed to treat.

A key feature of a system based on product separation is the need for parallel treatment; multiple treatment technologies may be required to operate simultaneously. For example, if urine, brownwater and greywater are separated using a urine-diverting flush toilet, the urine, urine precipitates, brownwater, brownwater sludge (which results after brownwater treatment), greywater and the greywater sludge and scum (which result from proper settling and separation), must all be treated. A thorough understanding of the products being treated is required to systematically design and optimize the treatment technologies required. Recognizing the need for parallel *treatment technologies* within this functional group is important to tailoring the treated output products for subsequent use and/or disposal.

16.3.5 Use and disposal

Use and disposal is the name of the functional group that includes the technologies or methods which distribute the products in a beneficial or benign way.

The opportunities for using faecal sludge have recently attracted increased attention. Sludge that has been dried can serve as a carbon-rich fuel source (e.g., the cement industry), applied directly to land as a soil amender (USEPA 1994), or processed further to release the limited, but valuable quantities of metals and nutrients (Hermann 2009). A system which aims to apply sludge directly must meet strict contaminant criteria and would therefore attempt to limit industrial or chemical inputs. The latter example underscores the important role of the *user interface* in generating a desirable output product. Conversely, the recovery of copper from incinerated sludge depends on non-excreta inputs to the system. The use of sludge as a fuel source is an emerging process, but clearly the sludge must be dry enough to burn, and this depends on the type and efficiency of the treatment technology used prior.

Figure 16.1a is a simplified schematic of how a series of functional groups would be linked with the input/intermediate/output products flowing between them. Of course, a real system does not consist of functional groups but rather of technologies, selected from the functional groups. Figure 16.1b shows the simple example of a system which handles only urine and faeces; note that some technologies process more than one product, while others are product-specific. In the case of eThekwini (shown) there is no dedicated *treatment technology* for the urine or faeces, but rather the treatment (dehydration and ureolysis) is carried out as a function of storage.

By thinking in terms of functional groups and the products which flow between them, the complexity of a multi-product system is revealed. The previous section attempted to illustrate how interdependent the functional groups become once the quality of the input and output products become important (as is one of the goals of source separation). In other words, technology replacements can be easily made within a functional group or by trading off between functional groups, but the implications for the type, number and quality of the subsequent or resultant output products will be affected. These trade-offs will also have significant consequences for the operation and management of the system as a whole.

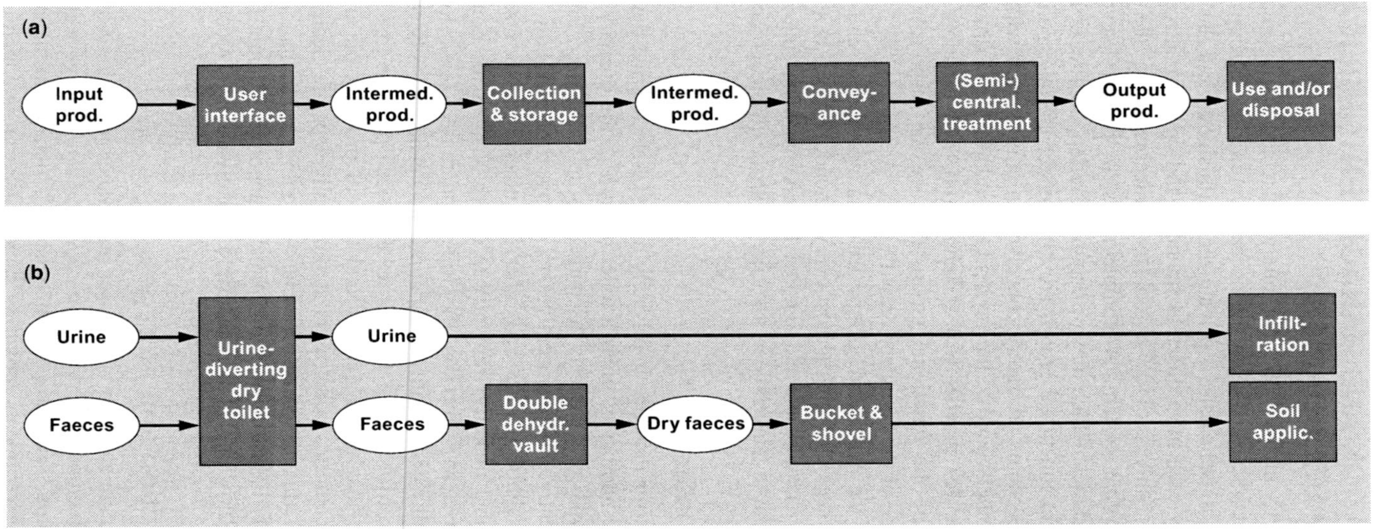

Figure 16.1 (a) Simplified schematic diagram of how input products are transformed into intermediate products and become output products as they travel through the functional groups. (b) System diagram for the eThekwini (Durban) model of source separation and urine infiltration showing the specific technologies used for each functional group (see also Lüthi and Panesar 2013).

16.4 OPERATION AND MANAGEMENT: IMPLICATIONS FOR SYSTEM BOUNDARIES

The system boundary includes all input, intermediate and output products, and the specific technologies selected from the necessary functional groups. Within one system boundary, households, the private sector (business) and the public sector (government), may each have different financial and operational responsibilities. Functional groups only provide a tool for conceptualizing the technologies needed for a gap-free system. A complete system requires several different management boundaries, which may or may not overlap, conflict or leave gaps.

Because different functional groups are often under the managerial domain of disparate sectors, systems are rarely conceived in consultation with all of those responsible. In other words, the system boundaries are drawn from the perspective of one (or maybe two) stakeholder(s), and therefore there is the potential to exclude the functional groups that are the responsibility of those stakeholders outside of the boundary drawn.

By looking at the functional groups from an operational and management perspective, it becomes clear that almost every functional group is simultaneously prioritized by one stakeholder, and ignored by another. The following sections briefly summarize the key stakeholders associated with the different functional groups.

16.4.1 User interface

By and large, the *user interface* is privately owned, selected and maintained. The operation (e.g., the number of times a toilet is flushed, the quantity of greywater generated or the types of chemicals discharged) is up to the user. Financially, the user has few incentives or disincentives to change how, or how often a *user interface* is used, although fines and other instruments can be used to control discharge quality and quantity from private enterprises. At the household level, regulation can have an effect on the toilets purchased (e.g., low-flush) but essentially, the quality and quantity of input products generated is beyond the scope of enforceable law. More often, it is norms and social pressure that play a strong regional role in regulating the quantity and type of products put into a system. In some southern countries, toilet users automatically separate cleansing materials (e.g., toilet paper) due to years of social marketing by sanitation authorities, while users in many industrialized cities remain unapologetic about flushing away unused chemicals knowing that there is little, and likely no, legal recourse.

16.4.2 Collection and storage

Currently, this functional group is of lesser importance in the industrialized world than it is in the developing world where *collection and storage* technologies are the main, if not only, sanitation option outside of rich, urban areas. Save for a few examples of governments providing sanitation services which centre around collection and storage technologies (e.g., ventilated improved pits, septic tanks), this type of technology is financed, operated and maintained by the homeowner. Unfortunately, when the word "on-site" technology is used, it further reinforces the idea of sanitation as a technology, thus exacerbating the problem of system gaps. Homeowners who elect to base their sanitation needs on *collection and storage* technologies often forget that products, for example sludge from a septic tank, must be transported to a subsequent location, by a third party, for a cost. Thus, the poorly named "on-site" technology, has expensive, labour-intensive "off-site" consequences that would not have been neglected had an overview of the full system boundary been presented.

As mentioned, collection and storage technologies are less common in industrialized countries, but as source-separation increases, so too will the need to examine the management needs associated with this functional group. The responsibilities for the regulation and transport of products stored at the household level remain unclear: the storage of urine, faeces, rainwater, or greywater remains in murky, undefined legal realms, especially in urban areas. If source separation is to become a large-scale reality, architects, engineers, planners and regulators must soon address the zoning, health and occupational issues associated with *collection and storage* technologies in our growing cities.

16.4.3 Conveyance

Sewers used to transport blackwater are owned and most commonly operated and maintained by public (government) utilities, whereas private enterprises are assuming an increasingly important role in the transport of other intermediate and output products.

Vacuum trucks (also known as honey-suckers, pump trucks, exhauster trucks, etc.) transport sludge between *collection and storage* and *treatment technologies*. In the industrialized world, private operators who empty septic tanks and transport the sludge are highly regulated and in many ways supported with transfer stations or disposal points, legal status (e.g., permits, insurance), and social acceptance. Entrepreneurs who transport sludge in the developing world are largely forced to work without permits, pay bribes, and generally tolerate ostracism (Eales 2005). In both cases, the operators run businesses which must be profitable; they must charge customers enough to cover the costs for operation and maintenance of the vehicles, permits and dumping fees- factors which are rarely accounted for in the design of a system. Figure 16.2a shows a system, which has management gaps: the entrepreneurs needed to remove the sludge from the household and the government-run facility for sludge treatment, have both been overlooked. Figure 16.2b shows the technologies that were missing and the responsibility boundaries. The relationships between the different groups, that is, between households and entrepreneurs, between entrepreneurs and government, are often logistically and financially complicated. Conceptualizing system and management boundaries cannot resolve inter-agent conflicts, but it can help to prepare for and prevent them.

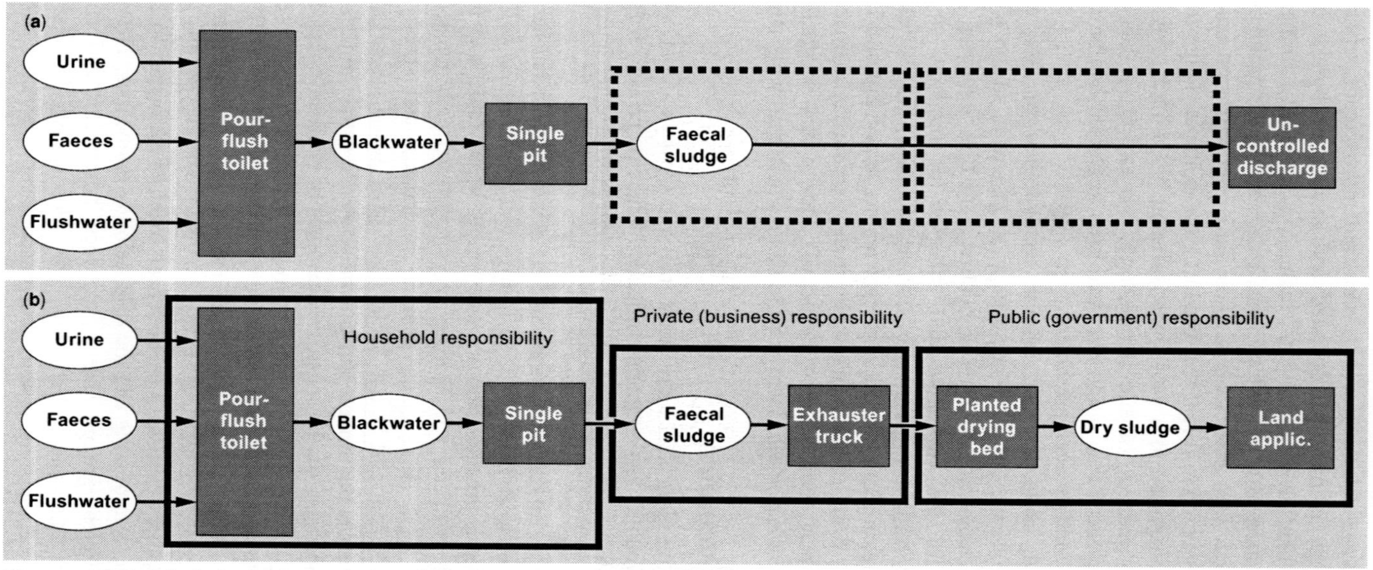

Figure 16.2 (a) A typical example of an incomplete pit-based system; missing conveyance, treatment and safe disposal technologies result in the uncontrolled discharged of untreated faecal sludge. (b) A completed pit-based system with no gaps – examples of appropriate technologies have been identified for each of the missing functional groups in Figure 16.2a.

16.4.4 Treatment

Though once the exclusive domain of the public sector, privately (or commercially) operated treatment plants have become increasingly common since at least the mid-1990s (Arnold 2009). Decentralized and product-specific *treatment technologies* go even further by providing opportunities for small enterprises, institutes and even households to operate and take responsibility for their own product treatment and use.

As mentioned above, the relationships between the agencies responsible for different functional groups in the system are complex. Shifting the responsibility of treatment to new agents may create new business opportunities, increase transparency, or reduce waste but at the cost of increased inter-agent complexity. The benefits of decentralized treatment are made abundantly clear in this edition, but the consequences of their management and connections to other functional groups within the system must be considered carefully in order that the system be sustainable.

16.4.5 Use and disposal

Water for irrigation, nutrients for fertilizer and energy for industrial and domestic use are opportunities that will dramatically affect the way that intermediate and output products are transported, treated, and ultimately, used. Public-private partnerships, as is the case with Ostara – a company that extracts nitrogen and phosphorus from digester supernatant at treatment plants – may well become a model for financial and environmental sustainability (Baur *et al.* 2009). Waste Enterprisers (WE), based in Accra, Ghana, is a business-minded social enterprise with a focus on nutrient recovery (in the form of fish) and good treatment (Waste Enterprisers 2010). The business offers clients two different models of cooperation: Implement-Own-Operate whereby WE assumes responsibility for treatment, but retains (most of) the profit, whereas the Implement-Train-Transfer model, sees the client retaining responsibility and profits after an initial upstart phase (ibid).

As the demand for sustainably sourced resources increases, so too will the need to develop these kinds of innovative business models. Models that integrate, or are well connected with, the treatment and resale market will be industry leaders. Sanitation has commonly been seen as a public-service which operates at a loss. Source separation and innovative *use and disposal* technologies will help to minimize operational losses, and may one day offset a system's costs to the point where it is cost neutral or even profitable.

This analysis of the management boundaries within the system boundary clearly points to the fact that decentralized solutions increasingly require careful consideration of the different responsibilities assumed by the growing number of agents involved in any one system.

16.5 CONCLUSIONS AND RECOMMENDATIONS

A systems approach, which clearly identifies not only the technologies and the linkages between functional groups, but the operational responsibilities for each, ensures a complete, sustainable sanitation system.

Product identification and characterization is essential: a careful analysis of the input, intermediate and output products will dictate the number and location of functional groups needed. Product-specific technologies should be chosen from the functional groups identified; a system is then made up of logically linked, product specific technologies, though effort should be taken to minimize redundancy, wherever necessary.

A system boundary defines the extent of the infrastructure; within the project boundary, the management boundaries must be drawn to delineate the responsibilities within the system. However, because each service provider operates over a limited range of technologies, several functional groups – specifically *conveyance* and *use and disposal* – "fall between the cracks," usually because these are the points at which management is transferred from one stakeholder to another. Technology development in each of the functional groups has been impressive, but the greatest challenge yet will lie in developing and implementing the policies and relationships between the stakeholders that are needed to ensure an efficient transfer of products between the functional groups.

With new business models will come new earning opportunities along with increased system complexity; negotiating the financial, logistical and legal relationships between the agents responsible for different functional groups within a system will be essential to the successful implementation and development of decentralized technologies and systems.

REFERENCES

Arnold C. A. (2009). Water privatization trends in the United States: human rights, national security, and public stewardship. *William & Mary Environmental Law and Policy Review*, **33**, 785–849.

Baur R., Prasad R. and Britton A. (2009). P-recovery from sewage sludge ash – technology transfer from prototype to industrial manufacturing facilities. In: *International Conference on Nutrient Recovery from Wastewater Streams*, D. Mavinic, K. Ashley and F. Koch (eds.), Vancouver, 10–13th May 2009, IWA Publishing, London, UK, pp. 351–360.

Eales K. (2005). Bringing pit emptying out of the darkness: A comparison of approaches in Durban, South Africa, and Kibera, Kenya. Sanitation partnership series, Building Partnerships for Development in Water and Sanitation, London, UK.

Friedler E., Butler D. and Alfiya Y. (2013). Wastewater composition. In: *Source Separation and Decentralization for Wastewater Management*, T. A. Larsen, K. M. Udert and J. Lienert (eds.), IWA Publishing, London, UK, pp. 241–257.

Heinss U., Larmie S. A. and Strauss M. (1999). Characteristics of faecal sludges and their solids–liquid separation. Eawag-Sandec unpublished intern. report, Swiss Federal Institute of Aquatic Science and Technology (Eawag), Dübendorf, Switzerland.

Hermann L. (2009). P-recovery from sewage sludge ash – technology transfer from prototype to industrial manufacturing facilities. In: *International Conference on Nutrient Recovery from Wastewater Streams*, D. Mavinic, K. Ashley and F. Koch (eds.), Vancouver, 10–13th May 2009, IWA Publishing, London, UK, pp. 403–416.

Ingallinella A. M., Sanguinetti G., Koottatep T., Montangero A. and Strauss M. (2002). The challenge of faecal sludge management in urban areas – strategies, regulations and treatment options. *Water Science and Technology*, **46**(10), 285–294.

Lüthi C. and Panesar A. (2013). Source separation in middle- and low-income countries. In: *Source Separation and Decentralization for Wastewater Management*, T. A. Larsen, K. M. Udert and J. Lienert (eds.), IWA Publishing, London, UK, pp. 455–461.

Siegrist H., Laureni M. and Udert K. M. (2013) Transfer into the gas phase: ammonia stripping. In: *Source Separation and Decentralization for Wastewater Management*, T. A. Larsen, K. M. Udert and J. Lienert (eds.), IWA Publishing, London, UK, pp. 337–350.

Udert K. M., Larsen T. A. and Gujer W. (2003). Biologically induced precipitation in urine-collecting systems. *Water Supply*, **3**(3), 71–78.

USEPA (1994). A plain English guide to the EPA part 503 biosolids rule. US Environmental Protection Agency, Office of Wastewater Management, Washington DC, USA.

Waste Enterprisers (2010). http://waste-enterprisers.com (accessed 19 July 2011).

World Health Organization (2006). Guidelines for the safe use of wastwater, excreta and greywater. Volume 4: Excreta and greywater use in agriculture. World Health Organization, Geneva, Switzerland.

Chapter 17

Wastewater composition

Eran Friedler, David Butler and Yuval Alfiya

17.1 INTRODUCTION

Domestic wastewater flows and quality strongly depend on the type of wastewater generating appliance (the source), on the type and mode of use which may differ from appliance to appliance and from user to user, and on the household tenants' habits and culture. Differences in reported data may also stem from geographic and socio-economic variations and the date of the study: for example the water demand of washing machines and dishwashers was reduced significantly in the last two decades, and the formulations of detergents changed during the same period. The aim of this chapter is to characterize and quantify the flows and pollutant loads of the different sub-streams of domestic wastewater.

17.2 DOMESTIC WASTEWATER FLOWS

Domestic wastewater is composed of a number of sub-streams. We distinguish their source in relation to domestic appliances as shown in Table 17.1, to help understand the composition and relevance of the stream. The principal distinction we make is between *blackwater* and *greywater* which are further divided to sub-streams having their individual flow characteristics.

Average daily domestic wastewater discharge stands at 148 $L \cdot p^{-1} \cdot d^{-1}$ in the eleven countries shown in Table 17.2, with relatively low standard deviation (33 $L \cdot p^{-1} \cdot d^{-1}$). By far the larger of the two sub-streams is greywater, which amounts to 73% of the total domestic wastewater discharge (108 $L \cdot p^{-1} \cdot d^{-1}$). This is subdivided almost equally between dark and light grey fractions. In developing countries, water consumption tends to be substantially lower than in the countries shown in Table 17.2. Consequently, wastewater production can be as low as 20–30 $L{\cdot}p^{-1}{\cdot}d^{-1}$ (Morel and Diener 2006). This will result in higher concentrations of pollutants together with lower volumes available for reuse. The most concentrated waste flows originate from toilets without flushing, such as waterless urinals or dry toilets, for example, urine-diverting dehydration toilets (UDDTs) or ventilated pit latrines (VIPs) (Tilley *et al.* 2008).

Table 17.1 Wastewater sub-streams and their sources.

Stream	Sub-stream	Source
Blackwater (BW)	Yellowwater	Urine
	Brownwater	Faeces and toilet paper
	Beigewater	Anal cleansing water
Greywater (GW)	Light greywater	Shower (S) Bath tube (B) Bathroom washbasin (WB)
	Dark greywater	Kitchen sink (KS) Dishwasher (DW) Washing machine, and laundry where applicable (WM)

Blackwater is generated via the water closet (WC) and consists of a combination of yellowwater (urine), brownwater (faeces plus anal cleanser – typically toilet paper) and in some countries beigewater (water used for anal cleansing), the latter two streams include the carrier flushwater. Friedler *et al.* (1996) defined four types of WC use: urine only, faeces and urine, faeces only, and other, with respective proportional use of 74, 21, 2 and 3%. The increasing application of dual-flush toilets has led to a decrease of the water used for urine flushing: after urination half flush volumes can be practiced (e.g., from 6 to 3 L per use). The development of these new appliances should also be borne in mind when source-separation of blackwater to its sub-streams is considered. Of the greywater stream, the bath and shower together (BT and SH) were signalled as the major contributor of wastewater, consisting 28% ($42\,L{\cdot}p^{-1}{\cdot}d^{-1}$) of the overall wastewater generation.

Looking further at the micro-scale, appliance discharges can be characterized by frequency of use, mode of use, duration/discharge of each single use-event, and by the load of pollutants associated with each use. Some appliances (bath, washbasin and kitchen sink) may be operated in two modes, fill-and-empty and run-to-waste, that differ in the volume of generated discharge. In the fill-and-empty mode the drainage point of the appliance is stoppered and opened towards the end of the use-event, while in the run-to-waste mode the drainage point is open and wastewater is drained continuously throughout the use-event. In this context,

Table 17.2 Proportional daily flow contribution of in-house wastewater sources in [%]. See Table 17.1 for abbreviations.

	Australia[1]	Brazil[2]	Denmark[3]	Israel[4]	Malta[5]	Netherlands[6]	Oman[7]	Portugal[8]	Switzerland[9]	UK[6]	USA[10]	Average
BW	21	34	25	36	31	37	6	21	30	32	31	**27**
BT and SH	33	23	38*	26	27	26	49	36	20	21	23	**28**
WB	8	17		9	9	5	5	15	13	9	9	**10**
KS**	8	21	22	16	16	16	32	14	15	16	9	**17**
WM	27	5	15	8	17	16	8	11	19	21	25	**16**
DW	3			3				3	2	1	2	**2**
Total [$L \cdot p^{-1} \cdot d^{-1}$]	155	151	112	153	95	130	171	134	158	144	224	**148**

* BT, SH and WB; ** When DW stream is not stated KS may include DW discharge.

References: [1] Loh and Coghlan (2003), [2] Ghisi and Ferreira (2007), [3] Donner *et al.* (2010), [4] Friedler (2008), [5] Butler *et al.* (1995), [6] Memon and Butler (2006), [7] Parthapar *et al.* (2005), [8] Vieira *et al.* (2007), [9] Helvetas (2005), [10] Roesner *et al.* (2006).

Friedler and Butler (1996) found that the bath is used 85% of the time in fill-and-empty mode, which produces significantly higher discharges than run-to-waste mode (average of 95 *vs.* 21 L per use, respectively). The kitchen sink discharges also strongly depend on its mode of use. The most frequent discharge range in run-to-waste mode, which consisted 53% of all kitchen sink uses, was 0.13–1.4 L per use (>30% of uses in this mode), while the modal range in fill-and-empty mode was 6.5–8.3 L per use (25% of uses in this mode). The toilet, washing machine and dishwasher have a pre-selected mode of use (usually fill-and-empty) that cannot be modified by the user. The shower, in contrast, is primarily used in run-to-waste mode the generated volume of which strongly depends on users' habits and environmental awareness.

The data in Table 17.2 of course represent snapshot values that will vary over time both increasing due to increased appliance ownership and greater luxury water use and decreasing due to water demand management measures (Memon and Butler, 2006). For example, washing machine ownership has now reached almost saturation levels (e.g., in the UK, 96% in 2009[1]), whilst at the same time their water use efficiency has increased from 150 L per cycle in the 1970s to less than 50 L per cycle today. Similarly, dishwasher efficiency increased from more than 50 L per cycle to less than 10 L per cycle over a similar period making them generally more water-efficient than manual dish washing. However, in this case, there is greater potential for overall increase as current ownership levels are lower (35–40% in Israel and the UK and 60–65% in Denmark and Germany in 2009[2]). Striking advances have been made to WC efficiency in recent years with many EU countries mandating low water-use models at 6 L per flush or less, and some countries (like Israel) mandating dual flush toilets of 6 and 3 L per flush. Technological developments are pushing flush volumes ever lower, with a 1.5 L air-flushed toilet under development (Littlewood *et al.* 2007) and vacuum toilets (0.8–2 L). This trend of increasing water use efficiency will result in lower specific water consumption on the one hand and more concentrated wastewater on the other.

17.3 WASTEWATER FLOW PATTERNS

Water uses within individual households are stochastic in nature and create an intermittent flow pattern of liquid wastes that vary widely in strength and volume. This high variability necessitates incorporation of equalisation tank (for flow, quality and temperature) in most on-site wastewater reuse schemes. Analyzing wastewater discharges from a large number of individual households statistically enables to derive characteristic diurnal flow patterns. These characteristic flow patterns, including both blackwater and greywater sub-streams, exhibit a

---

[1] www.statistics.gov.uk

[2] Friedler - Unpublished compilation of data

significant morning peak at 6:00–10:00 followed by smaller evening peaks (19:00–23:00). Low flow periods occur in the late morning and early afternoon and virtually no-flow period late at night (2:00–5:00, Figure 17.1). Blackwater typically contributes 60–90% to the total flow at night and 20–40% during the day. The exception to this pattern is greywater discharge from the washing machine that typically occurs during the late morning and late afternoon and continues until midnight (Butler 1991, Butler *et al.* 1995, Briks and Hills 2007, Ghunmi *et al.* 2008, Wheatley and Surendran 2008, Eriksson *et al.* 2009, Kim *et al.* 2009). Peak intensity and timing varies from one place to another with differences explained by different lifestyles and cultures. As an example, Butler *et al.* (1995) showed that in England the morning peak is 1.5 times higher than in Malta and occurs half an hour later (8:00 vs. 7:30).

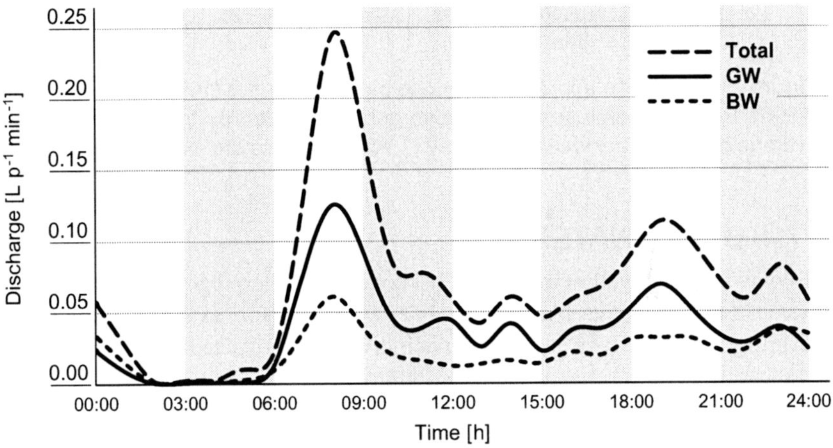

Figure 17.1 Diurnal flow pattern (weekdays) of domestic wastewater (Total), blackwater (BW) and greywater (GW) (based on Butler *et al.* 1995).

Domestic wastewater sub-streams exhibit variability also on a weekly scale. Differences between week days and weekend days are most pronounced, as many people tend to spend more time at home during the weekend. This is especially true for employed people, students, etc. For example, greywater from the washing machine exhibits the highest variability, with two days of peak use (1.5 on Saturday and Monday; Figure 17.2), while on Tuesday and Wednesday its use is significantly lower (0.42 and 0.49 respectively). Blackwater exhibits the lowest variability with slightly higher use during the weekend (1.1, and 1.2 on Saturday and Sunday). Regarding the diurnal pattern, its generation is more evenly distributed during weekend days exhibiting morning peaks about 20% lower than during weekdays appearing about an hour later (Friedler *et al.* 1996).

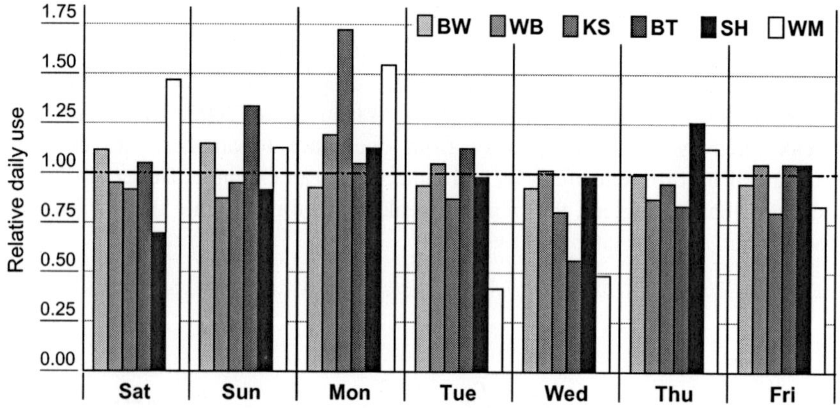

Figure 17.2 Relative appliance daily use based on Butler (1991) and Friedler *et al.* (1996). Relative daily use – no. of uses per day divided by the no. of uses on an average day (uses per week divided by 7). Abbreviations are explained in Table 17.1.

17.4 BLACKWATER

Brownwater and yellowwater can be collected either together in conventional systems and become blackwater, or separately (for example through urine diverting toilets), where each stream is individually collected and treated.

17.4.1 Yellowwater

Humans typically excrete 1,270 $g{\cdot}p^{-1}{\cdot}d^{-1}$ of urine, with a water content of 95% (Table 17.3). The remaining 5% are dissolved salts, which makes urine a saline solution. Fresh urine hardly contains any microorganisms; there are evidences that urine of healthy people may contain bacteria, but it is still uncertain whether these are viable or connected with bacterial infection (Wolfe *et al.* 2012). Faecal sterols measured in source-separated urine indicate cross-contamination with transmissible pathogens and other bacteria from faeces (Schönning *et al.* 2002). The organic load of yellowwater is relatively low, with a specific contribution of 13 $gCOD{\cdot}p^{-1}{\cdot}d^{-1}$, nevertheless, COD levels in source-separated urine are high ($\sim$10 $g{\cdot}L^{-1}$). This organic matter (measured as COD) is composed of organic acids, creatinine, amino acids and carbohydrates, which can be degraded by anaerobic bacteria such as fermenters during storage. The main dissolved compound in urine, urea (21 $\pm$ 3 $g{\cdot}L^{-1}$) is also organic, but does not present any COD (Udert *et al.* 2006).

Table 17.3 Compounds in urine and faeces.

		Urine			Faeces		
		Av.	**Std.**	**Range**	**Av.**	**Std.**	**Range**
Wet mass	$[g \cdot p^{-1} \cdot d^{-1}]$	1,270	363	1,010–1,530	200	100	100–350
Dry mass	$[g \cdot p^{-1} \cdot d^{-1}]$	58			41	10	31–53
Water cont.	[%]	95	1.4	94–96	77	7	65–85
pH	[–]	6.4	1.1	5.0–7.2			
EC	$[mS \cdot cm^{-1}]$	19	2.4	8.7–31			
TSS	$[g \cdot p^{-1} \cdot d^{-1}]$				22	23	6–60
$BOD_{5\,t}$	$[g \cdot p^{-1} \cdot d^{-1}]$	5.8	0.7	1.8–10	12	11	4.3–20
COD_t	$[g \cdot p^{-1} \cdot d^{-1}]$	13	2.6	5.0–24	31	41	2.6–63
TOC	$[g \cdot p^{-1} \cdot d^{-1}]$	4.3			3.5		
TAN	$[mg \cdot p^{-1} \cdot d^{-1}]$	550	36	320–880	301	408	13–590
N_{tot}	$[g \cdot p^{-1} \cdot d^{-1}]$	11	0.3	4–16	1.5	0.0	0.3–4.2
Urea	$[g \cdot p^{-1} \cdot d^{-1}]$	21	3				
P_t *	$[g \cdot p^{-1} \cdot d^{-1}]$	0.93		0.8–2.0	0.6		0.3–0.8
K	$[g \cdot p^{-1} \cdot d^{-1}]$	2.6	0.1	1.0–4.9	0.9	0.2	0.2–1.3
S	$[g \cdot p^{-1} \cdot d^{-1}]$	1.3		1.2–1.5	0.61	0.19	
Pb	$[\mu g \cdot p^{-1} \cdot d^{-1}]$	2.0			20		
Cu	$[\mu g \cdot p^{-1} \cdot d^{-1}]$	103			1,110		
Hg	$[\mu g \cdot p^{-1} \cdot d^{-1}]$	1.9	1.6	0.8–3.1	8.3		
Zn	$[\mu g \cdot p^{-1} \cdot d^{-1}]$	46			11		
FC	$[cell \cdot p^{-1} \cdot d^{-1}]$						10^8–10^{11}
FS	$[cell \cdot p^{-1} \cdot d^{-1}]$						10^7–10^{10}
Cps	$[cell \cdot p^{-1} \cdot d^{-1}]$						10^5–10^{12}

Compilation of data from: Siegrist *et al.* (1976), Ciba-Geiy (1977), Feachem *et al.* (1983), Fittschen and Hahn (1998), Almeida *et al.* (1999), Udert (2002), Udert *et al.* (2003), Vinnerås *et al.* (2006), Etter *et al.* (2011), Meinzinger and Oldenburg (2009). EC – Electrical conductivity; $_t$ – total; TAN – Total ammonium nitrogen; FC – Fecal coliforms; FS – Fecal Streptococci; Cps – *Clostridium perfringens* spores. * About 95% of P_t in urine is phosphate.

Urine is a major source of nitrogen, phosphorus and potassium in domestic wastewater (79, 47 and 71% respectively, Table 17.6), confirming its potential as a fertilizer. During storage and transport, urine is subject to processes that affect its characteristics such as urea hydrolysis, precipitation or volatilization. Bacteria in the collection system produce the enzyme urease which decomposes urea into ammonia and bicarbonate, thereby raising the pH value. This promotes precipitation of struvite ($MgNH_4PO_4 \cdot 6H_2O$), hydroxyapatite ($Ca_5(PO_4)_3(OH)$) and calcite ($CaCO_3$). Following urea hydrolysis 90% of the nitrogen is in the form of TAN (total ammonia nitrogen: $NH_3 + NH_4^+$), as opposed to fresh urine where 85% of the nitrogen is fixed as urea. Due to the pH increase (from 6.2 to 9.1), the fraction of ammonia (NH_3) of the TAN becomes high. Ammonia is very volatile (Henry's constant: 62 mol·L^{-1}·atm^{-1} at 25°C), therefore, it can be lost through volatilization during storage and transport of urine. Ammonia volatilization is not only a loss of a valuable nutrient, but it also can cause health and odour problems. Urea hydrolysis also increases the alkalinity strongly (Udert *et al.* 2006).

Urine contains micropollutants, such as hormones and pharmaceutical residues. Based on a quantitative screening of official pharmaceutical data in Switzerland, Lienert *et al.* (2007) calculated that 65 ± 27% of the active ingredients of pharmaceuticals consumed are excreted via urine, 42% as metabolites. As expected, the excretion rate of hormones differs between males and females: estrogens are excreted more by females and androgens more by males (Table 17.4).

Table 17.4 Excretion of selected natural hormones by males and females and concentrations in wastewater treatment plant (WWTP) influent (Based on Liu *et al.* 2009).

	Females [μg·p^{-1}·d^{-1}]			**Males [μg·p^{-1}·d^{-1}]**	**WWTP influent [ng·L^{-1}]**
	Post-m	**Pre-m**	**Pregnant**		
Estrone (E1)	5.0	11	1,190	3.9	n.d.-670
17β-estradiol (E2)	2.8	4.7	347	1.5	n.d.-162
Estriol (E3)	2.8	8.2	24,100	1.5	n.d.-660
Testosterone (T)		6.8		57	n.d.-95
Androsterone (AD)		1,570		3,340	5,700-14,400

Post-m – post-menopausal; Pre-m – pre-menopausal; n.d. – not detected.

17.4.2 Brownwater

Humans excrete some 200 g·p^{-1}·d^{-1} faeces, with a water content of 77% (Table 17.3). The amount of faeces excreted exhibits high variability and is dependent on diet, climate and state of health among other factors. Generally in developed countries, the amount of faeces lies in the range of 100–350 g·p^{-1}·d^{-1} (wet), while in developing countries it is 250–350 g·p^{-1}·d^{-1} due to the greater fibre and lower meat content in the diet. The higher the specific faeces excretion, the higher is its water content. Faeces, unlike urine, exhibit high organic load (31 gCOD·p^{-1}·d^{-1}), but lower nutrient loads (1.5 and 0.4 g·p^{-1}·d^{-1} N and P, respectively). Faeces are also characterized by very high levels of pathogens and indicator microorganisms. For example, faecal coliforms are common in the range of 10^6-10^9 cfu·$g_{faeces(wet)}^{-1}$ (10^8-10^{11} cfu·p^{-1}·d^{-1}).

Pollutant contribution of toilet paper is often overlooked although it contributes 11 and 8% of the TSS and COD loads, respectively (Table 17.6).

Friedler *et al.* (1996) found that average domestic toilet paper use is 12.4 sheets·p^{-1}·d^{-1} corresponding to 6.82 g·p^{-1}·d^{-1}. Toilet paper thus contributes 7.2 g·p^{-1}·d^{-1} of solids (of which 6.8 and 6.5 g·p^{-1}·d^{-1} are TSS and VSS respectively), 8.8 g·p^{-1}·d^{-1} of COD, while N and P contribution is negligible. The degradability of toilet paper is somewhat low since it consists mainly of cellulose.

17.5 GREYWATER

The quality characteristics of greywater (light and dark) are highly dependent on the behaviour of the dwellers, the appliances connected and the chemicals used. When greywater is considered for on-site reuse, it is advisable to identify the proportional volume and pollutant contribution of each generating appliance and select according to the reuse needs the streams with the lowest pollutant load. Friedler (2004) compared the greywater sub-streams with respect to their contribution to the main pollutant loads (Figure 17.3). The kitchen sink was noted as the major source of most pollutants. In general, dark greywater has higher pollutant loads than light greywater with the exception of pathogens (expressed as faecal coliforms). It should be noted that since the last stages of the washing machine and dishwasher wash cycles produce much less polluted greywater, they can be separated from the discharges of the first stages and included in the light greywater stream without significant addition of pollutants.

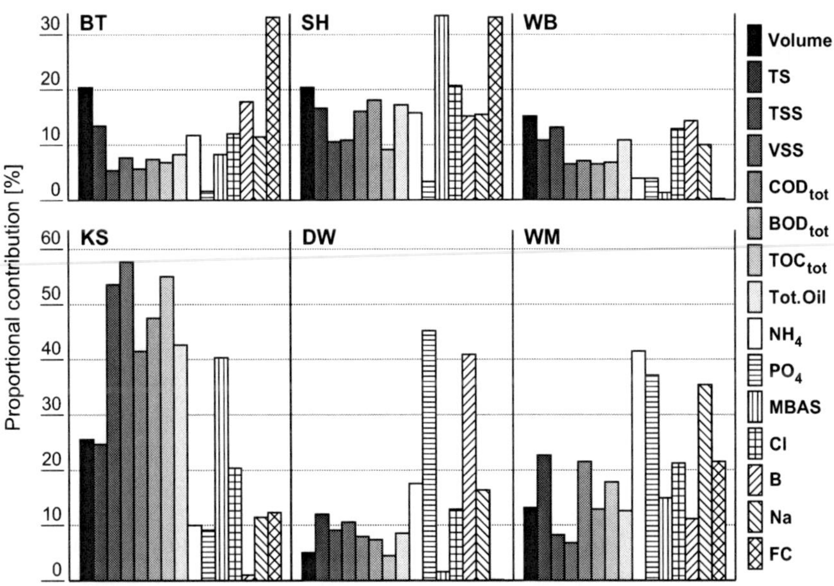

Figure 17.3 Proportional pollutant loads of individual greywater streams (based on Friedler 2004).

Organic matter. Greywater is the major contributor of organic substances (Table 17.6). Most of the organics in greywater originate from the kitchen sink (Figure 17.3). Kitchen sink and dishwasher greywater contains food residues, oils and fats, detergents, drain cleaners and bleaching agents. The kitchen sink contributes 40–60% of the major pollutant load in greywater (VSS, COD, BOD, total oil and methylene blue active substances, MBAS) (Almeida *et al.* 1999, Friedler 2004). The COD/BOD_5 ratio, which is a measure of biodegradability, ranges from 2–3.6 being higher in light greywater and indicating that the organic matter in the latter stream is somewhat less biodegradable (Jefferson *et al.* 2004, Morel and Diener 2006, Aizenchtadt *et al.* 2009).

Personal care products and detergents are found mainly in greywater, originating from household chemicals such as laundry and dishwasher detergents, shampoos, soaps, preservatives, perfumes, cosmetics and dyes. Eriksson *et al.* (2002) identified 900 organic chemicals in greywater, of which 66 were priority pollutants. In a later study, Eriksson *et al.* (2003) detected 200 xenobiotic organic compounds (XOC) in shower and washbasin greywater, generally at very low concentrations. Due to intensive use of detergents in washing machines and dishwashers, higher concentrations of surfactants are found in dark greywater while lower ones are expected in light greywater. Aizenchtadt *et al.* (2009) reported MBAS concentrations of 8 $mg{\cdot}L^{-1}$ in light greywater, while concentrations in combined greywater (light and dark) were in the range of 40–56 $mg{\cdot}L^{-1}$ (Table 17.5).

Hygiene. Light greywater is considered to be the least polluted wastewater stream (Figure 17.3), yet it can contain faeces-related pathogens from washing hands after excretion and bathing as well as skin and mucus tissue pathogens (Briks and Hills 2007). For example *Pseudomonas aeruginosa sp.* and *Staphylococcus aureus sp.* were detected at concentrations of $5–9{\cdot}10^3$ $cfu{\cdot}100mL^{-1}$. Faecal coliform levels of 10^6 $cfu{\cdot}100mL^{-1}$ have been recorded (Friedler *et al.* 2006), but care is needed in interpreting data due to the possibility of bacterial regrowth that may result in overestimation of the microbial risk.

Phosphorus. In places where washing machine and dishwasher detergents contain phosphorus, dark greywater can contribute significantly to the overall phosphorus load (Gray and Becker 2002). This contribution is expected to decrease due to the shift towards phosphorus-free detergents.

Heavy metals with substantial concentrations in domestic wastewater are zinc, copper and mercury. The main source of Zn and Cu has been identified as in-house plumbing, with greywater contributing about 90% (120 and 11.0 $mg{\cdot}p^{-1}{\cdot}d^{-1}$) of their loads to domestic wastewater (Gray and Becker 2002). The recent trend of shifting to plastic-based in-house plumbing is expected to reduce the concentrations of both metals. Dental amalgam (contributed by teeth brushing and by defecation) is believed to be the major source of mercury in domestic wastewater (>80%), while the rest is contributed by common household and toiletry products (AMSA 2000). The mercury load is expected to decrease as filling material is shifting from amalgam to composite resins.

Table 17.5 Pollutants in greywater.

	Loads				Concentrations			
		Av.	Std.	Range		Av./Gm.	Std.	Range
pH						7.2	0.52	6.4–10.0
EC					[$\mu S \cdot cm^{-1}$]	1.8	0.66	0.1–2.8
Turb					[NTU]	95	90	20–280
TSS	[$g \cdot p^{-1} \cdot d^{-1}$]	19	21	2–125	[$mg \cdot L^{-1}$]	216	270	2.0–1,070
VSS	[$g \cdot p^{-1} \cdot d^{-1}$]	6.9	8.6	1.6–20	[$mg \cdot L^{-1}$]	64		
COD_t	[$g \cdot p^{-1} \cdot d^{-1}$]	51	24	7.0–102	[$mg \cdot L^{-1}$]	620	707	7–2,570
TOC	[$g \cdot p^{-1} \cdot d^{-1}$]	8.4	6.4	1.8–16	[$mg \cdot L^{-1}$]	83	14	73–93
$BOD_{5\ t}$	[$g \cdot p^{-1} \cdot d^{-1}$]	19	19	1.0–63	[$mg \cdot L^{-1}$]	279	314	1–1,060
$BOD_{5\ d}$	[$g \cdot p^{-1} \cdot d^{-1}$]	7.6	7.7	1.9–19	[$mg \cdot L^{-1}$]	34	3.4	31–36
N_t	[$g \cdot p^{-1} \cdot d^{-1}$]	0.9	0.7	0.007–2.3	[$mg \cdot L^{-1}$]	23	38	0.1–128
TAN	[$g \cdot p^{-1} \cdot d^{-1}$]	0.32	0.49	0.11–1.3	[$mg \cdot L^{-1}$]	28	41	1–75
NO_3-N	[$g \cdot p^{-1} \cdot d^{-1}$]	0.64	1.3	0.007–2.5	[$mg \cdot L^{-1}$]	17		0.1–17
PO_4-P	[$g \cdot p^{-1} \cdot d^{-1}$]	0.14	0.19	0.017–0.41	[$mg \cdot L^{-1}$]	0.4		
P_t	[$g \cdot p^{-1} \cdot d^{-1}$]	0.5	0.4	0–2.2	[$mg \cdot L^{-1}$]	8.5	7.2	0.1–42
Cl^-	[$g \cdot p^{-1} \cdot d^{-1}$]	17	18	4.1–37	[$mg \cdot L^{-1}$]	181	37	9–227
K	[$g \cdot p^{-1} \cdot d^{-1}$]	0.3	0.6	0–4.1	[$mg \cdot L^{-1}$]	10	8.4	0.2–24
S	[$g \cdot p^{-1} \cdot d^{-1}$]	2.9		0.5–7.7	[$mg \cdot L^{-1}$]	38	49	0.5–72
MBAS	[$g \cdot p^{-1} \cdot d^{-1}$]	5.6	4.4	2.5–8.7	[$mg \cdot L^{-1}$]	48	12	40–56
B	[$g \cdot p^{-1} \cdot d^{-1}$]	0.16	0.1	0–0.22	[$mg \cdot L^{-1}$]	0.6		
Na	[$g \cdot p^{-1} \cdot d^{-1}$]	12	14	2.0–32	[$mg \cdot L^{-1}$]	148	29	7.4–480

(*Continued*)

Table 17.5 Pollutants in greywater (*Continued*).

	Loads				Concentrations			
		Av.	Std.	Range		Av./Gm.	Std.	Range
Hg	[$mg \cdot p^{-1} \cdot d^{-1}$]				[$\mu g \cdot L^{-1}$]			0.56–36
FC	[$cfu \cdot p^{-1} \cdot d^{-1}$]	$6.3 \cdot 10^7$			[$cfu \cdot L^{-1}$]	$8.2 \cdot 10^5$ *	$1.6 \cdot 10^8$	$2 \cdot 10^3 – 1 \cdot 10^9$
FS	[$cfu \cdot p^{-1} \cdot d^{-1}$]	$1.3 \cdot 10^6$			[$cfu \cdot 100\ mL^{-1}$]	$1.7 \cdot 10^3$ *		
Pa	[$cfu \cdot p^{-1} \cdot d^{-1}$]	$7.2 \cdot 10^6$			[$cfu \cdot 100\ mL^{-1}$]	$9.4 \cdot 10^3$ *	$1.7 \cdot 10^4$	$3 \cdot 10^3 – 3 \cdot 10^4$
Sa	[$cfu \cdot p^{-1} \cdot d^{-1}$]	$3.7 \cdot 10^6$			[$cfu \cdot 100\ mL^{-1}$]	$4.8 \cdot 10^3$ *	$5.3 \cdot 10^3$	$2 \cdot 10^3 – 1 \cdot 10^4$
FE	[$cfu \cdot p^{-1} \cdot d^{-1}$]	$1.9 \cdot 10^7$			[$cfu \cdot 100\ mL^{-1}$]	$2.5 \cdot 10^4$ *		$1 \cdot 10^3 – 1 \cdot 10^5$
Cp	[$cfu \cdot p^{-1} \cdot d^{-1}$]	$9.6 \cdot 10^5$			[$cfu \cdot 100\ mL^{-1}$]	$1.3 \cdot 10^3$ *		
SC	[$cfu \cdot p^{-1} \cdot d^{-1}$]			$0 – 1.5 \cdot 10^6$	[$cfu \cdot 100\ mL^{-1}$]	$2.0 \cdot 10^3$ *		$0 – 1 \cdot 10^4$

Compilation of: Fox *et al.* (2002), Ottoson and Stenström (2003), Jefferson *et al.* (2004), Friedler (2004), Gross *et al.* (2005), Palmquist and Hanaeus (2005), Friedler *et al.* (2006), Vinnerås *et al.* (2006), Briks and Hills (2007), Gilboa and Friedler (2008), Ghunmi *et al.* (2008), Jamrah *et al.* (2008), Winward *et al.* (2008), Meinzinger and Oldenburg (2009). GM – geometric mean (*); MBAS – methylene blue active substances; FS – Fecal Streptococci; Pa – *Pseudomonas aeruginosa*; Sa – *Staphylococcus aureus*; FE – Faecal Enterococci; Cp – *Clostridium perfringens*; SC – Somatic Coliphages; see Table 17.3 for further abbreviations.

Table 17.6 Nominal and proportional pollutant loads in greywater and blackwater (urine, faeces, toilet paper).

	Load [$g \cdot p^{-1} \cdot d^{-1}$]					Contribution [%]			
	GW	**Urine**	**Faeces**	**Toilet paper**	**Total**	**GW**	**Urine**	**Faeces**	**Toilet paper**
TSS	19	12	23	6.8	61	32	19	38	11
BOD_{5t}	19	5.8	12	n.a.	37	52	15	33	n.a.
COD_t	51	13	31	8.8	104	50	11	31	8
N_t	0.9	9.2	1.5	0.0	11.6	8	79	13	0
P_t	0.5	0.8	0.4	0.0	1.8	28	47	25	0
K	0.3	2.9	0.9	n.a.	4.1	7	71	22	n.a.

Based on data in Table 17.3 and 17.5. Toilet paper data taken from Friedler *et al.* (1996), n.a. – data not available

Salts. Sodium is added to washing powders as a counter ion ("filler,") boron as a whitening agent and chloride as a counter ion. High salinity can adversely affect soil structure and plant growth, high sodium levels can harm soil structure (esp. clay soils), while boron starts to be toxic to plants at relatively low concentrations. For example, during the 1990's laundry detergents contributed 7, 42 and 80–90% of the total chlorides, sodium and boron addition to municipal wastewater in Israel (Anonymous 2000). Since then, the maximum allowed specific loads of these ions in laundry detergents were reduced by regulations (SII 1999) in order to enable reuse in agriculture. In addition to the above, dishwashers use adds extra sodium chloride for regeneration of its internal ion-exchanger (water softening). The dishwasher is also a major contributor of boron with average concentrations of up to 4 $mg \cdot L^{-1}$ in its discharge (Friedler 2004). This may impair the possibility of reuse in irrigation due to inflicted damages on plants and soils. Lately, the government of Israel has developed regulations limiting boron concentration in dishwasher detergents (SII 2006).

Temperature. Some sub-streams of greywater may exhibit elevated temperatures and wide temperature fluctuations (e.g., bath and shower 15–40°C; washing machine typically 40° but can reach 80–85°C; dishwasher 50–70°C). These fluctuations should be considered when designing decentralized biological treatment, since medium to high temperatures can favour microbial growth and calcium carbonate precipitation, while very high temperatures or sharp temperature fluctuations may inhibit bio-degradation of organic matter.

17.6 PROPORTIONAL CONTRIBUTION OF NUTRIENTS AND ORGANICS

The loads contributed by each domestic wastewater source and its proportional contribution are presented in Table 17.6. Urine was found to be the major contributor of nitrogen (79%), phosphorus (47%) and potassium (71%). This strengthens the potential of using source-separated urine as a fertilizer. Urine and faeces together contribute most of the daily load of nitrogen, phosphorus, potassium and TSS (93, 72, 93 and 57%, respectively), and almost half the BOD and COD loads (48 and 42%, respectively). Dark greywater is the major source of BOD and COD in greywater.

17.7 DISCUSSION AND SIGNIFICANCE

The aim of the chapter has been to collate, summarise and interpret data from a wide variety of sources. In it we have discussed in detail the characteristics of wastewater in the context of its source and its potential for reuse. This is particularly important when considering the feasibility of separation opportunities. Only by understanding the characteristics of each sub-stream, dependent on the local context, can rational decisions be made on the degree and nature of separation that should be attempted. The current focus in many countries is on improving water use efficiency. This clearly has benefits in terms of enhancing the local water supply balance. However, when the water usage is reduced, the concentration of pollutant in the wastewater streams will increase, which can have an impact on wastewater treatment. As the scarcity of nutrients becomes even more pressing (esp. phosphorus), greater and more widespread attention has to be given to on- or off- site reuse of source-separated wastewater. Careful characterization has already shown some of the key challenges to separation, such as the presence of priority pollutants derived from personal care products and pharmaceuticals. The need for characterization of wastewater fractions will only become more important in the future as new challenges emerge and new goals take priority.

REFERENCES

Aizenchtadt E., Ingman D. and Friedler E. (2009). Analysis of the long-term performance of an on-site greywater treatment plant using novel statistical approaches. *Urban Water J.*, **6**(5), 341–354.

Almeida M. C., Butler D. and Friedler E. (1999). At-source domestic wastewater quality. *Urban Water*, **1**(1), 49–55.

Anonymous (2000). Reduction of salinity in Israel's municipal sewage. *Israel Env. Bull.*, **23**(4), 15–17 (*in Hebrew*).

Briks R. and Hills S. (2007). Characterisation of indicator organisms and pathogens in domestic greywater for recycling. *Environ. Mon. Ass.*, **129**, 61–69.

Butler D. (1991). A small-scale study of wastewater discharges from domestic appliances. *J. IWEM*, **5**, 178–185.

Butler D., Friedler E. and Gatt K. (1995). Characterising the quantity and quality of domestic wastewater inflows. *Water Sci. Technol.*, **31**(7), 13–24.

Ciba-Geigy (1977). Wissenschaftliche Tabellen Geigy, Teilband Körperflüssigkeiten (Scientific tables Geigy, volume body fluids). 8th edn., Basel, Switzerland.

Donner E., Eriksson E., Revitt D. M., Scholes L., Lützhøft H. C. H. and Ledin A. (2010). Presence and fate of priority substances in domestic greywater treatment and reuse systems. *Sci. Total Environ.*, **408**(12), 2444–2451.

Eriksson E., Andersen H. R., Madsen T. S. and Ledin A. (2009). Greywater pollution variability and loadings. *Ecol. Eng.*, **35**(5), 661–669.

Eriksson E., Auffarth K., Eilersen A. M., Henze M. and Ledin A. (2003). Household chemicals and personal care products as sources for xenobiotic organic compounds in grey wastewater. *Water SA*, **29**(2), 135–146.

Eriksson E., Auffarth K., Henze M. and Ledin A. (2002). Characteristics of grey wastewater. *Urban Water*, **4**(1), 85–104.

Etter B., Tilley E., Khadka R. and Udert K. M. (2011). Low-cost struvite production using source-separated urine in Nepal. *Water Res.*, **45**(2), 852–862.

Feachem R. G., Bradley D. J., Garelick H. and Mara D. D. (1983). Sanitation and disease: health aspects of excreta and wastewater management. John Wiley & Sons, Chichester, USA.

Fittschen I. and Hahn H. H. (1998). Characterization of the municipal wastewater part human urine and preliminary comparison with liquid cattle excretion. *Water Sci. Technol.*, **38**(6), 9–16.

Fox K., Cassani G., Facchi A., Schroder F., Poelloth C. and Holt M. (2002). Variation in boron loads reaching European sewage treatment works. *Chemosphere*, **47**(5), 499–505.

Friedler E. (2004). Quality of individual domestic greywater streams and its implication on on-site treatment and reuse possibilities. *Environ. Technol.*, **25**(9), 997–1008.

Friedler E. (2008). The water saving potential and the socio-economic feasibility of greywater reuse within the urban sector – Israel as a case study. *Int. J. Environ. Stud.*, **65**(1), 57–69.

Friedler E. and Butler D. (1996). Quantifying the inherent uncertainty in the quantity and quality of domestic wastewater. *Water Sci. Technol.*, **33**(2), 65–78.

Friedler E., Brown D. M. and Butler D. (1996). A study of WC derived sewer solids. *Water Sci. Technol.*, **33**(9), 17–24.

Friedler E., Kovalio R. and Ben-Zvi A. (2006). Comparative study of the microbial quality of greywater treated by three on-site treatment systems. *Environ. Technol.*, **27**(6), 653–663.

Ghisi E. and Ferreira D. F. (2007). Potential for potable water savings by using rainwater and greywater in a multi-storey residential building in southern Brazil. *Build. Environ.*, **42**(7), 2512–2522.

Ghunmi L. A., Zeeman G., van Lier J. and Fayyed M. (2008). Quantitative and qualitative characteristics of grey water for reuse requirements and treatment alternatives: the case of Jordan. *Water Sci. Technol.*, **58**(7), 1385–1396.

Gilboa Y. and Friedler E. (2008). UV disinfection of RBC treated light greywater effluent: kinetics, survival and regrowth of selected microorganisms. *Water Res.*, **42**(4–5), 1043–1050.

Gray S. R. and Becker N. S. C. (2002). Contaminant flows in urban residential water systems. *Urban Water*, **4**(4), 331–346.

Gross A., Azulai N., Oron G., Ronen Z., Arnold M. and Nejidat A. (2005). Environmental impact and health risks associated with greywater irrigation: a case study. *Water Sci. Technol.*, **52**(8),161–169.

Helvetas (2005). Wasserverbrauch in der Schweiz (Water consumption in Switzerland), Factsheet, Helvetas Swiss Intercooperation, Zurich, Switzerland.

Jamrah A., Al-Futaisi A., Parthapar S. and Al Harrasi A. (2008). Evaluating greywater potential for sustainable water resources management in Oman. *Environ. Mon. Ass.*, **137**, 315–327.

Jefferson B., Palmer A., Jeffrey P., Stuetz R. and Judd S. (2004). Grey water characterisation and its impact on the selection and operation of technologies for urban reuse. *Water Sci. Technol.*, **50**(2), 157–164.

Kim W. J., Managaki H., Furumai H. and Nakajima F. (2009). Diurnal fluctuation of indicator microorganisms and intestinal viruses in combined sewer system. *Water Sci. Technol.*, **60**(11), 2791–2801.

Lienert J., Bürki T. and Escher B. I. (2007). Reducing micropollutants with source control: Substance flow analysis of 212 pharmaceuticals in faeces and urine. *Water Sci. Technol.*, **56**(5), 87–96.

Littlewood K., Memon F. A. and Butler D. (2007). Downstream implications of ultra-low flush WCs. *Water Practice Technol.*, **2**(2), doi10.2166/wpt.2007.037.

Liu Z., Kanjo Y. and Mizutani S. (2009). Urinary excretion rates of natural estrogens and androgens from humans, and their occurrence and fate in the environment: A review. *Sci. Total Environ.*, **407**(18), 4975–4985.

Loh M. and Coghlan P. (2003). Domestic water use study in Perth, Western Australia 1998–2001. Water Corporation, Perth, Australia.

Meinzinger F. and Oldenburg M. (2009). Characteristics of source-separated household wastewater flows: a statistical assessment. *Water Sci. Technol.*, **59**(9), 1785–1791.

Memon F. A., Butler D. (2006). Domestic water consumption trends and techniques for demand forecasts. In: Water Demand Management, D. Butler, F. A. Memon (eds.), IWA Publishing, London, UK.

Morel A. and Diener S. (2006). Greywater management in low and middle-income countries, Review of different treatment systems for households or neighbourhoods. Swiss Federal Institute of Aquatic Science and Technology (Eawag). Dübendorf, Switzerland.

Ottoson J. and Stenström T. A. (2003). Faecal contamination of greywater and associated microbial risks. *Water Res.*, **37**(2), 645–655.

Palmquist H. and Hanaeus J. (2005). Hazardous substances in separately collected grey- and blackwater from ordinary Swedish households. *Sci. Total Environ.*, **348**(1-3), 151–163.

Prathapar S. A., Jamrah A., Ahmed M., Al Adawi S., Al Sidairi S. and Al Harassi A. (2005). Overcoming constraints in treated greywater reuse in Oman. *Desalination*, **186**(1-3), 177–186.

Roesner L., Qian Y., Criswell M., Stromberger M. and Klein S. (2006). Long-term effects of landscape irrigation using household graywater – Literature review and synthesis. Water Environment Research Foundation (WERF), Alexandria VA, USA.

Schönning C., Leeming R. and Stenström T. A. (2002). Faecal contamination of source-separated human urine based on the content of faecal sterols. *Water Res.*, **36**(8), 1965–1972.

Siegrist R., Witt M. and Boyle W. C. (1976). Characteristics of rural household wastewater. *J. Env. Eng. ASCE*, **102**(3), 533–548.

SII (1999). SI 438 – Cleaning powders, Environmental requirements and labeling requirements: Laundry powders. The Standards Institute of Israel (SII), (in Hebrew).

SII (2006). SI 1417 – Dishwashing powders for dishwashing machine: Environmental quality assurance requirements and labeling requirements. The Standards Institute of Israel (SII), (in Hebrew).

Tilley E., Lüthi C., Morel A., Zurbrügg C. and Schertenleib R. (2008). Compendium of sanitation systems and technologies. Swiss Federal Institute of Aquatic Science and Technology (Eawag), Dübendorf, Switzerland.

Udert K. M. (2002). The fate of phosphorus and nitrogen in source-separated urine. PhD thesis. Swiss Federal Institute of Technology, Zurich, Switzerland.

Udert K. M., Larsen T. A. and Gujer W. (2003). Estimating the precipitation potential in urine-collecting systems. *Water Res.*, **37**(11), 2667–2677.

Udert K. M., Larsen T. A. and Gujer W. (2006). Fate of major compounds in source-separated urine. *Water Sci. Technol.*, **54**(11–12), 413–420.

Vieira P., Almeida M. C., Baptista J. M. and Ribeiro R. (2007). Household water use: a Portuguese field study. *Water Sci. Technol.: Water Supply*, **7**(5–6), 193–202.

Vinnerås B., Palmquist H., Balmer P. and Jönsson H. (2006). The characteristics of household wastewater and biodegradable solid waste – A proposal for new Swedish design values. *Urban Water J.*, **3**(1), 3–11.

Wheatley A. D. and Surendran S. (2008). Greywater treatment and reuse results from a 50-person trial. *Urban Water J.*, **5**(3), 187–194.

Winward G. P., Avery L. M., Frazer-Williams R., Pidou M., Jeffrey P., Stephenson T. and Jefferson B. (2008). A study of the microbial quality of grey water and an evaluation of treatment technologies for reuse. *Ecol. Eng.*, **32**(2), 187–197.

Wolfe A. J., Toh E., Shibata N., Rong R., Kenton K., FitzGerald M., Mueller E. R., Schreckenberger P., Dong Q., Nelson D. E. and Brubaker L. (2012). Evidence of uncultivated bacteria in the adult female bladder. *J. Clinic. Microbiol*, **50**(4), 1376–1383.

Chapter 18

Treatment of the solid fraction

Ralf Otterpohl and Christopher Buzie

18.1 INTRODUCTION

This chapter presents processes which may be technically and economically appropriate for the treatment of faecal solids in decentralized treatment facilities or on-site at the point of solid generation. We focus on low-tech options, because they are currently the most suitable solutions for on-site and decentralized settings. One very common process for the treatment of faecal solids, namely anaerobic digestion, is not presented here, as it is discussed in detail by Zeeman and Kujawa-Roeleveld (2013).

For every treatment option, it is recommended that the design criteria and operational scheme be selected depending on whether the treated solid fraction is intended for integration into or exclusion from the material cycle. The design will focus essentially on minimizing external energy input and ensuring low or modest levels of mechanization as well as low capital and operating costs. Most importantly, a crucial step prior to the technology selection and designing activities is to establish a comprehensive faecal solid management scheme that details the treatment goals as well as the organizational, technical, institutional, legal and financial aspects of the sanitation system.

18.2 COMPOSITION OF FAECAL SOLIDS

The efficiency with which excreta can be treated is strongly influenced by its physical and biochemical characteristics – notably its water and ammonia content. When collected in toilets, faecal solids contain not only faeces and other solids such as toilet paper, but can also contain urine and flushing water.

produced in the anaerobic digester. However, this is an unlikely option for on-site treatment of faecal solids.

18.9 CONCLUSIONS AND OUTLOOK

Treatment of human excreta is challenging, irrespective of the option selected. Montangero and Strauss (2004) have pointed out that the design of treatment schemes cannot be based on standard characteristics because the composition of faecal sludge varies greatly. This is not only true for faecal sludge, but even more so for faecal solids in general. The composition of the faecal solids is influenced by nutrition, the addition of urine and flushing water, the environmental conditions such as temperature or groundwater and processes occurring during transport and storage. Due to the high variability of the characteristics of the faecal solids, even low-tech technologies such as presented in this chapter require careful maintenance and observation.

Today's treatment options for faecal solids are often labour-intensive and require considerable commitment, which is unsuitable for most sanitation users in industrialized countries. Technologies for on-site treatment can only be successful if they do not require a lot of maintenance. This has to be considered for the further development of processes for faecal solid treatment.

Many researchers and practitioners are currently working on the development of new technologies for on-site treatment. There is a strong focus on biological systems. As we have shown, such systems can be operated with simple reactor setups such as compost heaps. New biological technologies have recently been developed. Two examples were presented in this chapter: Terra Preta sanitation and vermicomposting. Another example is treatment with black soldier fly larvae, which convert faecal matter into protein-rich biomass (Diener *et al.* 2009).

However, physico-chemical processes can also be viable options for on-site treatment. The example of the dehydration vault shows that such processes do not always require sophisticated reactor setups. Nevertheless, modern process control options and electricity can make even complex process configurations interesting for on-site treatment. One example is the Swedish Cinderella toilet (Fritidstoa 2012), which incinerates the faecal matter directly in the toilet.

It is unclear as to which technology will be the gold standard for on-site treatment of faecal solids in the future. Actually, it is very likely that there will be a wide range of technologies to choose from. This will help us select the optimal reactor for a given setting.

REFERENCES

AFNOR (2002). *Norme Française NF U 44-095: Amendements organiques – Composts contenant des matières d'intérêt agronomique, issues du traitement des eaux*, (French standard NF U 44-095: Organic soil improver – Composts containing materials of agricultural interest, which originate from water treatment), AFNOR Editions, St Denis La Plaine, France.

Bajsa O., Nair J., Mathew K. and Ho G. E. (2004). Vermiculture as a tool for domestic wastewater management. *Water Science and Technology*, **48**(11–12), 125–132.

Brock T. D., Madigan M. T., Martinko J. M. and Parker J. (1994). Biology of Microorganisms, 7th ed. International ed, Prentice Hall, Englewood Cliffs, NJ, USA.

Brodie H. L., Carr L. E. and Condon P. (2000). A comparison of static pile and turned windrow methods for poultry litter compost production. *Compost Science and Utilization*, **8**, 178–189.

Butt K. R. (1993). Utilisation of solid paper-mill sludge and spent brewery yeast as feed for soil dwelling earthworms. *Bioresource Technology*, **44**(2), 105–107.

Buzie C. (2010). Development of a continuous single chamber vermicomposting toilet with urine diversion for onsite application. PhD thesis. Institute of Wastewater Management and Protection, Technical University Hamburg – Harburg, Germany.

CREPA (2004). Fact sheet on on-site sanitation latrines, Centre Régional pour l'Eau Potable et l'Assainissement à faible coût (CREPA), Burkina Faso.

De Bertoldi D. M. G., Vallini G. and Para A. (1983). The biology of composting: A review. *Waste Management and Research*, **1**(2), 157–176.

Diener S., Zurbrügg C. and Tockner K. (2009). Conversion of organic material by black soldier fly larvae: Establishing optimal feeding rates. *Waste Management and Research*, **27**(6), 603–610.

Dominguez J. and Edwards C. A. (2004). Vermicomposting organic wastes: A review. In: *Soil Zoology for Sustainable Development in the 21st Century*, S. H. Shakir and W. Z. A. Mikhail (eds). Self – Publisher, Cairo, Egypt. pp. 369–396.

El-jalil M. E., Zinedine A. and Faid M. (2001). Some microbiological and chemical properties of poultry wastes manure after lactic acid fermentation. *International Journal of Agriculture and Biology*, **10**(4), 405–411.

Factura H., Bettendorf T., Buzie C., Pieplow H. and Reckin J.and Otterpohl R. (2010). Terra Preta sanitation: re-discovered from an ancient Amazonian civilisation – integrating sanitation, bio-waste management and agriculture. *Water Science and Technology*, **61**(10), 2673–2679.

Friedler E., Butler D. and Alfiya Y. (2013). Wastewater composition. In: *Source Separation and Decentralization for Wastewater Management*, T. A. Larsen, K. M. Udert and J. Lienert (eds.), IWA Publishing, London, UK, pp. 241–257.

Fritidstoa (2012). Fritidstoa AB, Förbränningstoaletter (incineration toilets), http://www.fritidstoa.se (accessed 05 April 2012).

Haug R. T. (1980). Compost engineering: principles and practice. *Ann Arbor Science*, Ann Arbor, Michigan, USA.

Koné D. and Strauss M. (2004). Low-cost options for treating faecal sludges (FS) in developing countries – challenges and performance. In: A. Liénard and H. Burnett editors. *Proceedings of the 6th International Conference on Waste Stabilisation Pond and 9th International Conference on Wetland Systems*, Avignon (France), 28 Sep–1 Oct 2004.

Koné D., Cofie O., Zurbrügg C., Gallizzi K., Moser D., Drescher S. and Strauss M. (2007). Helminth eggs inactivation efficiency by faecal sludge dewatering and co-composting in tropical climates. *Water Research*, **41**(19), 4397–4402.

Lopez Zavala M. A., Funamizu N. and Takakuwa T. (2004). Temperature effect on aerobic biodegradation of feces using sawdust as a matrix. *Water Research*, **38**(9), 2405–2415.

Mitscherlich E. and Marth E. H. (1984). Microbial Survival in the Environment: Bacteria and Rickettsiae Important in Human and Animal Health. Springer, Berlin, Germany.

Montangero A. and Strauss M. (2004). Faecal Sludge Treatment. Department of Water and Sanitation in Developing Countries (Sandec), Swiss Federal Institute of Aquatic Science and Technology (Eawag), Dübendorf, Stwitzerland.

Nedgwa P. M. and Thompson S. A. (2001). Effects of C-to-N ratio on vermicomposting of biosolids. *Bioresource Technology*, **75**(1), 7–12.

Otterpohl R., Braun U. and Oldenburg M. (2004). Innovative technologies for decentralised water-, wastewater and biowaste management in urban and peri-urban areas. *Water Science and Technology*, **48**(11–12), 23–32.

Otterpohl R. and Buzie C. (2011). Wastewater: reuse-oriented wastewater systems-low- and high-tech approaches for urban areas. In: *Waste: A Handbook for Management*, T. M. Letcher and D. A. Vallero (eds.), Academic Press, Burlington, MA, USA, pp. 127–136.

Polprasert C. (2007). Organic waste recycling, Technology and Management. 3rd ed., IWA Publishing, London, UK.

Rynk R. (1992). On-farm composting handbook. Natural Resource, Agriculture, and Engineering Service, Ithaca, NY, USA.

Shalabi M. (2006). Vermicomposting of faecal matter as a component of source control sanitation. PhD thesis, Institute of Wastewater Management and Water Protection. Technical University Hamburg – Harburg, Germany.

Schönning C. and Stenström T. A. (2004). Guidelines for the safe use of urine and faeces in ecological sanitation systems. EcoSanRes Publication Series, Report 2004-1, Stockholm Environment Institute, Stockholm, Sweden.

Strauss M., Larmie S. A. and Heinss U. (1997). Treatment of sludges from on-site sanitation — Low-cost options. *Water Science and Technology*, **35**(6), 129–136.

Strauss M. and Montangero A.. (2002). FS Management – Review of Practices, Problems and Initiatives. Swiss Federal Institute of Aquatic Science and Technology (Eawag), Dübendorf, Switzerland.

Tchobanoglous G. and Burton F. L. (1991). Wastewater Engineering; Treatment, Disposal, and Reuse, 3rd ed., Metcalf & Eddy, McGraw – Hill, New York, USA.

Tchobanoglous G., Theisen H. and Vigil S. A. (1993). Integrated Solid Waste Management Engineering Principles and Management Issues. McGraw Hill, New York, USA.

Tilley E., Lüthi C., Morel A., Zurbrügg C. and Schertenleib R. (2008). Compendium of Sanitation Systems and Technologies. Swiss Federal Institute of Aquatic Science and Technology (Eawag), Dübendorf, Switzerland.

USCC (2008). Greenhouse Gases and the Role of Composting: A Primer for Compost Producers. The United States Composting Council (USCC), Ronkonkoma NY, USA.

U.S-EPA (2003). Control of Pathogens and Vector Attraction in Sewage Sludge, United States Environmental Protection Agency, National Risk Management Research Laboratory, Cincinnati, OH.

Veeken A., de Wilde V. and Hamelers B. (2002). Passively aerated composting of straw-rich pig manure: effect of compost bed porosity. *Compost Science and Utilization*, **10**(2), 114–128.

Vinnerås B. (2002). Possibilities for Sustainable Nutrient Recycling by faecal Separation Combined with Urine Diversion. PhD thesis, Department of Biometry & Engineering, Swedish University of Agricultural Sciences, Uppsala, Sweden.

WHO (2006). Water, Sanitation and Health Electronic Library. 4rd ed., World Health Organisation (WHO), Geneva, Switzerland.

Yadav K. D., Tare V. and Ahammed M. M. (2010). Vermicomposting of source-separated human faeces for nutrient recycling. *Waste Management*, **30**(1), 50–56.

Yadav K. D., Tare V. and Ahammed M. M. (2011).Vermicomposting of source-separated human faeces by *Eisenia fetida*: effect of stocking density on feed consumption rate, growth characteristics and vermicompost production. *Waste Management*, **31**(6), 1162–1168.

Yang S. Y., Ji K. S., Baik Y. H., Kwak W. S. and McCaskey T. A. (2006). Lactic acid fermentation of food waste for swine feed. *Bioresource Technology*, **97**(15), 1858–1864.

Zeeman G. and Kujawa-Roeleveld K. (2013). Anaerobic treatment of source-separated domestic wastewater. In: *Source Separation and Decentralization for Wastewater Management*, T. A. Larsen, K. M. Udert and J. Lienert (eds.), IWA Publishing, London, UK, pp. 307–320.

Chapter 19

Aerobic elimination of organics and pathogens: greywater treatment

Bruce Jefferson and Paul Jeffrey

19.1 INTRODUCTION

Aerobic biological processes have been used since the beginning of modern wastewater treatment to reduce organic load and pathogens for the purpose of safe environmental discharge and in more recent times water recycling. In the context of source-separated wastewater, these original goals are still central to a number of the possible waste streams (i.e., blackwater, greywater, etc.). Greywater derives its pollution load excluding sources of human excreta and as such is generally limited in solids and nutrients compared to urine, black- and brownwater (see Friedler *et al.* 2013). Consequently, the major treatment challenges are related to organics and pathogen removal. The latter become increasingly important if reuse is the treatment driver rather than safe discharge. Water quality standards around the world reflect the focus on organics and pathogens (Pidou *et al.* 2007). All reported regulations contain a coliform standard to ensure protection of human health with additional limits on residual chlorine, organics and solids in most cases. Whilst the exact standards vary widely between regulatory bodies, any technology, or treatment trains, that can deliver a biological oxygen demand (BOD) of less than 10 $mg{\cdot}L^{-1}$, a turbidity of less than 2 NTU and a non-detectable level of faecal coliforms is likely to be acceptable worldwide. However, less restrictive standards in certain regions and applications provide flexibility in technology selection and design that can be critical in ensuring economic viability.

The exclusion of excreta means that in comparison to the other source-separated waste streams greywater exhibits relatively low biodegradability, high concentrations of xenobiotic compounds and a high variability in concentrations. Such descriptions do not initially suggest the suitability of aerobic biological processes yet they represent by far the most common approach to treatment constituting almost 90% of all full-scale schemes. In fact aerobic biological treatment of greywater is currently the most successful application of decentralized treatment of source-separated wastewater. Consequently, the information in this chapter will be provided, in part, from a combination of pilot and full-scale case studies.

A series of comprehensive reviews exist on greywater characteristics (Friedler *et al.* 2013, Eriksson *et al.* 2002, Jefferson *et al.* 2004), reuse standards (Dixon *et al.* 1999, Maimon *et al.* 2010) and treatment technologies (Pidou *et al.* 2007, Li *et al.* 2009, Ghunmi *et al.* 2011) negating the need to replicate such information in the current chapter. The following chapter will therefore discuss the issues of aerobic elimination of organics and pathogens with a focus towards greywater that can then be extended to the discussion of other waste streams.

19.2 COMPOSITION AND TREATABILITY

19.2.1 Organic compounds

When reading the published studies and reviews the most significant characteristic to be consistently described concerns the variability of greywater. To illustrate, across all reported studies to date (Figure 19.1) the BOD of greywater ranges between 5 and 900 $mg{\cdot}L^{-1}$ (chemical oxygen demand (COD) between 23 and 1600 $mg{\cdot}L^{-1}$) with a median value of 150 $mg{\cdot}L^{-1}$ (COD: 240 $mg{\cdot}L^{-1}$). In comparison, the difference between a weak and strong sewage influent is 100 to 350 $mg{\cdot}L^{-1}$ BOD. Consequently, one of the biggest challenges in designing robust affordable treatment systems is the management of the load which varies by strength and flow. As such, a certain level of bespoke design is required to match the specific challenge being encountered, that is a technology suitably designed to treat a BOD of 900 $mg{\cdot}L^{-1}$ may not be suitable to treat a load of 50 $mg{\cdot}L^{-1}$. This restricts cost minimization, thereby reducing economic viability.

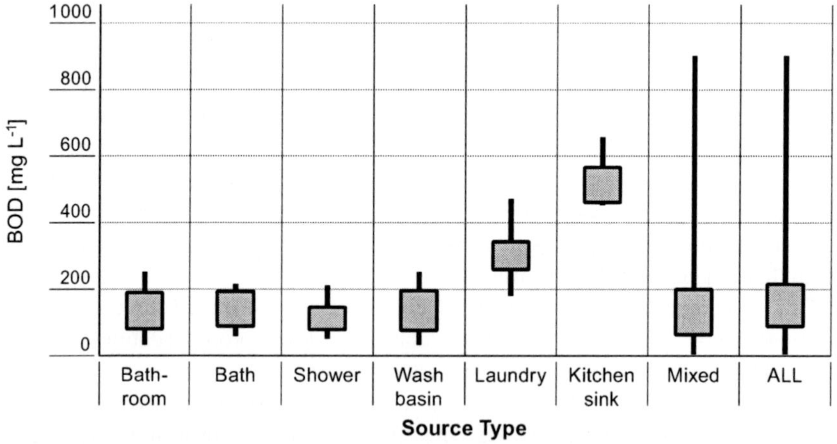

Figure 19.1 Reported ranges in organic strength for different greywater sources. The box shows the range between the 25 and 75 percentiles, the lines show the complete data range. Based on a comprehensive internal review of literature values.

Analysis at individual source level (Figure 19.1) indicates that BOD levels in different bathroom greywater sources is similar with median values of 100–130 $mg \cdot L^{-1}$ (COD: 170–280 $mg \cdot L^{-1}$). The major factor influencing variability is related to water use levels with a recent survey in the UK revealing that average shower times are around 10 minutes (Waterwise 2009). Age is correlated with showering times such that on average people aged 18–24 shower for five minutes longer than those aged over 55 years providing some level of predictability in terms of treatment needs.

Greywater is generally biodegradable with 80% of all schemes reporting a COD: BOD ratio below 3. However, in a few cases COD:BOD ratios as high as 15.8 have been observed in which case no substantial biodegradation can be expected. Whilst the organics may be amenable to biodegradation, a cocktail of other ingredients is required to ensure effective operation of a bioreactor including macronutrients (nitrogen and phosphorus) and micronutrients. The majority of nutrients are discharged via human excreta with, for instance, a survey in Sweden suggesting that 90% of nitrogen and 67% of phosphorus is associated with urine (Günther 2000). Median values reported in bathroom sources support such findings with ammonia (1.5 $mg \cdot L^{-1}$) and phosphorus (1 $mg \cdot L^{-1}$) levels well below typical concentrations found in domestic sewage. Consequently, a typical COD:N:P ratio for greywater is 100:2:1 which is below the recommended balance of 100:20:1 or 100:5:1 (depending on reference) suggesting greywater is likely to be deficient in nitrogen, phosphorus or both. Indeed, investigations into the benefits of nutrient dosing have shown that organic loading rates can be increased by up to 160% when nutrient levels are adjusted by dosing (Jefferson *et al.* 2001). In addition to macronutrient limitations greywater is potentially limited in a number of micronutrients such as trace metals including: iron, copper, aluminium, molybdenum, cobalt and zinc. The benefits of micronutrient balancing are similar to those of macronutrient adjustment with studies reporting a 140% increase in biomass respiration when the zinc content of greywater was balanced (Jefferson *et al.* 2001).

The variability in the levels of greywater constituents reflects differences in the choices and actions of people producing greywater as a consequence of geographical, social and economic drivers. To illustrate, recent measurements have shown that the addition of one millilitre of shampoo leads to greywater COD concentrations which vary between 5600 and 53,700 $mg \cdot L^{-1}$ depending on which product is chosen (Kadewa 2010). Similar variations are seen across all the product groups and in terms of all parameters. No obvious trend has been identified which matches load exerted and product descriptors such as cost or brand type (own brand, market leading or ecological) making predictions of greywater characteristics based on demographic information problematic (Knops *et al.* 2007). The same is true in terms of the toxicity and biodegradability of the products with the same study finding that environmentally labelled products do not exhibit reduced toxicity or greater biodegradability than leading brand alternatives. Consequently, unless use of specific brands can be defined (such as in hotels) source control of the characteristics of greywater are limited to reducing total usage amounts of the products rather than trying to influence the brand of product used.

One final consideration that must be taken into account is the risk of a toxic shock to biological processes through discharge of products not normally associated with greywater. Results from a recent survey revealed that the presence in bathroom greywater sources of bleach (32% of respondents), food (23%), cleaning products (66%) and washing powder (27%) was likely on a weekly basis (Knops *et al.* 2007). Interestingly, the perception of a substance being harmful to the environment did not significantly influence the likely discharge frequency. This was most pronounced with older respondents although in part this is due to the opportunity of discharge as much as behavioural choice. Estimations of likely exposure volumes and buffer capacity associated with greywater storage indicate that even in single household systems the risk is predominately limited to use of bleach where actual usage rates are likely to impact treatment performance. Accordingly, a level of education is required to ensure risks are minimized. This is especially important in buildings where cleaning is not undertaken by those responsible for the greywater system (e.g., in hotels, schools, halls of residence etc.).

19.2.2 Xenobiotics

Increasing interest is being shown beyond simple organic removal to consider the removal of specific individual chemicals of concern (xenobiotics – XOC) that could enter the environment and human population through the reduced treatment pathways commonly used with source-separated waste treatment. The XOC of greatest interest in relation to greywater are derived from household and personal care products (PCP) as pharmaceuticals and hormones are mainly related to urine and faeces (Friedler *et al.* 2013).

Reported studies have suggested over 900 possible substances which fit into 14 different compound groups (Eriksson *et al.* 2002) with the ones of greatest occurrence being fragrances, UV filters, biocides, preservatives and detergents (Eriksson *et al.* 2002, Leal *et al.* 2010). Whilst detailed information about levels of the different possible compounds is not readily available, usage rates are thought to be very high for some products. For instance, fragrance ingredients such as galaxolide and tonalide are common in PCP with European production of the chemicals reported to be between $1\text{–}5{\cdot}10^6\ \mathrm{kg{\cdot}y^{-1}}$ (Leal *et al.* 2010). The use of parabens as preservative agents and triclosan as a biocidal agent in PCP means that they can exist in concentrations of around $10\ \mathrm{mg{\cdot}g^{-1}}$ in some PCPs. Whilst many compounds are likely to exist at fairly consistent levels throughout the year, UV filters pose an additional treatment challenge in that their use is seasonal. Health awareness concerning skin cancer means that usage rates of the product are predicted to increase significantly. In addition, engineered nanoparticles are beginning to be used in PCPs, especially in UV filters, with very little currently known about the risks or their removal during treatment. Each individual chemical is likely to be found in greywater at a concentration within the $\mu g{\cdot}L^{-1}$ range with for instance levels of 10.7 $\mu g{\cdot}L^{-1}$ (galaxolide), 7.5 $\mu g{\cdot}L^{-1}$ (nonylphenol) and 15.6 $\mu g{\cdot}L^{-1}$ (triclosan) reported recently in a survey (Leal *et al.* 2010). However, the most reported XOC in relation to greywater treatment are surfactants with levels between 0.7 and 70 $mg{\cdot}L^{-1}$ as methylene blue active substances (MBAS; Shafran *et al.* 2006). This is substantially higher than in mixed wastewater.

19.2.3 Pathogens

Microbiological considerations are normally confined to indicator organisms such as total coliforms and faecal coliforms. Reported concentrations are similar across all greywater sources and range from non-detectable to 10^7 $CFU{\cdot}100mL^{-1}$ with a median value of 10^3 $CFU{\cdot}100mL^{-1}$. The similarity across sources reflects the fact that coliform levels are primarily related to demographic considerations. For instance, total coliform levels of $3.2{\cdot}10^5$ and 80 $CFU{\cdot}100mL^{-1}$ were found in greywater from households with and without children, respectively (Rose *et al.* 1991). The pathogenic population of greywater includes pathogens colonizing body surface and orifices such as the nose and mouth as well as residual traces of faeces. Estimates indicate that a faecal load of around 0.04 $g{\cdot}p^{-1}{\cdot}d^{-1}$ is discharged in greywater sources (Ottoson and Stenström 2003). In addition, greywater can be exposed to enteric pathogens washed from raw meat or vegetables such as *Campylobacter* and *Salmonella*. Whilst minimized by excluding kitchen sources this cannot be guaranteed as food preparation and cleaning practices vary widely. Overall, faecal coliform contamination is common in greywater with levels similar to those found in treated sewage effluent.

Identification of specific pathogens is limited due to the sample size and frequency required to ensure representation. However, pathogens identified in greywater include opportunistic pathogens (microorganisms that do not normally harm the host but can cause disease when the host's resistance is lowered) such as *Pseudomonas aeruginosa* and *Staphylococcus aureus* as well as pathogens transmitted with faeces such as *Salmonella* spp., *Giardia* spp. and *Cryptosporidium* spp. (Winward *et al.* 2008). A key difference between the two groups of pathogens is the frequency of detection. The opportunistic pathogens are regularly detected with one study finding *Pseudomonas aeruginosa* present in all the tested samples and another that found very high levels of *Staphylococcus aureus* from military camps with levels between 7.0–7.7 $\log_{10}{\cdot}100mL^{-1}$. In contrast, the pathogens transmitted with faeces are much less common as they require an affected individual contributing and thus are more likely in larger sample sizes and those containing children or the elderly. Environmental strains of *Legionella pneumophila* (serotypes 2–14) have also been recorded in wash basin greywater collected from a large facility at a concentration between 2.2–2.9 $\log_{10}\cdot 100\text{mL}^{-1}$. However, the strain most commonly associated with infection (serotype 1) has not been detected to date in greywater (Birks *et al.* 2004). Information concerning viruses is more restricted with no human viruses detected in the limited screening studies that have been reported. However, coliphages have been reported (Ottoson and Stenström 2003, Eriksson *et al.* 2002) demonstrating a potential transmission route. However, in another study three viral indicators (somatic phage, host: *Escherichia coli* CN13 and F-RNA phages, hosts: *E. coli* F^{+}_{amp}, *E. coli* K_{12}) were found not to be present suggesting a low risk (Gilboa and Friedler 2008).

Regrowth or die-off in greywater during storage also needs to be considered. Indicator organisms have been shown to increase by as much as 2 $\log_{10}$ units within 24–48 hours during storage whilst enteric pathogens such as *Salmonella* spp., *Shigella* spp. and *Campylobacter* spp. do not appear to regrow but do persist for days (Rose *et al.* 1991).

19.3 TECHNOLOGIES FOR AEROBIC TREATMENT

To date reports exist for around 100 different treatment schemes for the processing of greywater with an approximately even split between research and development as well as full-scale. Almost all full-scale schemes utilize an aerobic biological process with the exception of single house scale units in the UK where the lack of regulation has led to a growth of simple systems based on coarse filtration followed by high dose chlorine disinfection (Pidou *et al.* 2007).

Irrespective of the type of aerobic biological process being used there are a number of common associated features required to ensure the technology works effectively including storage/flow balancing, a pre-treatment step to remove large debris and a post-treatment step to disinfect the water and/or provide a residual disinfection level consistent with relevant regulations. At the scale of most operations the biggest challenge is flow balancing with peak flows typically occurring in the morning between 6 and 10 am and in the evening between 6 and 9 pm. Convention is to store the untreated water to minimize the size of treatment technology required and enable as near constant flow operation as possible. Detailed analysis reveals that for a single house the optimum storage size is 500 L beyond which the additional water savings do not offset the associated costs (Dixon *et al.* 1999). Upper limits of storage capacity of 32 days have been cited to avoid unnecessary increase in pathogenic species (Mustow and Grey 1997, Winward *et al.* 2009) although greywater has been found to degrade in storage rapidly leading to concerns about odour emission when storage exceeds two days (Jefferson *et al.* 2004).

19.3.1 Removal of organic compounds

The organic characteristics of greywater matched against the likely water quality standards tend to restrict technology choice to an aerobic biological process with 65–70% of all reported schemes and research trials based on intensive or extensive biological processes. Intensive biological processes aim to reduce the required reactor size by intensifying the biological processes by either increasing the concentration of retained biomass or increasing the rate of oxygen transfer. Examples are the activated sludge process, membrane bioreactors (MBRs) or biologically aerated filters (BAFs). In contrast, extensive processes operate in a more passive manner such that they do not tend to use active aeration or mixing. Examples are reed beds (constructed wetlands) or waste stabilization ponds.

To meet the most restrictive of the reuse standards (e.g., regional state standards within the USA, Pidou *et al.* 2007) a BOD removal of around 80–95% is required depending on the influent concentration. The average BOD removal reported (Pidou *et al.* 2007) for intensive biological processes is 88% compared to 86% for extensive systems indicating that both are likely to be suitable although polishing of the treated water may be required.

19.3.1.1 Intensive biological processes

A wide range of intensive biological processes are used for greywater treatment including: rotating biological contactors (RBC), anaerobic filters, sequencing batch reactors (SBR), BAFs and MBRs. In most cases, effluent solids can be an issue so a downstream separation process is usually integrated into the treatment system. The most common approach and arguably the market leaders are MBRs where a membrane separator (ultrafiltration mainly) is directly submerged into a bioreactor chamber. This combination provides the most effective level of sustainable treatment possible in a single unit and has become very popular in the USA and Japan. Twenty seven published reports have detailed the performance of MBRs for greywater treatment of which all but two reliably met a 10 $mg \cdot L^{-1}$ BOD standard (Pidou *et al.* 2007). Equivalent levels of performance can be achieved with other biological systems and although these require downstream filtration, often with sand filters, to meet the tighter consents, they can be used without them when working to meet less stringent solids consents.

In many cases the BOD of the greywater influent is relatively low and as such design tends to be hydraulically limited. Typical hydraulic retention times (HRT) in MBRs range between 8–12 h (equating to an average organic loading rate of 0.88 $kgCOD \cdot m^{-3} \cdot d^{-1}$) with effective treatment reported at HRTs as low as 2 hours. However, HRTs are more commonly 12 h or greater as running at slower retention times generates operational benefits which do not have excessive cost implications at the scales under consideration. Importantly, in the case of MBRs a more passive approach to running the plants has greatly decreased fouling problems and the need to regularly clean the membranes. Accordingly, flux rates are generally low at around 5–15 $L \cdot m^{-2} \cdot h^{-1}$ (Pidou *et al.* 2007) compared to those used in municipal MBRs which are typically >20 $L \cdot m^{-2} \cdot h^{-1}$ (Judd 2011). In part the loading rate is limited due to the nutrient imbalance of the feed water. Previous studies have shown nutrient balancing can enhance loading rates with one study increasing the loading rate for 90% BOD removal from 1 $kgCOD \cdot m^{-3} \cdot d^{-1}$without nutrient balancing to 1.7 $kgCOD \cdot m^{-3} \cdot d^{-1}$ with nutrient addition (Jefferson *et al.* 2001) although cost limitation may restrict inclusion of such an approach in small decentralized systems.

19.3.1.2 Extensive biological processes

Reed beds are the most common extensive biological processes and benefit from positive public perception value as they are seen to be a more sustainable method of purification and a more "natural" process. The technology is easily set up and consequently many self-build systems exist. The design is based on either horizontal subsurface flow (anaerobic) or vertical flow (aerobic) configuration utilizing a range of media from gravel to soil. Improved removal is normally reported for vertical flow systems due to their aerobic degradation pathway. HRTs for the systems average around 4.5 d although much more compact systems have been demonstrated by utilizing a series of shallow troughs to maximize the aerobic environment (Kadewa *et al.* 2010) leading to total HRTs as low as 1 h. A variety of plant species have been used with *Phragmites australis* being the most common. To date no evidence has been provided that demonstrates the advantage of using specific species so local preference tends to strongly influence choice. However, care is required not to use species whose root zone is too dense or growth rate too excessive as this can cause operational problems. In some cases systems have been built without plants and so operating as percolation fields. In all cases the unplanted versions' performance was similar to that of the planted equivalent enabling very low physical footprint versions to be considered (stacked trays) although care is required to use covered systems to reduce the risk of clogging from excessive algal growth. Common problems with reed beds include uneven distribution and insufficient pre-treatment to remove clogging potential which can be normally resolved during initial set-up and maintenance scheduling.

19.3.1.3 Most effective processes

Overall, vertical flow reed beds and MBRs provide the most effective removal rates and can readily meet even the tightest water recycling standards in terms of BOD or COD. The small scale of these technologies and the variable nature of the flows and loads that require treatment mean that engineering sufficient resilience into the treatment performance is the key challenge facing engineers. Comparative analysis of the resilience of a range of technologies treating greywater reveals the beneficial impact of incorporating membranes which sustain much greater treatment resilience than non-barrier processes. The difference is illustrated in Figure 19.2 where the variation in effluent quality is plotted as a cumulative curve during prolonged trials under steady state conditions. Processes that exhibit near vertical curves around the median point with little or no tail at the top are considered the most resilient. For instance, in Figure 19.2 the MBR is able to deliver an effluent BOD of below 1 mg·L^{-1} consistently whereas the vertical flow reed bed (VFRB) can deliver a BOD below 2 mg·L^{-1} for only 65% of the time beyond which the effluent quality varies considerably. The trade-off is in terms of hydraulics where the opposite is true such that water quality is maintained but treated flow diminishes as the permeability of the membrane decreases. Consequently flow buffering is more important in MBRs and operating fluxes tend to be reduced compared to more traditional applications.

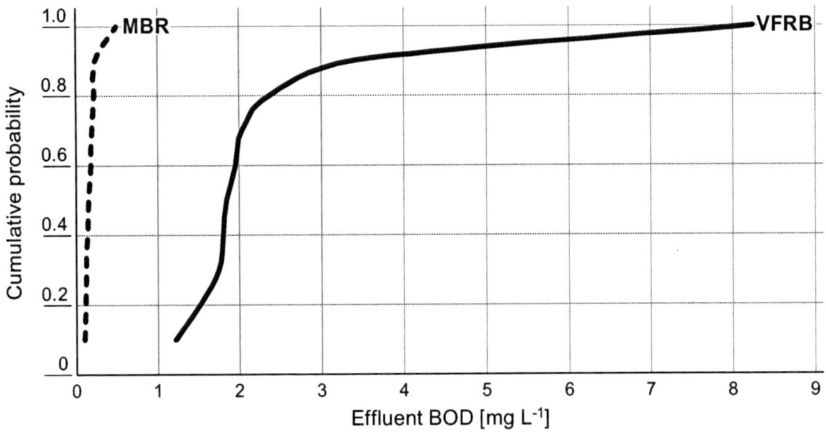

Figure 19.2 Typical resilience curves for an intensive (membrane bioreactor – MBR) and an extensive (vertical flow reed bed – VFRB) biological process treating greywater.

19.3.2 Xenobiotics removal

In relation to aerobic biological processes the main treatment pathways are thought to be combinations of adsorption and biodegradation with some additional removal possible due to stripping, chemical oxidation and photocatalysis (Caliman and Gavrilescu 2009). Detailed descriptions of the removal pathways and expected levels are summarized elsewhere (Virkutyte *et al.* 2010) revealing a very complex picture concerning expected removal and preferred treatment strategy. However, in general adsorption is maximized under short sludge retention times where fresh biomass is generated at its quickest providing additional sorption sites.

The required level of pathogen removal represents the main difference between treatment and reuse with the latter requiring very low levels in the final treated water. In such cases the role of the aerobic biological processes becomes more concerned with pre-treatment to maximize the efficacy of the downstream disinfection process (chlorine, UV etc.). This relates principally to the levels or residual organics and particles at the point of disinfection rather than direct pathogen removal across the aerobic biological process. Importantly, both chlorine and UV systems are negatively affected by the presence of organics and as such the efficiency of the upstream aerobic process is critical. For instance, in the case of chlorine disinfection the organic content accounts for 99% of the total chlorine demand. It is common to use fixed dose delivery systems for reuse of source-separated wastewater and so confidence in the consistency of pathogen removal is dictated by the stability of the residual organic levels leaving the aerobic biological process. The other major influence on disinfection efficacy is the presence of particles that act to shield microorganisms against the influence of the chosen disinfection process. Estimates suggest effective penetration levels of 40 and 2 μm for chlorine and UV respectively, limiting the ability of either to remove all microorganisms (Winward *et al.* 2008). Consequently, aerobic processes that produce large effluent solids will inhibit disinfection and reduce overall pathogen removal if a suitable physical process is not included. To illustrate the significance of these points reported trials suggested that a residual of around $1\ \log_{10}\cdot 100\text{mL}^{-1}$ is likely when directly disinfecting raw greywater but the level will be reduced to a non-detectable level if applied after pre-processing through a reed bed.

Heterotrophic plate count analysis has shown regrowth to be statistically greater in untreated compared to aerobically treated greywater demonstrating the benefits of removal. Application of UV at doses below those required for complete disinfection result in regrowth of UV resistant bacteria due to decreased competition from the more susceptible species (Gilboa and Friedler 2008). However, pre-treatment in either MBRs or RBCs has been shown to enable no significant regrowth for up to a week if the downstream UV system initially disinfects the treated water (Friedler *et al.* 2006).

19.4 CONCLUSIONS

The aim of the chapter has been to discuss the role of aerobic biological processes for the elimination of organics and pathogens from source-separated wastewater flows. The literature clearly demonstrates the importance of incorporating an aerobic stage irrespective of whether an intensive or extensive technology is used. The evidence has been based on greywater as it represents the most successful example of source-separated treatment. Analysis of full-scale systems reveals that aerobic biological processes dominate the selection choices of designers and engineers. This is because aerobic processes are most suited to tackle the key challenges that greywater flows present: high variability of the organic load and the high contribution of XOCs within the COD fraction. This extends to pathogen removal where aerobic processes are a more appropriate match to downstream disinfection systems such as UV and chlorine due to their ability to reach low organic residuals.

Translation of such evidence to other source-separated streams is more difficult as inclusion of faecal excreta provides considerably more organic carbon enabling anaerobic processes to become potentially more viable. However, the variability and scale of source-separated treatment means that aerobic processes are likely to remain at the heart of organics and pathogen elimination in source-separated flows for the considerable future. The key challenge is to deliver resilience of treatment even for aerobic systems against the challenges of scale and variability. Currently this requires a level of bespoke design and operation and with it a restriction in the level of cost minimization achieved. Ultimately this will need to be overcome if the opportunity for source-separated treatment is to be realized.

REFERENCES

Birks R. and Hill S. (2007). Characterisation of indicator organisms and pathogens in domestic greywater for recycling. *Environmental Monitoring and Assessment*, **129**, 61–69.

Caliman F. A. and Gavrilescu M. (2009). Pharmaceuticals, personal care products and endocrine disruption agents in the environment – a review. *Clean*, **37**, 277–303.

Dixon A. M., Butler D., Feweks A. and Robinson M. (1999). Measurement and modelling of quality changes in stored untreated greywater. *Urban Water*, **1**, 293–306.

Donner E., Eriksson E., Revitt D. M., Scholes L., Holten Lutzhoft H-C. and Ledin A. (2010). Presence and fate of priority substances in domestic greywater treatment and reuse systems. *Science of the Total Environment*, **408**, 2444–2451.

Eriksson E., Auffarth K.., Henze M. and Ledin A. (2002). Characteristics of grey wastewater. *Urban Water*, **4**, 85–104.

Eriksson E., Donner E. and Ledin A. (2010). Presence of selected priority and personal care substances in an onsite bathroom greywater treatment facility. *Water Science and Technology*, **62**, 2889–2898.

Friedler E., Butler D. and Alfiya Y. (2013). Wastewater composition. In: *Source Separation and Decentralization for Wastewater Management*. T. A. Larsen, K. M. Udert and J. Lienert (eds.), IWA Publishing, London, UK, pp. 241–257.

Friedler E., Kovalio R. and Ben-Zvi A. (2006). Comparative study of the microbial quality of greywater treated by three on-site treatment systems. *Environmental Technology*, **27**, 653–663.

Ghunmi L. A., Zeeman G., Fayyad M. and van Lier J. B. (2011). Grey water treatment systems: A review. *Critical Reviews in Environmental Science and Technology*, **41**, 657–698.

Gilboa Y. and Friedler E. (2008). UV disinfection of RBC treated light greywater effluent: Kinetics, survival and regrowth of selected microorganisms. *Water Research*, **42**, 1043–1050.

50% in greywater (values based on own experiences). The estimations suggest that biomass integration can remove only a small fraction of the nitrogen in urine and blackwater (5% and 10%, respectively), but all of the nitrogen in greywater. In fact, Jefferson and Jeffrey (2013) report that macronutrients (nitrogen and phosphorus) and micronutrients have to be dosed to some greywater in order to ensure effective biodegradation of the organic substances.

20.2.2 Nitrification

The term nitrification is used to describe the biological oxidation of ammonia (NH_3 and NH_4^+) to nitrate (NO_3^-). Two microbial groups are involved in nitrification: ammonia oxidizing bacteria (AOB) mediate the oxidation of ammonia to nitrite, while nitrite oxidizing bacteria (NOB) mediate the oxidation of nitrite to nitrate. The first step is also called nitritation. Both bacterial groups, AOB and NOB, are chemo-litho-autotrophs. Ammonia oxidizing archaea (AOA) have also been found in wastewater treatment plants, but recent measurements suggest that they are of minor importance for nitrification (Wells *et al.* 2009).

Equations 20.1 and 20.2 show the stoichiometry of the reactions in AOB (Poughon *et al.* 2001). In a first step, the enzyme ammonia monooxygenase (AMO) catalyzes the oxidation of ammonia to hydroxylamine. Elemental oxygen (O_2) provides the oxygen atom. In the second step, the short lived hydroxylamine is oxidized by the enzyme hydroxylamine oxidoreductase (HAO) to NO_2^-. In this reaction, a water molecule is the oxygen donor.

$$NH_3 + 0.5\,O_2 \rightarrow NH_2OH \tag{20.1}$$

$$NH_2OH + H_2O + O_2 \rightarrow NO_2^- + H^+ + 2\,H_2O \tag{20.2}$$

In the nitrite oxidation reaction (Equation 20.3) NO_2^- or rather its acid HNO_2 (nitrous acid) is oxidized to NO_3^- with water as oxygen donor (Poughon *et al.* 2001).

$$HNO_2 + H_2O + 0.5\,O_2 \rightarrow NO_3^- + H^+ + H_2O \tag{20.3}$$

Suzuki *et al.* (1974) demonstrated that free ammonia (NH_3) is the substrate of AOB and Hunik *et al.* (1993) presented evidence that nitrous acid (HNO_2) is the actual substrate of NOB. The dependency on NH_3 and HNO_2 can explain the strong influence of the pH value on AOB and NOB activity. However, high concentrations of the substrates NH_3 and HNO_2 are not necessarily beneficial. Anthonisen *et al.* (1976) determined that AOB and NOB are inhibited by high concentrations of NH_3 and HNO_2. AOB can tolerate substantially higher NH_3 concentrations than NOB, but both bacterial groups have similar 50% inhibition levels for HNO_2. The complex pH influence is usually negligible in the treatment of municipal wastewater due to the low ammonia concentrations (around

40 mgN·L^{-1}), but reliable pH control can be crucial in the treatment of highly concentrated solutions such as digester supernatant, urine or blackwater.

Nitrite accumulation is often used for novel nitrogen removal processes such as nitritation/denitrification or nitritation/anammox. There are several possibilities to ensure nitrite accumulation in high-strength solutions. Hellinga *et al.* (1998) demonstrated that NOB can be selectively washed out at high pH values, because the growth rate of NOB diminishes with rising pH, while the AOB growth rate increases. This effect is due to combined limitation and inhibition by NH_3 and HNO_2. Another factor that fosters nitrite accumulation is high temperature: the AOB growth rate increases faster with temperature than the NOB growth rate. A third possibility to favor AOB over NOB is low oxygen concentration (Blackburne *et al.* 2008): NOB have a lower oxygen affinity than AOB.

Ammonia oxidation lowers the pH value, because two moles of alkalinity are consumed when one mole of ammonia is oxidized (Equations 20.1 and 20.2). Since common ammonia oxidation stops at a pH value of about 5.5 (Udert *et al.* 2005), the alkalinity controls, how much ammonia can be oxidized. In stored urine and digester supernatant only about 50% of the ammonia can be nitrified, because both solutions contain ammonia and alkalinity at a molar ratio of 1.

The presence of organic substances can impede nitrification. Some organic substances directly inhibit nitritation, for example high concentrations of acetate, propionate, butyrate, phenol or cresol (Eilersen *et al.* 1994, Gomez *et al.* 2000, Texier and Gomez 2007). An indirect effect on nitrification is the competition by organo-heterotrophic bacteria, which consume oxygen or nitrite while degrading organic compounds.

Nitrification is a possible source of the atmospheric pollutants nitric oxide (NO) and nitrous oxide (N_2O) (Schreiber *et al.* 2012). NO and N_2O can be produced by AOB in a process called nitrifier denitrification, which is the reduction of nitrite to NO and N_2O. The probable purpose of this process is detoxification, when nitrite levels are high. A second pathway is the oxidation of hydroxylamine by HAO to NO, which in turn is reduced to N_2O by a yet unidentified enzyme. NOB were proposed to form NO and N_2O, when they use organic substances (for example pyruvate or glycerol) for the reduction of nitrate to nitrite under anoxic conditions.

20.2.3 Heterotrophic denitrification

Organo-heterotrophic bacteria (HET) use organic substances for energy generation and biomass synthesis. HET gain the most energy, when they use oxygen as electron acceptor. However, some HET can also use nitrite or nitrate during denitrification. Heterotrophic denitrifiers are a heterogeneous group of bacteria with various substrate utilization patterns (Kuenen and Robertson 1988). The complete denitrification pathway consists of a sequence of several intermediates. The first intermediate after NO_3^- reduction is NO_2^-, followed by NO and N_2O. The final product is dinitrogen gas (N_2). Denitrifiers can perform the entire or only parts of the denitrification pathway, have different affinities to the intermediates of the denitrification pathway and they differ in their response to inhibitory substances such as nitrite. Imbalances in the microbial community, substrate limitations or inhibition effects can result in the accumulation of denitrification intermediates.

Equations 20.4 and 20.5 exemplify the stoichiometry of denitrification with NO_2^- or NO_3^- as electron acceptors and acetic acid as electron donor

$$3\,CH_3COOH + 8\,NO_2^- + 8\,H^+ \rightarrow 4\,N_2 + 6\,CO_2 + 10\,H_2O \tag{20.4}$$

$$5\,CH_3COOH + 8\,NO_3^- + 8\,H^+ \rightarrow 4\,N_2 + 10\,CO_2 + 14\,H_2O \tag{20.5}$$

According to this stoichiometry, denitrification with NO_3^- requires 2.86 $gCOD{\cdot}gN^{-1}$, while denitrification with nitrite requires 1.71 $gCOD{\cdot}gN^{-1}$. In wastewater treatment plants, the actual consumption of organic substances during denitrification with nitrate or nitrite can be considerably higher, because some of the organic substrate is required for biomass synthesis and some may be lost to aerobic respiration. Nevertheless, nitrogen removal via nitrite requires substantially fewer resources (organic substrate, oxygen) than nitrogen removal via nitrate. Since nitrite accumulation in high-strength wastewaters can be achieved with some simple operational measures, the combination of nitritation and denitrification of nitrite has been optimized for nitrogen removal from digester supernatant (Hellinga *et al.* 1998, Fux *et al.* 2006). A critical point about denitrification of nitrite solutions is NO and N_2O production. High nitrite concentrations can lead to NO and N_2O emissions not only during anoxic phases (denitrification) but also aerobic phases (nitritation) (Fux *et al.* 2006, Kampschreur *et al.* 2008).

While nitrification consumes alkalinity, denitrification results in an alkalinity increase: one mole of protons is consumed per mole of NO_3^- reduced or NO_2^- reduced (Equations 20.4 and 20.5). This effect can be used to provide additional alkalinity for solutions, which do not have sufficient alkalinity to oxidize all ammonia to nitrite or nitrate (Hellinga *et al.* 1998).

20.2.4 Anaerobic ammonium oxidation (Anammox)

Denitrification is also possible without organic substrate. The process is mediated by a group of chemo-litho-autotrophic bacteria called anammox bacteria (anaerobic ammonium oxidizing bacteria, AMX). These bacteria use ammonia as electron donor and nitrite as electron acceptor. The stoichiometry for AMX metabolism has been determined by Strous *et al.* (1998)

$$\begin{aligned} &NH_4^+ + 1.32\,NO_2^- + 0.066\,HCO_3^- + 0.13\,H^+ \\ &\quad \rightarrow 1.02\,N_2 + 0.26\,NO_3^- + 0.066\,CH_2O_{0.5}N_{0.15} + 2.03\,H_2O \end{aligned} \tag{20.6}$$

According to this stoichiometry AMX require about 30% more nitrite than ammonia and they oxidize 11% of the initial total nitrogen to nitrate. AMX need very specific environmental conditions: they are highly sensitive to oxygen

(Strous *et al.* 1997), irreversibly inhibited by elevated nitrite concentrations (Strous *et al.* 1999) and even under optimal conditions, their growth is very slow (minimum doubling time of 11 days at 32°C, Strous *et al.* 1998).

Although chemo-litho-autotrophs, AMX are influenced by the presence of organic substances. Güven *et al.* (2005) observed that some organic compounds directly inhibit anammox bacteria (for example alcohols). In general, high concentrations of biodegradable organic compounds impede AMX activity by fostering the growth of HET (Chamchoi *et al.* 2008, Molinuevo *et al.* 2009). However, organic substances can also have a beneficial effect on AMX: Kartal *et al.* (2007, 2008) reported that AMX can oxidize acetic acid and propionic acid and thereby enhance their competitiveness.

20.3 NITROGEN STABILIZATION IN URINE

20.3.1 Conditions in stored urine

Urine is a highly concentrated solution containing about 80% of the excreted nitrogen (Friedler *et al.* 2013). In toilets, pipes and storage tanks, urea, which is the main nitrogen compound, is hydrolyzed to NH_3 and CO_2 by the ubiquitous enzyme urease (Udert *et al.* 2003a). After urea hydrolysis, 90% of the total nitrogen is ammonia (approx. 8000 $mgN{\cdot}L^{-1}$), the alkalinity is around 500 $mmol{\cdot}L^{-1}$ and the pH value is approximately 9 (Udert *et al.* 2006). These concentrations are given for a theoretical urine composition based on medical data. In stored urine, which has been collected in existing NoMix systems, the concentrations are usually lower due to dilution with flushing water or ammonia volatilization (Siegrist *et al.* 2013).

Urea hydrolysis makes source-separated urine an unstable solution, because ammonia is easily lost by volatilization. The loss of ammonia causes smell problems and can have negative effects on the environment. Furthermore, it lowers the fertilizer value of urine. Source-separated urine can be stabilized by adding high amounts of acids, but nitrification is a more resource-efficient alternative to acid dosage (Udert and Wächter 2012).

20.3.2 Nitrification without base dosage

Without base addition about half of the ammonia can be oxidized to nitrate, because the molar ratio of alkalinity to ammonia is 1. During nitrification, the pH decreases to values around 6 and 90% of the COD is degraded. Stable nitrification of urine was described by Udert *et al.* (2003b) and Udert and Wächter (2012). In both studies a CSTR with biofilm carriers (moving bed biofilm reactor, MBBR) was used to ensure that the NOB were not washed out. Due to the high ammonia concentrations and the high pH value of the influent (first study: 7100 $mgN{\cdot}L^{-1}$ and pH 9.2; second study: 2400 $mgN{\cdot}L^{-1}$ and pH 8.7) inhibition effects by NH_3 and, during process instabilities, HNO_2 had to be considered. When the influent load was constant a high nitrification rate was reached (380 $gN{\cdot}m^{-3}{\cdot}d^{-1}$

at pH 6.3, Udert *et al.* 2003b), but a sudden rise of the pH value due to an inflow increase would cause nitrite accumulation and inhibition of NOB (Figure 20.1). Therefore, Udert and Wächter (2012) kept the pH in a narrow range by controlling the inflow: the pH dropped due to nitrification during phases with no inflow and it rose, when the inflow was switched on.

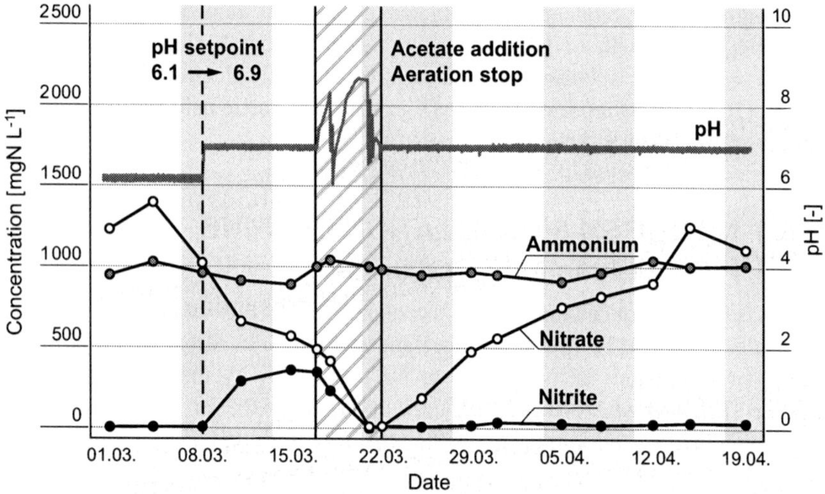

Figure 20.1 Nitrite accumulation in a moving bed biofilm reactor (MBBR) for urine nitrification. A sudden increase of the pH set-point from 6.1 to 6.9 caused the accumulation of nitrite and thereby the inhibition of NOB. The accumulated nitrite was removed with heterotrophic denitrification (aeration stop and acetate addition). From Udert and Wächter (2012).

While a sudden pH increase can have detrimental effects, a pH drop is less problematic. At a pH value close to 5.5, nitrification stops completely, but can be restarted safely by slowly increasing the alkalinity. The low pH limit of 5.5 is probably due to physiological limitations of AOB. However, if aeration continues for several weeks, acid-tolerant AOB will grow in and will cause a further pH decrease to values below 3 (Udert *et al.* 2005). At this low pH NOB cannot grow, but nitrite is completely oxidized by chemical reactions. The main product of chemical nitrite oxidation is nitrate but some intermediates (HNO_2, NO or N_2O) can volatilize. A combination of biological nitritation and chemical nitrite oxidation after acid dosage is an interesting alternative to pure biological processes and was patented by Verhave (2007).

20.3.3 Complete ammonia oxidation with base dosage

In order to oxidize all ammonia to nitrate the alkalinity in urine has to be increased. Oosterhuis and van Loosdrecht (2009) used sodium hydroxide (NaOH) to keep the pH at 7, Feng *et al.* (2008) dosed sodium carbonate (Na_2CO_3) to achieve a constant pH value of 8 and Jiang *et al.* (2011) used the same base (Na_2CO_3) to keep the pH value between 7.3 and 7.6. The influent of all three studies was diluted urine (total nitrogen concentrations between 700 and 1700 $mgN{\cdot}L^{-1}$). Various reactor types were used: Jiang *et al.* (2011) operated a SBR and prevented process instabilities by dosing the influent in three charges during one cycle. By this, the pH was kept between 7.3 and 7.6. Oosterhuis and van Loosdrecht (2009) used a suspended biomass CSTR with some biofilm carrier material attached to the wall and Feng *et al.* (2008) operated a packed-bed biofilm reactor in batch mode and in continuous mode. High nitrification rates (and ammonia oxidation efficiencies) were achieved by Jiang *et al.* (2011) and Oosterhuis and van Loosdrecht (2009): 1100 $mgN{\cdot}L^{-1}{\cdot}d^{-1}$ (100%) and 390 $mgN{\cdot}L^{-1}{\cdot}d^{-1}$ (97%), respectively.

20.3.4 Use of nitrified urine

Stabilized urine is a potential fertilizer, but the high water content makes transportation costly. Udert and Wächter (2012) demonstrated that distillation could be used to concentrate nearly all nutrients in a solid, if urine had been nitrified. They estimated that a small treatment facility consisting of a MBBR for nitrification and a distillation apparatus for water removal (80% energy recovery) would require about 0.5 $kWh{\cdot}L_{urine}^{-1}$ (primary energy). One critical point is the thermal stability of the solid ammonium nitrate product. The ammonium nitrate content in the fertilizer product is below the EU safety limits for fertilizers, but nevertheless, the product should not be exposed to very high temperatures. One way to increase the thermal stability is complete nitrification.

Nitrified urine can also be used to prevent corrosion in sewers. Corrosion by sulfide oxidation is a common problem in sewers with long wastewater retention times (for example in the Netherlands) or in sewers, which contain a high fraction of seawater (for example in Hong Kong). Nitrate dosage can prevent corrosion because nitrate instead of sulfate is reduced and therefore no sulfide is formed. Oosterhuis and van Loosdrecht (2009) and Jiang *et al.* (2011) demonstrated that decentralized nitrification of source-separated urine can be a suitable approach to alleviate sewer corrosion. In-sewer denitrification of nitrified urine would also reduce the need for enhanced nitrogen removal in wastewater treatment plants.

20.4 NITROGEN REMOVAL FROM URINE

In urine, the ratio of biodegradable COD to nitrogen is too low for complete nitrogen removal via heterotrophic denitrification. The ratio is only between 1.1 $gCOD{\cdot}gN^{-1}$ in fresh urine (Friedler *et al.* 2013) and 1.5 $gCOD{\cdot}gN^{-1}$ in stored urine after ammonia volatilization (Bürgmann *et al.* 2011). A viable alternative is nitritation/anammox, either in two separate reactors or combined in one single reactor.

20.4.1 Nitritation/anammox in a two-reactor set-up

Nitritation of urine can be achieved in SBRs and CSTRs. In a study by Udert *et al.* (2003b) an SBR was operated for more than 3.5 years with stable nitrite production. The strong pH changes (between 8.8 and 6.0) and the accumulation of nitrite suppressed NOB. The cycle duration was 2 d but ammonia oxidation was completed after less than 0.5 d. Despite the long idle time, NOB did not grow in and the nitrate concentration was always negligibly low. The ammonia oxidation rate was 280 $mgN{\cdot}L^{-1}{\cdot}d^{-1}$ but could have been increased to 1100 $mgN{\cdot}L^{-1}{\cdot}d^{-1}$ by shortening the cycle to 0.5 d. Stable nitrite production was also achieved in a heated CSTR (30°C) with suspended biomass (SRT 4.8 d). The nitritation rate was 790 $mgN{\cdot}L^{-1}{\cdot}d^{-1}$. Nitrogen removal via anammox was tested by dosing the effluent of the nitritation CSTR to a batch reactor containing AMX from a reactor operated on digester supernatant (total COD 11.4 $g{\cdot}L^{-1}$). In this batch experiment a nitrogen removal rate of 1000 $mgN{\cdot}L^{-1}{\cdot}d^{-1}$ was achieved. The batch experiment showed that nitrogen removal in a two-step nitritation/anammox process is essentially possible. However, further long-term experiments are needed for a sufficient proof.

20.4.2 Nitritation/anammox in a single reactor

Several studies have shown that nitritation and anammox can be operated simultaneously in a single reactor (Joss *et al.* 2009, Vlaeminck *et al.* 2009). One challenge for simultaneous nitritation/anammox is competition by heterotrophic bacteria, when biodegradable organic substances are available. Urine does not contain sufficient organic substances for complete nitrogen removal with heterotrophic denitrification, but the organic content is so high that HET will be competitors of AOB and AMX. In order to elucidate the effect of organic substances, Udert *et al.* (2008) compared nitritation/anammox treatment of digester supernatant and urine in a SBR with intermittent aeration. Stored urine from the Eawag collection tank was diluted five times to reach a similar ammonia concentration as digester supernatant (610 and 640 $mgN{\cdot}L^{-1}$ in digester supernatant and urine, respectively). The SBR cycle length was either fixed or controlled with a low pH set-point. With both solutions, the nitrogen removal rate (200 $mgN{\cdot}L^{-1}{\cdot}d^{-1}$ for urine) and the ammonia removal efficiency (78% for urine) were about the same. Since only 78% of the ammonia was removed in urine, the ratio of biodegraded COD to removed ammonia was sufficient (1.9 $gCOD{\cdot}gN^{-1}$) for heterotrophic denitrification with nitrite. Nevertheless, AMX must have contributed to nitrogen removal, because HET probably oxidized some organic substrate with oxygen (see section 20.5.2).

In the same reactor a 94% nitrogen removal (at 100 $mgN{\cdot}L^{-1}{\cdot}d^{-1}$) was achieved by optimizing the pH set-point, the length of the anoxic phase and the airflow (Bürgmann *et al.* 2011). Under these conditions, the ratio of biodegraded COD to removed ammonia was 1.5 $gCOD{\cdot}gN^{-1}$, indicating that AMX contributed substantially to the overall nitrogen removal. By raising the pH set-point and increasing the aeration rate, a maximum nitrogen removal rate of

430 mgN·L^{-1}·d^{-1} was reached (90% nitrogen removal). Unfortunately, the high performance could not be sustained and was followed by a drop of the nitrogen removal rate to values around 170 mgN·L^{-1}·d^{-1} (nitrogen removal >80%). Concomitantly, NO_2^-, NO and N_2O accumulated. Although AMX were still present at considerable numbers, the performance could not be improved anymore by adjusting the control parameters. An evaluation of physico-chemical and microbial data revealed that a microbial regime shift had happened. Based on an ordination analysis, the researchers hypothesized that the concomitant change of an important process control parameter (pH set-point), a strongly influential environmental parameter (temperature) and a reinforcing response parameter (N_2O and NO) caused a fundamental change of the microbial community. The influence of N_2O or NO still has to be clarified. To our knowledge, N_2O does not have any adverse or stimulating effects on AMX. However, NO is a necessary intermediate of AMX metabolism and AMX can tolerate very high concentrations (Kartal *et al.* 2010). Therefore, it is likely that NO did not inhibit AMX. Instead, AMX could have reduced their nitrite consumption, because NO was available. Anyway, this study shows that nitritation/anammox in a single reactor requires careful process control to prevent detrimental changes of the microbial community.

20.5 NITROGEN REMOVAL FROM BLACKWATER

20.5.1 Blackwater collected in vacuum toilets

Vlaeminck *et al.* (2009) and de Graaff *et al.* (2010, 2011) investigated nitrogen removal from blackwater collected with vacuum toilets at the demonstration site in Sneek (The Netherlands). The organic content of the blackwater was too low for heterotrophic denitrification, because most of the organics were removed in an anaerobic digester before nitrogen treatment. As a consequence both research groups chose nitritation/anammox as nitrogen removal process. An overview of the influent concentrations and performance data is given in Table 20.1.

Vlaeminck *et al.* (2009) used a rotating biological contactor (RBC) for simultaneous nitritation and anammox (one-stage oxygen-limited autotrophic nitrification/denitrification, OLAND). The reactor was started with a synthetic solution and after steady state had been reached, the effluent was slowly changed to digested blackwater. The temperature was kept at 26°C. With digested blackwater the oxygen concentration decreased strongly and as a consequence the ammonia oxidation rate dropped. To raise the oxygen concentration, about 15% of the biomass was removed. However, the resulting oxygen concentration (0.7 mg·L^{-1}) supported NOB growth so that nitrate accumulated. By raising the pH value with base ($NaHCO_3$) a sufficiently high NH_3 concentration (3 mgN·L^{-1}) was reached to inhibit NOB. The final nitrogen removal rate was 700 mgN·L^{-1}·d^{-1} and the nitrogen removal efficiency was 76%. It should be noted that a lower oxygen concentration would have suppressed NOB as well, but the nitrogen removal rate would have been lower.

Table 20.1 Performance data of reactors for nitrogen removal from blackwater.

Study	Wastewater	Reactor	Oxygen reactor [$mg \cdot L^{-1}$]	Ammonia influent [$mgN \cdot L^{-1}$]	TKN influent [$mgN \cdot L^{-1}$]	Biodegr. COD to TKN, influent [$gCOD \cdot gN^{-1}$]	N elimination rate [$mgN \cdot L^{-1} \cdot d^{-1}$]	N removal efficiency [%]
Vlaeminck *et al.* (2009)	Blackwater (vac.) anaerob. digested	RBC (OLAND)	0.7	1,100	1,100	(1) 0.11	700	76
de Graaff *et al.* (2010, 2011)	Blackwater (vac.) anaerob. digested	CSTR (nitritation) SBR (anammox)	aerobic anoxic	1,000		(1) 0.60	(2) 250	87
Hocaoglu *et al.* (2011)	Blackwater, solids screened	MBR operated as SBR	0.15–0.2 0.3–0.35 0.5–0.55	130	160	(1) 4.8	140 126 63	73 70 36
Knerr *et al.* (2011)	Blackwater, solids screened	MBR with recirculation	aerobic and anoxic compartments	230	280	(1) 2.1	(1) 49	42
Boehler *et al.* (2007)	Blackwater	MBR with recirculation	aerobic and anoxic compartments	130			(1) 33	nearly 100
Luostarinen *et al.* (2006)	Blackwater and kitchen waste anaerob. digested	MBBR operated as SBR or CSTR	intermittent aeration	26	N_{tot}: 31	(1) 1.2	(1) 7.4	49

(1) Value was derived from the literature data. (2) CSTR and SBR combined.

Note: Abbreviations are explained in the text.

De Graaff et al. (2010) used a two-step process for nitritation and anammox. An airlift reactor was chosen for nitritation. The reactor was started with a synthetic solution and later slowly accustomed to digester supernatant. In a first phase, the temperature was kept at 34°C, but later lowered to 25°C. At 34°C acid (HCl) had to be dosed to keep the pH value below 7.5 in order to prevent AOB inhibition. However, acid dosage was not necessary at 25°C and a higher nitrite to ammonia ratio was reached (1.3 instead of 1.0 $gN{\cdot}gN^{-1}$). The researchers argued that strong fluctuations of the sludge retention time (1 to 17 d) and of the pH in the reactor (6.3 to 7.7) were responsible for the suppression of NOB. Due to the high nitrite concentrations, N_2O emissions were significant (between 0.6 and 2.6% of the total nitrogen load). The nitrogen load was 1100 $mgN{\cdot}L^{-1}{\cdot}d^{-1}$ and 520 $mgN{\cdot}L^{-1}{\cdot}d^{-1}$ at 34°C and 25°C, respectively.

Two parallel SBRs containing AMX granules were fed with the effluent of the nitritation reactor (de Graaff *et al.* 2011). Although the two SBRs were operated at different temperatures (34°C and 25°C), they had a similar nitrogen removal rate (500 $mgN{\cdot}L^{-1}{\cdot}d^{-1}$). In both reactors, $CaCl_2{\cdot}2H_2O$ had to be dosed to ensure granulation of the AMX biomass, otherwise nitrogen removal rates dropped rapidly. Additionally, both reactors were flushed with N_2 to ensure anoxic conditions, and with CO_2 for pH control. This might not be possible nor necessary in full-scale applications. N_2O emissions were low, accounting for 0.02 to 0.1% of the nitrogen load.

Based on these first studies, an RBC system seems to be more suitable for on-site applications due to its easier maintenance and operation. However, the resilience of both processes still has to be tested at pilot-scale.

20.5.2 Conventionally collected blackwater

Table 20.1 shows performance data of four on-site reactors treating blackwater that has been collected in conventional toilets (Hocaoglu et al. 2011, Knerr et al. 2011, Boehler et al. 2007 and Luostarinen et al. 2006). Three of the presented four studies used MBR systems, only one study (Luostarinen *et al.* 2006) used an alternative system, which consisted of anaerobic pretreatment in an UASB and subsequent nitrogen removal in a moving bed biofilm reactor (MBBR). The MBBR was operated either as SBR or as CSTR. In all four examples the concentrations were considerably lower than in the vacuum-collected black-water: the total nitrogen concentration (or total Kjeldahl nitrogen, TKN) was between 31 and 280 $mgN{\cdot}L^{-1}$. The blackwater in the study of Luostrainen *et al.* (2006) also contained kitchen waste to boost the methane production in an UASB. Boehler *et al.* (2007) recycled nearly 100% of the treated effluent for toilet flushing, because the reactor was situated in a cable car station on 3300 m above sea level, where water availability was limited.

Theoretically, the biodegradable COD content of untreated blackwater is sufficiently high for complete nitrogen removal via heterotrophic denitrification.

However, the aeration regime has to be carefully adjusted to minimize aerobic COD degradation. Furthermore, the particulate organic material including the activated sludge produced in the reactor has to be used as organic substrate as well to provide sufficient electron acceptors. For example, Boehler *et al.* (2007) could achieve nearly complete nitrogen removal, when the activated sludge was recirculated to the equalization tank, where the organic substances where hydrolyzed. Anaerobic pretreatment (Luostarinen *et al.* 2006) and removal of particulate material can reduce the organic content to levels, which are not sufficient for complete nitrogen removal. In the study by Knerr *et al.* (2011), about 75% of the COD was removed with a rotation screen (0.1 mm). Consequently, the ratio of biodegradable COD to TKN was reduced from 8.6 $gCOD{\cdot}gN^{-1}$ to 2.1 $gCOD{\cdot}gN^{-1}$ (assuming 80% biodegradability of COD).

Hocaoglu *et al.* (2011a) demonstrated that nitrogen removal from blackwater is possible with continuous aeration, but oxygen concentrations have to be kept low. At oxygen concentrations between 0.15–0.2 $mg{\cdot}L^{-1}$ the nitrogen removal efficiency and nitrogen removal rate were 50% and 55% higher than at an oxygen concentration of 0.5 $mg{\cdot}L^{-1}$.

20.6 CONCLUSIONS

The choice of a biological nitrogen removal process depends on the availability of biodegradable organic substrate. Nitritation/anammox is recommended for nitrogen removal from urine and anaerobically digested blackwater, because both solutions have too little organic substrate for complete heterotrophic denitrification. However, good process control is necessary to prevent process failures, such as nitrite accumulation or NO and N_2O emissions.

Heterotrophic denitrification of blackwater is a suitable process if the organic substrates are not used for energy recovery. A proven reactor set-up is an MBR with internal recirculation between anoxic and aerobic compartments. Still, heterotrophic denitrification can become limited by organic substrate, if particulate organic material is screened in front of the denitrification reactor or if the aeration is too high.

Nitrification is a suitable process to stabilize nitrogen in urine. The alkalinity is sufficiently high to nitrify about half of the ammonia. However, to achieve complete nitrification, the alkalinity has to be increased by at least a factor of two.

In nitrification processes, dosage of acid or base can be necessary to prevent inhibition of nitrifiers. In the case of urine nitrification, the pH and thereby the nitrifier activity can be stabilized by controlling the influent rate.

REFERENCES

Anthonisen A. C., Loehr R. C., Prakasam T. B. S. and Srinath E. G. (1976). Inhibition of nitrification by ammonia and nitrous acid. *J. WPCF.*, **48**(5), 835–852.

Blackburne R., Yuan Z. G. and Keller J. (2008). Partial nitrification to nitrite using low dissolved oxygen concentration as the main selection factor. *Biodegradation*, **19**(2), 303–312.

Boehler M., Joss A., Buetzer S., Holzapfel M., Mooser H. and Siegrist H. (2007). Treatment of toilet wastewater for reuse in a membrane bioreactor. *Water Sci. Technol.*, **56**(5), 63–70.

Bürgmann H., Jenni S., Vazquez F. and Udert K. M. (2011). Regime shift and microbial dynamics in a sequencing batch reactor for nitrification and anammox treatment of urine. *Appl. Environ. Microbiol.*, **77**(17), 5897–5907.

Chamchoi N., Nitisoravut S. and Schmidt J. E. (2008). Inactivation of anammox communities under concurrent operation of anaerobic ammonium oxidation (anammox) and denitrification. *Bioresource Technol.*, **99**(9), 3331–3336.

de Graaff M. S., Zeeman G., Temmink H., van Loosdrecht M. C. M. and Buisman C. J. N. (2009). Long term partial nitritation of anaerobically treated black water and the emission of nitrous oxide. *Water Res.*, **44**(7), 2171–2178.

de Graaff M. S., Temmink H., Zeeman G., van Loosdrecht M. C. M. and Buisman C. J. N. (2011). Autotrophic nitrogen removal from black water: calcium addition as a requirement for settleability. *Water Res.*, **45**(1), 63–74.

Eilersen A. M., Henze M. and Kløft L. (1994). Effect of volatile fatty acids and trimethylamine on nitrification in activated sludge. *Water Res.*, **28**(6), 1329–1336.

Feng D., Wu Z. and Xu S. (2008). Nitrification of human urine for its stabilization and nutrient recycling. *Bioresource Technol.*, **99**(14), 6299–6304.

Friedler E., Butler D. and Alfiya Y. (2013). Wastewater composition. In: *Source Separation and Decentralization for Wastewater Management*, T. A. Larsen, K. M. Udert and J. Lienert (eds.), IWA Publishing, London, UK, pp. 241–257.

Fux C., Velten S., Carozzi V., Solley D. and Keller J. (2006). Efficient and stable nitritation and denitritation of ammonium-rich sludge dewatering liquor using an SBR with continuous loading. *Water Res.*, **40**(14), 2765–2775.

Gomez J., Mendez R. and Lema J. M. (2000). Kinetic study of addition of volatile organic compounds to a nitrifying sludge. *Appl. Biochem. Biotechnol.*, **87**, 189–202.

Gujer W., Henze M., Mino T. and van Loosdrecht M. (1999). Activated sludge model No. 3. *Water Sci. Technol.*, **39**(1), 183–193.

Güven D., Dapena A., Kartal B., Schmid M. C., Maas B., van de Pas-Schoonen K., Sozen S., Mendez R., Op den Camp H. J. M., Jetten M. S. M., Strous M. and Schmidt I. (2005). Propionate oxidation by and methanol inhibition of anaerobic ammonium-oxidizing bacteria. *Appl. Environ. Microbiol.*, **71**(2), 1066–1071.

Hellinga C., Schellen A. A. J. C., Mulder J. W., van Loosdrecht M. C. M. and Heijnen J. J. (1998). The SHARON process: An innovative method for nitrogen removal from ammonium-rich waste water. *Water Sci. Technol.*, **37**(9), 135–142.

Hocaoglu S. M., Insel G., Cokgor E. U. and Orhon D. (2011a). Effect of low dissolved oxygen on simultaneous nitrification and denitrification in a membrane bioreactor treating black water. *Bioresource Technol.*, **102**(6), 4333–4340.

Hunik J. H., Meijer H. J. G. and Tramper J. (1993). Kinetics of Nitrobacter agilis at Extreme Substrate, Product and Salt Concentrations. *Appl. Environ. Microbiol.*, **40**(2–3), 442–448.

Jefferson B. and Jeffrey P. (2013). Aerobic elimination of organics and pathogens: greywater treatment. In: *Source Separation and Decentralization for Wastewater Management*,

T. A. Larsen, K. M. Udert and J. Lienert (eds.), IWA Publishing, London, UK, pp. 275–289.

Jiang F., Chen Y., Mackey H. R., Chen G. H. and van Loosdrecht M. C. M. (2011). Urine nitrification and sewer discharge to realize in-sewer denitrification to simplify sewage treatment in Hong Kong. *Water Sci. Technol.*, **64**(3), 618–626.

Joss A., Salzgeber D., Eugster J., König R., Rottermann K., Burger S., Fabijan P., Leumann S., Mohn J. and Siegrist H. R. (2009). Full-scale nitrogen removal from digester liquid with partial nitritation and anammox in one SBR. *Environ. Sci. Technol.*, **43**(14), 5301–5306.

Luostarinen S., Luste S., Valentín L. and Rintala J. (2006). Nitrogen removal from on-site treated anaerobic effluents using intermittently aerated moving bed biofilm reactors at low temperatures. *Water Res.*, **40**(8), 1607–1615.

Kampschreur M. J., van der Star W. R. L., Wielders H. A., Mulder J. W., Jetten M. S. M. and van Loosdrecht M. C. M. (2008). Dynamics of nitric oxide and nitrous oxide emission during full-scale reject water treatment. *Water Res.*, **42**(3), 812–826.

Kampschreur M. J., Temmink H., Kleerebezem R., Jetten M. S. M. and van Loosdrecht M. C. M. (2009). Nitrous oxide emission during wastewater treatment. *Water Res.*, **43**(17), 4093–4103.

Kartal B., Rattray J., van Niftrik L. A., van de Vossenberg J., Schmid M. C., Webb R. I., Schouten S., Fuerst J. A., Damsté J. S., Jetten M. S. M. and Strous M. (2007). Candidatus "Anammoxoglobus propionicus" a new propionate oxidizing species of anaerobic ammonium oxidizing bacteria. *Syst. Appl. Microbiol.*, **30**(1), 39–49.

Kartal B., Van Niftrik L., Rattray J., van de Vossenberg J. L. C. M., Schmid M. C., Damsté J. S., Jetten M. S. M. and Strous M. (2008). *Candidatus* 'Brocadia fulgida': an autofluorescent anaerobic ammonium oxidizing bacterium. *FEMS Microbiol. Ecol.*, **63**, 46–55.

Kartal B., Tan N. C. G., van Biezen E. D., Kampschreur M. J., van Loosdrecht M. C. M. and Jetten M. S. M. (2010). Effect of nitric oxide on anammox bacteria. *Appl. Environ. Microbiol.*, **76**(18), 6304–6306.

Knerr H., Rechenburg A., Kistemann T. and Schmitt T. G. (2011). Performance of a MBR for the treatment of blackwater. *Water Sci. Technol.*, **63**(6), 1247–1254.

Kuenen J. G. and Robertson L. A. (1988). Ecology of nitrification and denitrification. In: *The Nitrogen and Sulphur Cycles*, J. A. Cole and S. J. Ferguson (eds), Cambridge University Press, Cambridge, pp. 161–218.

Molinuevo B., García M. C., Karakashev D. and Angelidaki I. (2009). Anammox for ammonia removal from pig manure effluents: Effect of organic matter content on process performance. *Bioresource Technol.*, **100**(7), 2171–2175.

Oosterhuis M. and van Loosdrecht M. C. M. (2009). Nitrification of urine for H_2S control in pressure sewers. *Water Practise Technol.*, **4**(3).

Poughon L., Dussap C.-G. and Bros J.-B. (2001). Energy model and metabolic flus analysis for autotrophic nitrifiers. *Biotechnol. Bioeng.*, **72**(4), 416–433.

Schreiber F., Wunderlin P., Udert K. M. and Wells G. F. (2012) Nitric oxide and nitrous oxide turnover in natural and engineered microbial communities: Biological pathways, chemical reactions and novel technologies. *Frontiers Microbiol.*, **3**, article 372.

Siegrist H., Laureni M. and Udert K. M. (2013). Transfer into the gas phase: ammonia stripping. In: *Source Separation and Decentralization for Wastewater Management*,

T. A. Larsen, K. M. Udert and J. Lienert (eds.), IWA Publishing, London, UK, pp. 337–350.

Strous M., Heijnen J. J., Kuenen J. G. and Jetten M. S. M. (1998). The sequencing batch reactor as a powerful tool for the study of slowly growing anaerobic ammonium-oxidizing microorganisms. *Appl. Microbiol. Biotechnol.*, **50**(5), 589–596.

Strous M., Gerven E., Kuenen J. G. and Jetten M. S. M. (1997). Effects of aerobic and microaerobic conditions on anaerobic ammonium-oxidizing (Anammox) sludge. *Appl. Environ. Microbiol.*, **63**(6), 2446–2448.

Strous M., Kuenen J. G. and Jetten M. S. M. (1999). Key physiology of anaerobic ammonium oxidation. *Appl. Environ. Microbiol.*, **65**(7), 3248–3250.

Suzuki I., Dular U. and Kwok S. C. (1974). Ammonia or ammonium ion as substrate for oxidation by Nitrosomonas europaea cells and extracts. *J. Bacteriol.*, **120**(1), 556–558.

Texier A. C. and Gomez J. (2007). Simultaneous nitrification and p-cresol oxidation in a nitrifying sequencing batch reactor. *Water Res.*, **41**(2), 315–322.

Udert K. M., Larsen T. A., Biebow M. and Gujer W. (2003a). Urea hydrolysis and precipitation dynamics in a urine-collecting system. *Water Res.*, **37**(11), 2571–2582.

Udert K. M., Fux C., Münster M., Larsen T. A., Siegrist H. and Gujer W. (2003b). Nitrification and autotrophic denitrification of source-separated urine. *Water Sci. Technol.*, **48**(1), 119–130.

Udert K. M., Larsen T. A. and Gujer W. (2005). Chemical nitrite oxidation in acid solutions as a consequence of microbial ammonium oxidation. *Environ. Sci. Technol.*, **39**(11), 4066–4075.

Udert K. M., Larsen T. A. and Gujer W. (2006). Fate of major compounds in source-separated urine. *Water Sci. Technol.*, **54**(11–12), 413–420.

Udert K. M., Kind E., Teunissen M., Jenni S. and Larsen T. A. (2008). Effect of heterotrophic growth on nitritation/anammox in a single sequencing batch reactor. *Water Sci. Technol.*, **58**(2), 277–284.

Udert K. M. and Wächter M. (2012). Complete nutrient recovery from source-separated urine by nitrification and distillation. *Water Res.*, **46**(2), 453–464.

Verhave W. A. (2007). Process for the conversion of liquid waste biomass into a fertilizer product. Patent application, WO 2007/097612 A2.

Vlaeminck S. E., Terada A., Smets B. F., Van Der Linden D., Boon N., Verstraete W. and Carballa M. (2009). Nitrogen removal from digested black water by one-stage partial nitritation and anammox. *Environ. Sci. Technol.*, **43**(13), 5035–5041.

Wells G. F., Park H. D., Yeung C. H., Eggleston B., Francis C. A. and Criddle C. S. (2009). Ammonia-oxidizing communities in a highly aerated full-scale activated sludge bioreactor: Betaproteobacterial dynamics and low relative abundance of Crenarchaea. *Environ. Microbiol.*, **11**(9), 2310–2328.

Chapter 21

Anaerobic treatment of source-separated domestic wastewater

Grietje Zeeman and Katarzyna Kujawa-Roeleveld

21.1 INTRODUCTION

Anaerobic treatment refers to biological processes that degrade organic waste into useful compounds such as energy-rich methane gas (CH_4). It is particularly suitable for waste streams with a high content of organic substances. New sanitation concepts based on source separation can provide such waste streams. The present chapter will focus mainly on the anaerobic treatment of (i) blackwater (faeces and urine), and (ii) brownwater produced by urine diversion. Different dilutions are considered for both wastewater streams depending on the toilet systems in use. Such different dilutions determine the type of anaerobic reactor system that can be applied. When urine and faeces are collected separately, a third waste stream remains, namely greywater which contains the wastewater from the shower, bath, kitchen and laundry. The concentration of organic substances in greywater is low, but its large volume means that its organic load is similar to that in blackwater (Kujawa-Roeleveld *et al.* 2006). Despite its strong dilution, greywater can also be considered for anaerobic treatment due to its high temperature (Hernandez *et al.* 2010a, Abu Ghunmi *et al.* 2011). This chapter will briefly discuss the state of the art in anaerobic greywater treatment.

Separate anaerobic treatment of urine for energy production will not be discussed here, because urine contains only a minor fraction (*ca.* 10%) of the energy-rich organic substances in household wastewater.

This chapter only addresses anaerobic methane production, although other energy sources, such as bio-hydrogen and electricity using microbial fuel cells (MFC) might also be produced from source-separate domestic wastewaters in future. As short-term applications are not expected for complex wastewaters such as black-, brown- and greywater, the latter technologies will not be further discussed here.

21.2 THE ANAEROBIC CONVERSION PROCESS

Anaerobic microbiological decomposition is a process in which micro organisms derive energy and grow by metabolizing organic material in an oxygen-free environment. The anaerobic digestion process can be subdivided into four main phases, each requiring its own characteristic group of micro-organisms. These are:

(1) Hydrolysis: enzymes excreted by acidogenic bacteria convert bio-polymers into soluble monomers and dimers
(2) Acidogenic bacteria convert soluble organic compounds to volatile fatty acids (VFA) and carbon dioxide (CO_2)
(3) Acetogenic bacteria convert VFA to acetate and hydrogen (H_2) and CO_2
(4) Methanogenetic bacteria use acetic acid or H_2 and CO_2 to produce CH_4

The final product of this series of microbial conversion processes is biogas, consisting of CH_4 and CO_2, in addition to the biomass. For stable digestion, it is vital that various biological conversions remain well balanced in order to prevent the accumulation of intermediate compounds leading to process instability. Hence, depending on the buffer capacity, an accumulation of volatile fatty acids can result in a pH decrease that inhibits methanogenesis, thus further decreasing the pH.

21.3 REACTOR TYPES

Two basic approaches can be distinguished in the anaerobic treatment of waste streams. In the first approach, the waste stream and the active biomass have the same retention time in the reactor, that is the hydraulic retention time (HRT) and the sludge retention time (SRT) are equal. The second type of reactors allows a long retention of the biomass, so that the SRT significantly exceeds the HRT. The second system is advised for anaerobic treatment of wastewaters, while systems without biomass retention are more appropriate for slurries or solid wastes.

21.3.1 Reactors without enhanced biomass retention

21.3.1.1 Continuous Stirred-Tank Reactor (CSTR)

The most frequently used anaerobic system without sludge/biomass retention is the CSTR. This reactor type has three main characteristics. Firstly it is fed continuously. Secondly, liquid and suspended solids, including biomass, are completely mixed. In other words, the sludge retention time (SRT) equals the hydraulic retention time (HRT). Thirdly, the concentration and composition of the reactor content is equal to that of the effluent. The CSTR is mainly used for anaerobic treatment of highly concentrated waste streams with volatile suspended solid (VSS) contents of 50 $gVSS{\cdot}L^{-1}$ or more. Thus primary and secondary sludge from wastewater treatment plants (WWTPs) or animal manure are frequently digested in a CSTR. Concentrated blackwater could also be treated in a CSTR. When a CSTR reactor is being designed, the slowest conversion process will determine its size. In general, hydrolysis is the rate-limiting step in the anaerobic conversion of complex substrates, which usually have a high fraction of suspended organics, consisting of lipids, carbohydrates and proteins. The hydrolysis rate will therefore determine the size of the anaerobic reactor system. Anaerobic hydrolysis can be described by first-order kinetics with respect to the biodegradable substrate (Sanders 2001). The following formula is based on a mass balance and can be used to calculate the necessary HRT(d):

$$HRT = \frac{C_{S,\text{in}} - C_S}{k_h \cdot C_S} \tag{21.1}$$

where C_S is the total concentration of the organic substrate ($kgCOD{\cdot}m^{-3}$) in the reactor, $C_{S,in}$ the total concentration of the organic substrate ($kgCOD{\cdot}m^{-3}$) in the influent and k_h the first-order hydrolysis constant (d^{-1}).

The "Chinese dome" digester, used on-site in large numbers in rural areas in China (Chen *et al.* 2010), can be considered a CSTR system, although it might not always ensure complete mixing. Whereas conventional CSTRs use forced mixing, either mechanically or via gas or liquid recirculation, Chinese dome digesters are mixed by the variation in gas pressure occurring naturally due to the daily pattern of gas generation and consumption. A schematic of a Chinese dome digester is presented in Figure 21.1.

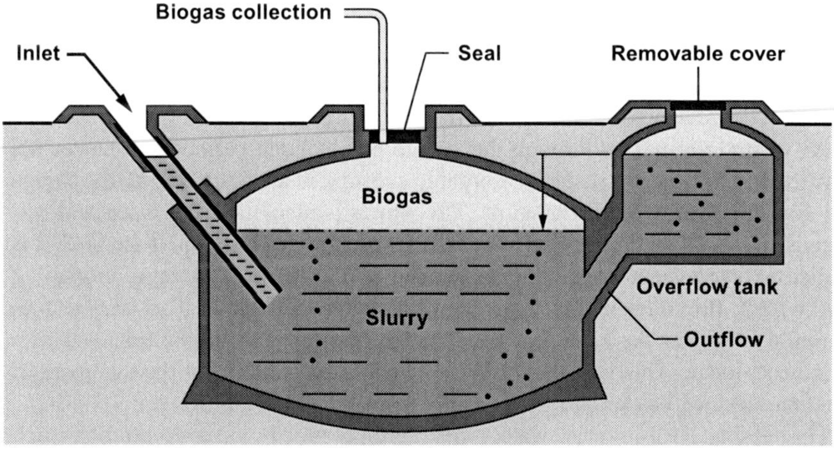

Figure 21.1 Schematic of a Chinese dome digester (from ISIS 2006, with permission).

Other, less frequently used anaerobic systems without sludge retention are plug flow, batch and accumulation (=fed-batch) systems (de Mes *et al.* 2003). The latter is discussed below.

21.3.1.2 Accumulation (AC) system

An accumulation (AC) system has the advantage of combining digestion and storage in a single reactor volume (Kujawa-Roeleveld *et al.* 2006). An AC system can be characterized by having inflow without outflow; the volume of the reactor content is therefore variable. Elmitwalli *et al.* (2011) developed a dynamic model based on the ADM1 model for predicting the process performance of AC systems treating blackwater or blackwaste (mainly faeces). The performance of the system is mainly determined by the filling time, followed by the process temperature (Elmitwalli *et al.* 2011).

21.3.2 Reactors with enhanced biomass retention

The SRT can be increased far beyond the HRT if the biomass is retained in the reactor by using internal settler systems or external settlers with sludge recycling or fixation of biomass on support material. Anaerobic sludge bed reactors are undoubtedly by far the most popular anaerobic treatment system to date (Franklin 2001). Several types of anaerobic sludge bed reactors are currently in use, but the UASB and the UASB septic tank are probably the most suitable for the treatment of black-, brown- or greywater (Zeeman and Lettinga 1999).

21.3.2.1 Upflow Anaerobic Sludge Blanket (UASB)

The UASB reactor concept was developed in the late 1960s at Wageningen University in the Netherlands (Lettinga *et al.* 1980). Wastewater enters the bottom of the reactor through an inlet distribution system and passes upward through a dense anaerobic sludge bed (Figure 21.2).

As wastewater passes through the sludge bed, soluble organic substances are converted to biogas and suspended organic substances are entrapped in the sludge bed for subsequent biodegradation. The upward circulation of water and gas establishes a well-settleable sludge consisting of granules or flocs. If the UASB is applied to wastewater containing suspended solids (such as sewage, brown- or blackwater), flocculent rather than granular sludge will grow. The former does not settle as well as the latter, but good sludge retention can still be achieved with flocculent sludge. This was shown by de Graaff *et al.* (2010) for the treatment of vacuum-collected blackwater.

The required HRT for wastewaters with high suspended solid concentrations such as brown- and blackwater can be calculated using Equation 21.2 (Zeeman and Lettinga 1999):

$$HRT = \frac{f_{X_S,\text{in}} \cdot C_{S,\text{in}}}{X_S} \cdot f_R \cdot (1 - f_H) \cdot SRT \tag{21.2}$$

where $C_{S,\text{in}}$ kgCOD·m^{-3} is the total concentration of organic compounds in the influent, $f_{Xs,\text{in}}$ kgCOD·kgCOD^{-1} the fraction of suspended organic solids in the influent ($X_{S,\text{in}}$), X_S (kgCOD·m^{-3}) the sludge concentration in the reactor, f_R the fraction of $X_{S,\text{in}}$ that is removed from the water stream (i.e., entrapped in the sludge bed) and f_H the hydrolyzed fraction of the removed solids.

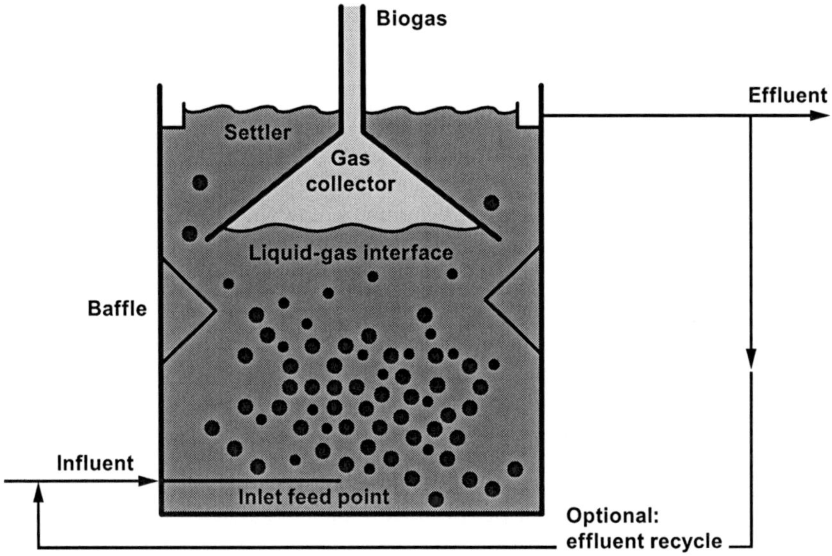

Figure 21.2 Schematic of an upflow anaerobic sludge blanket reactor (UASB).

21.3.2.2 UASB septic tank

The most important difference between a conventional UASB and a UASB septic tank is that the latter system is also designed for the accumulation of sludge. In contrast to a conventional septic tank, the UASB version is operated in upflow mode, resulting in both improved physical removal of suspended solids and biological conversion. The height of the sludge bed in this tank increases with time, while that of a UASB is kept at a steady level. Whereas a UASB needs regular sludge disposal, the sludge in a UASB septic tank needs to be disposed of only once every few years. This tank can therefore be attractive for use in small communities, where minimal maintenance is required.

21.3.2.3 *Anaerobic MBR*

In anaerobic membrane bioreactors (AnMBR), a membrane decouples the retention of solids and liquids. Several applications of AnMBRs have been investigated over the last couple of years and reviewed by Liao *et al.* (2006). They conclude that AnMBRs should be well suited for the treatment of highly concentrated particulate waste streams. The application of AnMBRs to blackwater was reported by van Voorthuizen *et al.* (2008).

21.4 ANAEROBIC TREATMENT OF BLACK- AND BROWNWATER

21.4.1 Maximum process temperature

The concentration of black- and brownwater is determined by the type of toilet used, and predominantly by the volume of water per flush. The higher the concentration and biodegradability of the organic compounds, the more CH_4 and thereby energy can be recovered per m^3 of wastewater. However, the net recovery of energy depends on more factors than just the amount of CH_4 produced. The temperature and concentration of the wastewater determines how much energy has to be invested to bring the wastewater to the necessary operating temperature for anaerobic treatment. The heat loss in the reactor must also be considered when attempting to increase the influent temperature. The highest permissible reactor temperature (T_R), whereby all the methane produced is used for heating, can be calculated with Equation 21.3:

$$(1-Y)\cdot f_C \cdot C_{S,\text{in}} \cdot Q \cdot H_S \cdot \eta_{\text{heat}} = \rho \cdot c_p \cdot (T_R - T_W) \cdot Q + A \cdot h \cdot (T_R - T_A) \qquad (21.3)$$

where Y is the sludge produced per organics removed ($kgCOD \cdot kgCOD^{-1}$), f_C the fraction of organics removed per influent organics ($kgCOD \cdot kgCOD^{-1}$), $C_{S,\text{in}}$ the concentration of the total organics in the influent ($kgCOD \cdot m^{-3}$) and Q the inflow ($m^3 \cdot h^{-1}$). Moreover, H_S is the energy content of the organic substrate ($3320\ kcal \cdot kgCOD^{-1}$ assuming that the organic substance is converted completely to methane, that is $0.35\ m^3\ CH_4 \cdot kgCOD^{-1}$), η_{heat} the efficiency of heat generation, ρ the density of wastewater (approx. $1000\ kg \cdot m^{-3}$), c_p the heat capacity of water (approx. $1\ kcal \cdot kg^{-1} \cdot °C^{-1}$), T_R the reactor temperature (°C), T_w the wastewater temperature (°C), T_A the ambient temperature (°C), A the heat exchange area (m^2) and h the total heat transfer coefficient ($kcal \cdot m^{-2} \cdot h^{-1} \cdot °C^{-1}$).

Elmitwalli *et al.* (2011) give an overview of data on the composition and flow ($L \cdot p^{-1} \cdot d^{-1}$) of brown- and blackwater as a function of different toilet types. These data can be used as a source for estimating the maximum permissible reactor temperature.

If the operating temperature is low, a high SRT is required for sufficient anaerobic degradation. Under such conditions, reactors with enhanced biomass retention should be chosen. Otherwise, the required reactor volume will be very large. Besides their large volume, low temperature reactors have further disadvantages. More CH_4, a potent greenhouse gas, will leave the anaerobic reactor dissolved in the effluent and will later escape to the atmosphere. In contrast, high temperatures, that is thermophilic treatment, increase pathogen reduction (Bendixen 1994) and extend the possibilities of safe sludge reuse.

21.4.2 Removal of organic substances and methane recovery for different waste streams

21.4.2.1 Vacuum-collected blackwater

The use of four types of reactors for anaerobic treatment of vacuum-collected blackwater under different conditions is reported in the literature, namely a CSTR, an AC system, a UASB septic tank and a UASB. The results of these various research efforts are summarized in Table 21.1.

Table 21.1 Performance of different anaerobic treatment systems for the treatment of vacuum-collected blackwater (adapted from de Graaff *et al.* 2010).

	[(1)]CSTR	[(2)]AC system	[(2)]UASB septic tank	[(3)]UASB septic tank	[(4)]UASB
Temperature [°C]	37	20	25	25	25
Reactor volume [L]	10	200	200	7,200	50
Av. infl. conc. [$gCOD \cdot L^{-1}$]	8.7	10.0	12.3	[(7)] 16.1	[(8)]7.7–9.8
COD_{tot} removal [%]	61	80	78	87	73
HRT [d]	20	150	29	30	8.7
SRT [d]	20	150	>365	>365	254
Methane [$NL \cdot p^{-1} \cdot d^{-1}$]	[(5)]9.0	10.2	14	13	10
Methane [$Nm^3 \cdot m^{-3}_{Influent}$]	[(5)]1.9	2.1	2.1	2.0	1.8
Methanization [(5)] [%]	60	58	60	57	54
Loading rate [$kgCOD \cdot m^{-3} \cdot d^{-1}$]	0.44	[(6)]0.3	0.42	0.36	1.0

[(1)]Wendland *et al.* (2007); [(2)]Kujawa-Roeleveld *et al.* (2006); [(3)]Meulman *et al.* (2008); [(4)]de Graaff *et al.* (2010); [(5)]calculated based on obtained methane production and influent load; [(6)]calculated based on the total volume after the feeding was stopped; [(7)] with an inflow of 5 $L \cdot p^{-1} \cdot d^{-1}$; [(8)]coarse filtered.

The removal efficiencies in the different anaerobic systems vary between 61 and 87%. The highest removal of 87% was achieved in a pilot UASB septic tank system. This system was operated at a low organic loading rate and hence with a large reactor volume (Meulman *et al.* 2008). The UASB reported by de Graaff *et al.* (2010) was operated for more than 900 d and showed very stable performance. A mass balance over the entire period illustrated a sludge production of 15% of the influent load. In the UASB reactors of de Graaff *et al.* (2010) the HRT could be maintained at 8.7 d at a temperature of 25°C, while the CSTR required a longer HRT (20 d) and a higher temperature (37°C) at a similar methane yield. The mean sludge concentration in the UASB of de Graaff *et al.* was 34 $gCOD{\cdot}L^{-1}$.

The anaerobic treatment of vacuum-collected blackwater in a UASB has been extensively researched and demonstrated and is currently applied at three full-scale locations, two offices and one new housing estate of 250 houses in The Netherlands and in a school in Ukraine (Zeeman 2011). A future (yet to be developed) double-switch vacuum toilet concept (urine is collected water-free and faeces with a minimum of one litre of flush water) could provide a high concentrated input stream for highly efficient anaerobic treatment. Bio-centres comprising the anaerobic treatment of concentrated black waste in connection with toilet blocks have recently been successfully introduced in slum areas in some developing countries (Esipisu 2008). Sufficient data on their reactor performance are still lacking.

21.4.2.2 Conventionally collected blackwater

When blackwater is collected via conventional toilets, its dilution is substantially higher than that collected via a vacuum. It is then recommended to treat it in a reactor with enhanced biomass retention such as a UASB reactor or a UASB septic tank. An AC system is not advisable due to its low loading rates and large reactor volumes.

Under tropical conditions, the anaerobic treatment of diluted streams can be performed at a relative low HRT. In contrast, temperate countries require longer HRTs as the biogas production is too low for sufficient heating of the diluted stream. Van Voorthuizen *et al.* (2008) used three combinations of biological treatment and membrane filtration to diluted blackwater collected from school toilets using approximately 5 L per flush. A UASB followed by effluent membrane filtration, an anaerobic MBR and an aerobic MBR were used for comparison. Anaerobic membrane treatment guaranteed an effluent free of suspended and colloidal matter but with a substantial concentration of soluble organic compounds. The process temperature in the anaerobic system was 37°C, which was too high to be cost-effective. The HRT in the UASB and the anaerobic MBR was 12 h and the average influent concentration amounted to 1139 $mgCOD{\cdot}L^{-1}$. The measured methane production in the UASB and in the anaerobic MBR was only 0.27 and 0.35 gCH_4-COD per gram blackwater COD respectively. However, it is assumed that the actual gas production could have been higher, since the mass balances could not be closed. Removal efficiencies of total COD were high, namely 91 and 86%, respectively.

Luostarinen *et al.* (2007) investigated the performance of a pilot-scale UASB septic tank of 1.2 m^3 which had been in operation for 13 y and could be assumed to be fully adapted to temperate conditions. It was fed with medium-strength blackwater with a mean concentration of 2897 $mgCOD{\cdot}L^{-1}$ and the ambient temperatures were 14–19°C. At an organic loading rate (OLR) of 4.1 $kgCOD{\cdot}m^{-3}{\cdot}d^{-1}$ and an HRT of 4.1 d, this system achieved a mean removal efficiency of 70%.

21.4.2.3 Black waste collected with minimum water

Chaggu (2004) used an accumulation system (IMPLWUS = improved pit latrine without urine separation) for the anaerobic storage and treatment of faeces, urine and small amounts of cleansing water under tropical ambient temperatures (25–30°C). The filling period was 380 d. Even after 380 d, full stabilization of the black waste was not achieved. In the first half year, methanogenesis and hydrolysis were inhibited by ammonia concentrations, although adaptation of the biomass was subsequently observed. Unfortunately no second filling period was implemented.

The results of research with AC systems for animal manure showed that improved reactor performance can be expected during the second operational period, as the biomass is then fully adapted to the new conditions (Zeeman 1991). Van Velsen observed already in 1979 that methanogenic biomass can adapt to ammonia concentrations as high as 5 $gNH_4^+{-}N{\cdot}L^{-1}$ (van Velsen 1979). Vasconcelos Fernandes (2010) showed that hydrolysis is unaffected by high ammonia concentrations if the biomass is well adapted. Elmitwalli *et al.* (2011) concluded from model calculations that highly concentrated brown or black waste, such as dry collected faeces or faeces separated from blackwater in a filter bag, is suited for treatment in an AC system providing that the filling period is longer than 150 d.

21.4.3 Boosting energy production by adding kitchen waste

Adding kitchen waste to blackwater can substantially increase the methane production. The theoretical organic load of blackwater and kitchen waste per person is approximately 122 $gCOD{\cdot}p^{-1}{\cdot}d^{-1}$, considering a production of faeces, urine and kitchen waste of 50, 12 and 60 $gCOD{\cdot}p^{-1}{\cdot}d^{-1}$ respectively (Kujawa-Roeleveld *et al.* 2006). At a typical methanization rate of 60%, the methane production is approximately 28 $L{\cdot}p^{-1}{\cdot}d^{-1}$. However, batch tests indicated that a mixture of blackwater and kitchen refuse showed a higher anaerobic biodegradability (70–80%). Therefore, a fully optimized digestion process may lead to a CH_4 production of approximately 35 $L{\cdot}p^{-1}{\cdot}d^{-1}$ (Kujawa-Roeleveld *et al.* 2006). Wendland *et al.* (2007) report similar values of 32 to 33 $L\ p^{-1}{\cdot}d^{-1}$.

21.4.4 Nutrient recovery and removal

The theoretically produced amounts of nitrogen, phosphorus and potassium from blackwater and kitchen waste are 4.47, 0.58 and 1.41 $kg \cdot p^{-1} \cdot y^{-1}$, respectively (Kujawa-Roeleveld and Zeeman 2006). Elzinga *et al.* (2009) presented results of a demonstration-scale plant treating the vacuum-collected blackwater from 32 houses. The anaerobic treatment of the concentrated blackwater was performed in a UASB septic tank. The effluent from the anaerobic reactor contained 7.6 $gN \cdot p^{-1} \cdot d^{-1}$ and 0.63 $gP \cdot p^{-1} \cdot d^{-1}$, amounting to 62% and 40% of the theoretically produced N and P (Zeeman and Kujawa-Roeleveld 2011). De Graaff *et al.* (2011c) performed a phosphorus mass balance for a UASB reactor running for 957 d with blackwater treatment. They found that 61% of the influent phosphorus ended up in the effluent. They also showed that 90% of dissolved phosphate can be removed from the effluent by struvite precipitation if magnesium is added at a molar ratio of 1.5 mol $Mg \cdot mol\ P^{-1}$ at pH of 8.0 or if the pH is adjusted to 9.0 and magnesium is dosed at a molar ratio of 1.3 mol $Mg \cdot mol\ P^{-1}$ (see also Kabdaşlı *et al.* 2013).

Nitrogen recovery from wastewater has to compete with the relatively energy-efficient ammonia production from atmospheric nitrogen. Both Vlaeminck *et al.* (2009) and de Graaff *et al.* (2011a) considered currently available nitrogen recovery techniques and concluded that at the prevailing ammonia concentrations in vacuum-collected and anaerobically treated blackwater (1 to 1.5 $gN \cdot L^{-1}$), biological removal of nitrogen including an anammox process step is most efficient (see Udert and Jenni 2013).

21.4.5 Removal of pharmaceuticals and hormones

Kujawa-Roeleveld *et al.* (2008) investigated the biodegradation potential of a number of selected pharmaceuticals in batch tests under various redox conditions: acetylsalicylic acid, diclofenac, ibuprofen, carbamazepine, metoprolol, clofibric acid, bezafibrate and fenofibrate. Under anaerobic conditions, only three out of the eight compounds, namely acetylsalicylic acid, ibuprofen and fenofibrate, were removed after a long batch digestion time of 30 days. Under aerobic conditions, the biodegradation rate of the same three compounds was shown to be much higher. De Mes (2007) investigated the fate of estrone (E1), 17ß-estradiol (E2) and 17a-ethynylestradiol (EE2) during the treatment of vacuum-collected blackwater in a UASB septic tank with micro-aerobic post-treatment. The temperature in the tank was 25°C, the HRT 49 d and the SRT 164 d. The concentrations of the natural estrogens E1 and E2 in the effluent were 4.02 $\mu g \cdot L^{-1}$ and 18.69 $\mu g \cdot L^{-1}$. No EE2 was detected. More than 70% of the E1 and 80% of the E2 were in their conjugated form. Conjugated compounds are usually more soluble and inactive forms of the parent compounds formed in the human body. Faeces contain enzymes that can hydrolyze conjugates back to their original and therefore active form (Ternes *et al.* 1999). The post-treatment effluent contained E1 and E2 in concentrations of 1.37 μg·L^{-1} and 0.65 μg·L^{-1} respectively. More than 77% of the measured unconjugated E1 and 82% of the E2 were associated with particles (>1.2 μm) in the final effluent, implying high sorption affinity of both compounds. De Mes (2007) proposes an additional physico-chemical treatment to reduce the emission of estrogens to surface waters, as effluent concentrations were still in the μg·L^{-1} range. The removal of pharmaceutical residues in vacuum-collected blackwater treated in a UASB followed by a nitritation and anammox reactor was reported by de Graaff *et al.* (2011b). Paracetamol, tetracycline and doxycycline were not detected in the anammox effluent. The other selected compounds were still present in the effluent of the anammox reactor at μg·L^{-1} levels. This reactor set-up also showed low removal of micropollutants, so that a chemico-physical post-treatment step is again required.

21.5 ANAEROBIC TREATMENT OF GREYWATER

The concentration of organic substances in greywater is low compared to blackwater, but the organic load is considerable due to the large volume. Different greywater treatment systems were recently reviewed by Abu Ghunmi *et al.* (2011). For minimum energy consumption, the review suggests a sequence of anaerobic and aerobic processes for greywater treatment with additional disinfection. A UASB or anaerobic filter is proposed for anaerobic treatment (Abu Ghunmi *et al.* 2010). The results of a few researches on the anaerobic treatment of greywater have been extensively reported (Elmitwalli and Otterpohl 2007, Elmitwalli and Otterpohl 2011, Abu Ghunmi *et al.* 2010 and Hernandez Leal *et al.* 2010a). Elmitwalli and Otterpohl (2007) reported COD removal of 52% at 30°C at an HRT of 6 and 10 h, and of 64% at 16 hours during UASB treatment of greywater. The SRT were 22 to 338 d, 64 to 377 d and 93 to 481d respectively. Between 38 and 51% of the organic substances were converted to CH_4. Hernandez Leal *et al.* (2010a) reported similar results. In their experiments, the organics removal was 51% at an HRT of 12 h, a temperature of 32°C and an SRT of 392 d. At a shorter HRT of 7 h, only 39% of the organics were removed (SRT = 97 d). Between 25 and 32% of the organics were converted to methane. Elmitwalli and Otterpohl (2011b) report greywater treatment in a UASB reactor at low ambient temperatures (14–25°C). The organics removal was low (31 to 41%) at an HRT and SRT range of 8–20 h and 330–432 d respectively. The maximum SRT is calculated the suspended solids in the effluent were not considered as biomass. In all three studies, the organics removal was low, which could be attributed to the limited removal of colloidal organic substances (Elmitwalli *et al.* 2000).

In a recent publication, Hernandez Leal *et al.* (2010b) show how a large fraction of the organics in greywater can be concentrated by applying bio-flocculation in a high-loaded MBR. The sludge produced has high biodegradability and adding it to a blackwater UASB reactor allows energy to be recovered from the greywater organics without installing a second UASB reactor. Hernandez Leal *et al.* (2010b) reported that the integration of bio-flocculation of greywater into decentralized sanitation concepts may increase the overall production of methane by 50%. Further research will show whether this co-digestion impedes the reuse of the sludge by adding micropollutants from personal care products. More studies are also needed to determine whether the addition of bio-flocculation sludge will affect the sludge retention in the UASB reactor over the long term.

21.6 CONCLUSIONS

The choice of reactor type for the anaerobic treatment of source-separated domestic waste streams depends mainly on the waste stream concentration. The concentrations of black- and brownwater are determined by the type of toilet.

The combined organics concentration and the wastewater temperature determine the maximum permissible temperature of the reactor content. Low-temperature reactors may be used, but they require large volumes, tend to lose more methane and have a low pathogen inactivation.

Pharmaceuticals and hormones are insufficiently removed during anaerobic treatment. Future research will focus on the development of effective post-treatment methods.

Anaerobic treatment of vacuum-collected blackwater in a UASB at a temperature of 25–35°C is a proven technology and is run at full-scale. At minimal dilution, this treatment becomes very similar to the anaerobic treatment of animal manure used worldwide on both small and large scales.

Anaerobic treatment of greywater in a UASB is possible, but the removal of organic substances and biogas production are low.

Bio-flocculation of greywater coupled with anaerobic treatment of the sludge in the blackwater reactor might be an attractive option for increased biogas production.

REFERENCES

Abu Ghunmi L., Zeeman G., Fayyad M. and van Lier J. B. (2010). Grey water treatment in a series anaerobic – aerobic system for irrigation. *Bioresource Technology*, **101**(1), 41–50.

Abu Ghunmi L., Zeeman G., Fayyad M. and van Lier J. B. (2011). Grey water treatment systems: A review. *Critical Reviews in Environmental Science and Technology*, **41**(7), 657–698.

Bendixen H. J. (1994). Safeguards against pathogens in Danish biogas plants. *Water Science and Technology*, **30**(12), 171–180.

Chaggu E. J. (2004). *Sustainable environmental protection using modified pit-latrines*. PhD thesis, Environmental Technology, University of Wageningen, The Netherlands.

Chen Y., Yang G., Sweeney S. and Feng Y. (2010). Household biogas use in rural China: A study of opportunities and constraints. *Renewable and Sustainable Energy Reviews*, **14**(1), 545–549.

de Graaff M. S., Temmink H., Zeeman G. and Buisman C. J. N. (2010). Anaerobic Treatment of Concentrated Black Water in a UASB Reactor at a Short HRT. *Water*, **2**(1), 101–119.

de Graaff M. S., Temmink H., Zeeman G., van Loosdrecht M. C. M. and Buisman C. J. N. (2011a). Autotrophic nitrogen removal from black water: Calcium addition as a requirement for settleability . *Water Research*, **45**(1), 63–74.

de Graaff M. S., Vieno N. M., Kujawa-Roeleveld K., Zeeman G., Temmink H. and Buisman C. J. N. (2011b). Fate of hormones and pharmaceuticals during combined anaerobic treatment and nitrogen removal by nitritation-anammox in vacuum collected black water. *Water Research*, **45**(1), 375–383.

de Graaff M. S., Temmink H., Zeeman G. and Buisman C. J. N. (2011c). Energy and phosphorus recovery from black water. *Water Science and Technology*, **63**(11), 2759–2765.

de Mes T. Z. D., Stams A. J. M. and Zeeman G. (2003). Methane production by anaerobic digestion of wastewater and solid wastes. In: *Biomethane and Biohydrogen. Status and Perspectives of Biological Methane and Hydrogen Production*. J. H. Reith, R. H. Wijffels and H. Barten (eds.), Dutch Biological Hydrogen Foundation, Petten, The Netherlands. pp. 58–94.

de Mes T. Z. D. (2007). *Fate of estrogens in biological treatment of concentrated black water.* PhD thesis, Environmental Technology, University of Wageningen, The Netherlands.

Elmitwalli T. and Otterpohl R. (2007). Anaerobic biodegradability and treatment of grey water in upflow anaerobic sludge blanket (UASB) reactor. *Water Research*, **41**(6), 1379–1387.

Elmitwalli T. and Otterpohl R. (2011). Grey water anaerobic treatment in upflow anaerobic sludge blanket (UASB) reactor at different temperatures. *Water Science and Technology*, **64**(3), 610–619.

Elmitwalli T. A., van Dun M., Bruning H., Zeeman G. and Lettinga G. (2000). The role of filter media in removing suspended and colloidal particles in an anaerobic reactor treating domestic sewage. *Bioresource Technology*, **72**(3), 235–242.

Elmitwalli T., Zeeman G. and Otterpohl R. (2011). Modelling anaerobic digestion of concentrated black water and faecal matter in accumulation system. *Water Science and Technology*, **63**(9), 2039–2045.

Esipisu E. (2008). *Human poop on firing line*. Daily Nation, HORIZON II, March 6, 2008. Nairobi, Kenya.

Franklin R. J. (2001). Full scale experience with anaerobic treatment of industrial wastewater. *Water Science and Technology*, **44**(8), 1–6.

Hernandez Leal L., Temmink H., Zeeman G. and Buisman C. J. N. (2010a). Comparison of three systems for biological grey water treatment. *Water*, **2**(2), 155–169.

Hernandez Leal L., Temmink H., Zeeman G. and Buisman C. J. N. (2010b). Bioflocculation of grey water for improved energy recovery within decentralized sanitation concepts. *Bioresource Technology*, **101**(23), 9065–9070.

ISIS (2006). *Biogas China*. Report 02/10/06. Institute for Science in Society (ISIS), http://www.i-sis.org.uk/BiogasChina.php (accessed on 06 April 2012).

Kabdaşlı I., Tünay O. and Udert K. M. (2013). Transfer into the solid phase. In: *Source Separation and Decentralization for Wastewater Management*, T. A. Larsen, K. M. Udert and J. Lienert (eds.), IWA Publishing, London, UK, pp. 351–366.

Kujawa-Roeleveld K. and Zeeman G. (2006). Anaerobic treatment in decentralized and source-separation-based sanitation concepts. *Reviews in Environmental Science and Biotechnology*, **5**(1), 115–139.

Kujawa-Roeleveld K., Elmitwalli T. and Zeeman G. (2006). Enhanced primary treatment of concentrated black water and kitchen residues within DESAR concept using two types of anaerobic digesters. *Water Science and Technology*, **53**(9), 159–168.

Kujawa-Roeleveld K., Schuman E., Grotenhuis T., Kragic D., Mels A. and Zeeman G. (2008). Biodegradability of human pharmaceutically active compounds (PhAC) in biological systems treating source separated wastewater streams. Conference proceedings, "Sanitation Challenge", Wageningen, The Netherlands, 19–21 May 2008.

Lettinga G., van Velsen A. F. M., Hobma S. W., de Zeeuw W. J. and Klapwijk A. (1980). Use of the upflow sludge blanket (USB) reactor. *Biotechnology Bioengineering*, **22**, 699–734.

Liao B. -Q., Kraemer J. T. and Bagley D. M. (2006). Anaerobic membrane bioreactors: Applications and research directions. *Critical Reviews in Environmental Science and Technology*, **36**(6), 489–530.

Luostarinen S., Sanders W., Kujawa-Roeleveld K. and Zeeman G., (2007). Effect of temperature on anaerobic treatment of black water in UASB-septic tank systems. *Bioresource Technology*, **98**(5), 980–986.

Sanders W. T. M. (2001). *Anaerobic hydrolysis during digestion of complex substrates*. PhD thesis, Environmental Technology, University of Wageningen, The Netherlands.

Ternes T. A., Kreckel P. and Mueller J. (1999). Behavior and occurrence of estrogens in municipal sewage treatment plants–II. Aerobic batch experiments with activated sludge. *The Science of the Total Environment*, **225**, 91–99.

Udert K. M and Jenni S. (2013). Biological nitrogen conversion processes. In: *Source Separation and Decentralization for Wastewater Management*, T. A. Larsen, K. M. Udert and J. Lienert (eds.), IWA Publishing, London, UK, pp. 291–305.

Vasconcelos Fernandes T. (2010). *Hydrolysis inhibition of complex biowaste*. PhD thesis, Environmental Technology, University of Wageningen, The Netherlands.

van Velsen (1979). Adaptation of methanogenic sludge to high ammonia concentrations. *Water Research*, **13**(10), 995–999.

van Voorthuizen E., Zwijnenburg A., van der Meer W. and Temmink H. (2008). Biological black water treatment combined with membrane separation. *Water Research*, **42**(16), 4334–4340.

Vlaeminck S. E., Terada A., Smets B. F., van der Linden D., Boon N., Verstraete W. and Carballa M. (2009). Nitrogen removal from digested black water by one-stage partial nitritation and anammox. *Environmental Science and Technology*, **43**(13), 5035–5041.

Wendland C., Deegener S., Behrendt J., Toshev P. and Otterpohl R. (2007). Anaerobic digestion of blackwater from vacuum toilets and kitchen refuse in a continuous stirred tank reactor (CSTR). *Water Science and Technology*, **55**(7), 187–194.

Zeeman G. (1991). *Mesophilic and psychrophilic digestion of liquid manure*. PhD thesis, Environmental Technology, Wageningen University, The Netherlands.

Zeeman G. (2011) Interview in: Full resource recovery in new approaches to sanitation. Edit. Bill McCann, Water **21**, 45–46.

Zeeman G. and Lettinga G. (1999). The role of anaerobic digestion of domestic sewage in closing the water and nutrient cycle at community level. *Water Science and Technology*, **39**(5), 187–194.

Zeeman G. and Kujawa-Roeleveld K. (2011). Resource recovery from source separated domestic waste(water) streams; full scale results. *Water Science and Technology*, **64**(10), 1987–1992.

Chapter 22

Electrochemical systems

Kai M. Udert, Shelley Brown-Malker and Jürg Keller

22.1 INTRODUCTION

Electrochemical processes offer many possibilities for decentralized wastewater treatment. Besides the direct oxidation and reduction of major pollutants they can also be used for coagulation and precipitation processes, disinfection or targeted micropollutant removal. While those processes require energy, wastewater can also be used as a direct source for electricity, either in purely chemical fuel cells or in bioelectrochemical systems. Several processes have been tested for source-separated waste streams, especially for urine. However, no pilot plants or long-term experiences have been reported so far. In this chapter, we present basic concepts of electrochemical treatment, possible applications for source-separated waste streams and we summarize and discuss existing data especially with respect to energy.

22.2 ELECTROCHEMICAL OXIDATION AND REDUCTION PROCESSES

In nearly all biological systems energy is gained by withdrawing electrons from a molecule with low reduction potential (e.g., acetate) and transferring them to a molecule with high reduction potential (e.g., oxygen). In biological processes, the two half reactions oxidation and reduction occur on a very small spatial scale, inside of a biological cell. In electrochemical processes the two half-reactions are clearly separated. A chemical substance (electron donor) is oxidized at the anode, the electrons flow via an electric connection to the cathode, where they are used to reduce another chemical compound (electron acceptor). Anode and cathode are typically located in two separate chambers. To ensure the charge balance, ions have to be exchanged between the two chambers, for example by a salt bridge or an ion exchange membrane.

A basic requirement for electrochemical processes is that the two electrodes have different potentials. The potential difference (voltage) can be due to the different potentials of the chemicals in the anode and in the cathode chamber. This is the case in a fuel cell: the anode catalyzes the oxidation of a reduced compound (e.g., hydrogen), while the cathode transfers the electron to an oxidized compound (e.g., oxygen). While fuel cells produce electricity, electrolytic reactors require electric energy. In electrolysis, the potential difference is established by an external power source.

The spatial separation of oxidation and reduction and the use of electric current offer unique opportunities for water treatment: no or only small amounts of chemicals have to be added; the reaction can be easily started, regulated and terminated via power control; a wide range of compounds can be oxidized or reduced and last but not least, the electric parameters current and voltage can be easily measured and used for process automation and control.

22.3 OXIDATION AND REDUCTION OF POLLUTANTS

This section will focus on pure electrochemical oxidation and reduction, while bioelectrochemical reactions will be covered in detail in Section 22.5 below.

22.3.1 Suitable anode materials

The stability requirements for anodes are typically high in electrochemical processes, while cathodes are protected by the electron supply against corrosion. Optimal anodes not only have to withstand corrosion, they also have to be selective for the target compounds (Comninellis and Chen 2010). For a long time, platinum was the most commonly used material for anodes. However, the high material costs prohibited the application of electrolysis for municipal wastewater treatment (Marinčić and Leitz 1978). In the last decades so-called dimensionally stable anodes (DSA) were developed (Hinden *et al.* 1982). DSA are characterized by a thin active coating deposited on a base metal, such as titanium. The coatings consist of noble metals and their oxides, for example IrO_2, RuO_2 or PtOx. Another new type of electrodes with very high chemical inertness and structural stability are boron-doped diamond (BDD) electrodes (Comninellis and Chen 2010). Recently, nickel and nickel hydroxide electrodes have been proposed for ammonia and urea oxidation (Kapałka *et al.* 2010a, Boggs *et al.* 2009). Nickel is less stable than DSA or BDD, but its structural stability and the electrochemical properties can be improved, for example by substituting some nickel with cobalt (Yan *et al.* 2012).

22.3.2 Electrochemical ammonia oxidation

The most common electrochemical process for ammonia removal is the indirect oxidation via chlorine. Chlorine is a strong oxidant, which is formed in chloride containing solutions at high voltages. Chlorine is in equilibrium with hypochlorous acid, which oxidizes ammonia to molecular nitrogen, a process also known as breakpoint chlorination (Kapałka *et al.* 2010b). Indirect oxidation via chlorine is a fast process, but it is very unspecific and can produce unwanted by-products such as halogenated organic compounds.

Ammonia can also be directly oxidized on various electrodes. The reaction is well understood for platinum and some other noble metals and their alloys (Bunce and Bejan 2011). Free ammonia (NH_3) is the actual reactant, therefore the kinetics depend on the pH of the solution. At low anode potentials, nearly all ammonia is oxidized to N_2, while at higher anode potentials considerable amounts of nitrate and some nitrite can be produced. High anode potentials also lead to electrode poisoning by nitrogen adsorption. Direct ammonia oxidation was also observed at novel electrodes such as BDD (Kapałka *et al.* 2010c), $Ni/Ni(OH)_2$ (Kapałka *et al.* 2010a) or DSA such as IrO_2 (Kapałka *et al.* 2009) or Pt-Ir (Boggs and Botte 2009).

22.3.3 Electrochemical urea degradation

Electrochemical urea degradation has been studied to a lesser extent than ammonia oxidation. Simka *et al.* (2007) reported that N_2 and CO_2 are the main products of indirect urea oxidation with nitrate as by-product. Indirect oxidation of urea in fresh urine was also reported by Ikematsu *et al.* (2006). They used two Pt-Ir DSA as anode and iron as cathode at a current density of 40 mA cm^{-2} and diluted the urine samples at least two times with 200 $mol{\cdot}L^{-1}$ NaCl. 95% of the total nitrogen was removed, the residual nitrogen was nitrate.

Recently, urea has received some attention as a possible hydrogen source, because the theoretical cell potential for hydrogen production from urea is only 0.37 V compared to 1.23 V for water electrolysis. Boggs *et al.* (2009) reported that urea was degraded at nickel electrodes, when fresh urine was dosed to a 1 $mol{\cdot}L^{-1}$ KOH solution. For these preliminary tests, they used cyclic voltammetry, a fast method to identify electrochemical reactions.

Lan *et al.* (2010) investigated direct electricity production from urea in alkaline anion-exchange membrane fuel cells. With power curve measurements in fresh urine, they achieved power densities as high as 1.1 $mW{\cdot}cm^{-2}$ at 20°C and 4.2 $mW{\cdot}cm^{-2}$ at 60°C. The anode consisted of nano-sized nickel and the cathode was made of MnO_2 and carbon. The cathode chamber was filled with humidified air (Lan and Tao 2011).

22.3.4 Influence of urea hydrolysis

So far, electrochemical urine treatment has been mainly studied with fresh urine. However, in urine-collecting systems urea is rapidly hydrolyzed by bacterial urease to ammonia and carbon dioxide (Udert *et al.* 2003). This spontaneous process changes the urine composition significantly. Fresh urine has an average pH of 6.2 and contains about 7.7 $gN{\cdot}L^{-1}$ urea and 0.48 $gN{\cdot}L^{-1}$ ammonia. In stored urine all urea is degraded to ammonia, the pH value is 9.1, the total carbonate concentration increased from nearly zero to 3.2 $gC{\cdot}L^{-1}$ and the alkalinity is approximately 0.5 $mol{\cdot}L^{-1}$ (Udert *et al.* 2006). The above values are based on medical literature. Real stored urine can have lower concentrations due to dilution and ammonia volatilization (Siegrist *et al.* 2013).

Amstutz *et al.* (2012) performed batch experiments to compare the nitrogen removal from synthetic solutions of fresh and stored urine. They used two zirconium cathodes and one Ti/IrO_2 anode and applied a constant current density of 15 ± 0.3 $mA{\cdot}cm^{-2}$ with the aim to produce chlorine for indirect oxidation. In the experiment with fresh urine 72% of the initial nitrogen was removed and 24% was converted to nitrate. In the experiments with stored urine, hardly any ammonia was oxidized. Two reasons were proposed: (i) competition between chloride and carbonate oxidation at the anode and (ii) removal of active chlorine by chlorate formation.

22.3.5 Removal of organic pollutants and pathogens

Indirect electrochemical oxidation can also be used to remove organic compounds, colour and microorganisms (Anglada *et al.* 2009). However, when using high anode potentials for the production of chlorine unwanted by-products such as halogenated organic compounds, chlorate and perchlorate can be formed (Kapałka *et al.* 2010c).

Direct electrochemical oxidation is a possible post-treatment for the removal of pharmaceutical residues, pesticides and disinfection by-products. Many micropollutants can be removed by direct oxidation at the anode (Radjenovic *et al.* 2011), while reduction at the cathode is a possible treatment for halogenated organic compounds such as trihalomethanes (Radjenovic *et al.* 2012).

22.4 ELECTROCHEMICAL DISSOLUTION OF METALS

22.4.1 Electrocoagulation

Some metal anodes are deliberately oxidized and dissolved, when applying an appropriately high anode potential. This effect can be used to dose metal ions. Common so-called sacrificial electrodes are iron and aluminium (Chen 2004). This process is often called electrocoagulation, because the dissolved cations can facilitate the aggregation of negatively charged colloids and particles. However, electrochemical metal dissolution is also used for other purposes: the dissolved cations can react to (hydr)oxides, which adsorb pollutants such as arsenic (Kumar *et al.* 2004) or they form precipitates with the target compound, for example iron phosphates (Zheng *et al.* 2009). The electrocoagulation process is used for water treatment in decentralized reactors (Holt 2005).

22.4.2 Electrochemical precipitation of phosphate from urine

Iron, aluminium and magnesium have been used for electrochemical phosphate precipitation from urine. The first study was reported by Ikematsu *et al.* (2006), who tested a combination of electrolysis and electrochemical precipitation for the treatment of fresh urine. The reactor consisted of two DSA and one iron electrode. By changing the current direction, the reactor was switched from nitrogen oxidation on DSA to iron dissolution for iron phosphate precipitation. Zheng *et al.* (2009, 2010) used synthetic and real fresh urine for their experiments with iron and aluminium electrodes. With both types of electrodes, complete phosphate removal was possible. At a current density of 40 $mA \cdot cm^{-2}$, 2.4 $gFe \cdot gP^{-1}$ (1.3 mol $Fe \cdot mol\ P^{-1}$) had to be dosed to remove 98% of the phosphate. This value was calculated based on the assumption that all of the current was used for the release of Fe (i.e., 100% current efficiency). The energy demand increased with the current density from 3.4 $Wh \cdot gP^{-1}$ at 10 $mA \cdot cm^{-2}$ to 8.7 $Wh \cdot gP^{-1}$ at 40 $mA \cdot cm^{-2}$ (electrode gap 5 mm). Another energy relevant factor was the gap between the electrodes: increasing the gap from 5 mm to 40 mm raised the energy demand by a factor of three. In general, the energy demand was slightly higher for aluminium electrodes (4.6 $Wh \cdot gP^{-1}$ at 10 $mA \cdot cm^{-2}$ and 5 mm electrode gap) than for iron electrodes.

Hug and Udert (2012) used electrochemical dissolution of metallic magnesium to precipitate struvite ($MgNH_4PO_4 \cdot 6H_2O$), a slow-release phosphate fertilizer, from stored urine. A minimum anode potential of −0.9 V vs. NHE (normal hydrogen electrode) was necessary to overcome passivation by hydroxide films. The current efficiency was always above 100% probably due to microgalvanic corrosion or the release of Mg^+ instead of Mg^{2+}. For example at −0.6 V vs. NHE the current efficiency was 120% and the current density 6.6 $mA \cdot cm^{-2}$. Current densities above 100% are due to autocatalytic dissolution of magnesium. The energy demand at this anode potential was 1.7 $Wh \cdot gP^{-1}$ (55 mm electrode gap). At a dosage of 1.0 mol $Mg \cdot mol\ P^{-1}$ 90% of the phosphate was precipitated. Phosphate recovery was limited by crystallization kinetics and higher phosphate recovery (up to 96%) could be achieved by storing the sample for some hours.

22.5 BIOELECTROCHEMICAL PROCESSES

22.5.1 Electroactive bacteria

The discovery that some bacteria can transport electrons across the cell wall onto solid electron acceptors has led to the development of bioelectrochemical systems (BES) for wastewater treatment. A wide range of electroactive bacteria are known (e.g., *Geobacter sulfurreducens, Shewanella oneidensis* or *Pseudomonas aeruginosa*), which grow on conductive surfaces such as graphite or carbon fibres, and can mediate anodic (oxidative) or cathodic (reductive) electrochemical processes. The biomass yield is low (Logan 2008), which allows the recovery of much of the chemical energy contained in the organic substances as electric current. In most BES the electroactive bacteria grow on the anode. Here, the bacteria oxidize the organics and transfer the electrons to the anode through either (i) direct contact via membrane-bound cytochromes or nanowires (El-Naggar *et al.* 2010), or (ii) endogenous soluble mediators (Logan 2008). There is an

(−100 mV vs. NHE). Butyric acid was only partly removed (39%) under these conditions, but largely oxidized at higher potentials. At the maximum potential (+400 mV vs. NHE), excellent VFA removal (up to 90%) was achieved at a removal rate of 2.1 $gCOD{\cdot}L^{-1}{\cdot}d^{-1}$ and a volumetric current density of 150 $A{\cdot}m^{-3}$. The coulombic efficiency (42–54%) was similar at all anode potentials.

When feeding the actual fermenter effluent, it was mixed with centrifuged anaerobic digester effluent at different ratios to provide sufficient buffer capacity. Influents with a 50:50 and 75:25 fermenter effluent to anaerobic digester effluent ratio gave even better volumetric current densities (260 $A{\cdot}m^{-3}$ for both influents) and coulombic efficiencies (75% and 103%, respectively) than the synthetic feed. The VFA removal rate was similar to the results with synthetic influent. The performance of the system was very stable over a period of several months. Nevertheless, the need to have a source of alkalinity could limit the application of this technology for brown- or blackwater.

22.7 BIOELECTROCHEMICAL URINE TREATMENT

22.7.1 Influence of urine composition

Source-separated urine is interesting for BES due to the high concentration of dissolved organic compounds (up to 10 g $COD{\cdot}L^{-1}$, Udert *et al.* 2006) and the high electric conductivity (around 40 $mS{\cdot}cm^{-1}$, Ronteltap *et al.* 2010). Removal of organic substances in an upstream BES can be an option for the optimization of nitrogen conversion processes such as nitritation/anammox or nitrification (Udert and Jenni 2013). However, it has to be kept in mind that fresh urine is an unstable solution (see Section 22.3.4). Since urease-producing bacteria will likely grow inside a BES, it can be expected that urea will also be degraded in a BES fed with fresh urine. As a consequence, phosphate minerals will precipitate (Udert *et al.* 2003) and will clog the narrow flow channels.

Stored urine is a more suitable influent for BES than fresh urine. Due to urea hydrolysis, the phosphate minerals have already precipitated, and the pH and the alkalinity are high (Udert *et al.* 2006). Furthermore, the COD is better available for electroactive bacteria: in fresh urine, long-chain organic acids, creatinine, amino acids and carbohydrates are the main organic compounds (based on COD, Udert *et al.* 2006). In stored urine most of these compounds are already broken down by fermentation. In a sample taken from the men's urine storage tank at Eawag 47% of the dissolved COD was acetate, 4% propionate and 6% butyrate (total COD 5.6 $g{\cdot}L^{-1}$, 7% particulate, unpublished data). Since acetate is the main substrate of electroactive bacteria, it can be expected that at least half of the COD in stored urine can be degraded in a BES. In fact, Marti (2010) reported a maximum COD removal of 50% in a continuously fed MEC (undiluted stored urine) and Zang *et al.* (2012) found 62.4% COD removal in an MFC operated in batch mode (five times diluted stored urine).

22.7.2 Ammonium exchange

In BES with cation exchange membranes, the charge balance is established by cation migration. The most concentrated cation in stored urine is NH_4^+ (0.39 mol·L^{-1}), followed by Na^+ (0.11 mol·L^{-1}) and K^+ (0.06 mol·L^{-1}, all values for undiluted urine, Udert *et al.* 2006). Therefore, a substantial part of the NH_4^+ will migrate from the anode chamber to the cathode chamber. Kuntke *et al.* (2012) utilized this effect in an MFC with a cation-exchange membrane and air-cathode to recover ammonia from urine. In their set-up, the ammonia was transferred from the anode chamber into the air stream at the cathode and then absorbed in an acid trap. NH_4^+ migration accounted for about 30% of the positive charge flow. However, a large fraction of the total ammonia transport did not depend on the electric current: diffusion of uncharged NH_3 contributed at least 42% to the total ammonia transport and was the main ammonia transport mechanism at low current densities. The authors estimated that 11.4% of the total ammonia could be removed with their MFC set-up. However, this removal is low compared to the ammonia losses that can occur in urine-collecting systems due to NH_3 volatilization (close to 50%, Siegrist *et al.* 2013).

22.7.3 Inhibition by ammonia

Fast increases of the ammonia concentration in synthetic ammonium acetate solutions (Clauwaert *et al.* 2008) or synthetic stored urine (Zöllig 2008) inhibit electroactive bacteria. However, the bacteria can be accustomed to high ammonia levels, if the ammonia increase is slow. Winkler (2009) started an MFC with a phosphate buffer (pH 7) containing high acetate concentrations (10 gCOD·L^{-1}) and low ammonia concentrations (0.87 gN·L^{-1}). After reaching steady state, the ammonia concentration in the influent was increased to values typical for stored urine (8.1 gN·L^{-1}), though the pH was kept at 7. In the next phase, the influent pH was increased by 0.5 pH units per day to a maximum value of 9.0. The bacteria were not inhibited and the final current density (0.15 A m^{-2}) was even three times higher than at pH 7. The observed lack of inhibition is in accordance with the study of Kuntke *et al.* (2012), who did not observe any ammonia inhibition in their MFC with stored urine (pH 8.85).

22.7.4 Sulfate removal

Marti (2010) observed sulfate removal in the anode chamber of an MEC running on stored urine. At an anode potential of 0.06 V vs. NHE, the sulfate removal was 1.48 gS·m^{-2}·d^{-1}. Most of the sulfate was reduced to sulfide, but 0.35 gS · m^{-2}·d^{-1} was completely removed. The most likely reason for the sulfate removal was electrochemical oxidation of sulfide to solid sulfur at the anode surface (Dutta *et al.* 2008). Marti calculated that 11.2% of the current could be attributed to sulfide oxidation. It should be mentioned that stored urine already contains a substantial amount of sulfide: in Marti's experiments the stored urine contained 39 mgS·L^{-1} sulfide, 239 mgS·L^{-1} sulfate and 4800 mg·L^{-1} COD. The experimental results showed that BES are a possible technology for the removal of sulfide and sulfate from stored urine. However, the results also demonstrated that sulfate reducers compete with electroactive bacteria in urine-fed BES.

22.7.5 Micropollutant removal

De Gusseme *et al.* (2012) presented a possible application of biocathodes for the removal of micropollutants. They used a biocathode containing *Shewanella oneidenis* and Pd catalysts to treat hospital wastewater and fresh urine containing the iodinated X-ray contrast compound diatrizoate. *Shewanella oneidenis* mediates the dehalogenation of diatrizoate with Pd as catalyst and hydrogen as electron donor. The removal of diatrizoate was less efficient in urine (54%) than in hospital wastewater (85%), which might be due to interferences with sulfur compounds. Sulfide is known to cause poisoning of the Pd catalyst. Direct dehalogenation of the same compound by biocatalyzed cathodic reduction was reported by Mu *et al.* (2011), with promising process rates and efficiencies for such persistent micropollutants as diatrizoate.

22.8 ENERGY CONVERSION AND DEGRADATION RATES

The estimated energy conversion values and degradation rates in Table 22.1 and 22.2 can be used to discuss possible applications of electrochemical systems for on-site reactors. The data on power production (Table 22.1) show that energy recovery from urine or brownwater cannot contribute significantly to the overall energy consumption of a society (4690 W·p^{-1} in Switzerland in 2009, BFE 2010). However, the energy produced in fuel cells may be sufficient for process monitoring and it can supply some energy for reactor operation.

The COD removal rates of MFC and MEC are comparable to efficient conventional biofilm systems (Morgenroth 2008). Possible advantages of BES are better process control and energy efficiency, but BES are more complex than simple biofilm reactors and are yet to be demonstrated in practical applications.

Electrochemical dissolution of magnesium for struvite production requires little energy input and the removal rates are high. According to Table 22.2 45 mW would be sufficient for the removal of one person's daily phosphate load in urine. The limiting factor for this process could be the price of metallic magnesium. Hug and Udert (2012) estimated that the material costs for metallic magnesium is similar to $MgCl_2$ and $MgSO_4$ but substantially higher than for MgO.

The rates of electrochemical nitrogen removal are much higher than for biological systems (Morgenroth 2008). However, the energy demand is also high, especially for indirect oxidation and the electrodes for direct oxidation can be expensive: Boggs and Botte (2009) used an anode consisting of carbon fibre paper, titanium foil and deposited platinum and iridium. Further research should focus on cheaper anode materials, even if the degradation rates are lower.

Table 22.1 Estimated energy production and degradation rates of fuel cell processes.

Process	Tarpget	Assumptions			Calculations	
		Coulombic efficiency [%]	Current density [A·m^{-2}]	Voltage [V]	Energy prod. [W·p^{-1}]	Degradation rate [g·m^{-2}·d^{-1}]
Fuel cell [(1)]	Urea as N (fresh urine)	90	33	0.33	0.68	140
MFC[(2)]	VFA as COD (stored urine)	10	0.50	0.50	0.044	36
MFC	VFA as COD (brownwater)	[(3)]50	[(2)]0.50	[(2)]0.50	0.54	7.2

Assumptions: [(1)] Lan and Tao (2011), Figure 22.5b, nano-sized nickel electrode, arbitrary coulombic efficiency; [(2)] Kuntke *et al.* (2012), Figure 22.4, air-cathode; [(3)] this work. Loads (based on Udert *et al.* 2006 and Friedler *et al.* 2013): urea in fresh urine: 9.6 gN·p^{-1}·d^{-1}, VFA in stored urine: 6.3 gCOD·p^{-1}·d^{-1} (50% of organic load. this work), total ammonia in stored urine: 10 gN·p^{-1}·d^{-1}, phosphate in stored urine: 0.68 gP·p^{-1}·d^{-1}, VFA in brownwater after fermentation: 16 gCOD·p^{-1}·d^{-1} (50% of organic load, this work).

Table 22.2 Estimated energy demand and degradation rates of electrolysis and electrochemical dissolution processes.

Process	Target	Assumptions			Calculations	
		Current efficiency [%]	Current density [A·m^{-2}]	Voltage [V]	Energy demand [W·p^{-1}]	Degradation rate [g·m^{-2}·d^{-1}]
Direct electrolysis [(1)]	Ammonia as N (stored urine)	100	190	0.52	1.3	790
Indirect electrolysis[(2)]	Urea as N (fresh urine)	N_2 38 NO_3^- 32	150	8.6	28	360
Precipitation[(3)]	Phosphate as P (stored urine)	120	6.6	1.1	0.045	110
MEC[(4)]	VFA as COD (stored urine)	280	0.84	0.80	0.25	17
MEC[(4)]	VFA as COD (brownwater)	280	0.84	0.80	0.6	17

Assumptions from: [(1)] Boggs and Botte (2009) Table 1 first case, Pt-Ir anode; [(2)] Amstutz *et al.* (2012), IrO_2 anode, voltage unpublished; [(3)] Hug and Udert (2012) −0.6 V anode potential, Mg:P dosage: 1.0 mol·mol^{-1}; [(4)] Marti (2008) stored urine. Loads, see Table 22.1.

22.9 CONCLUSIONS

Electrochemical systems are potentially very interesting for small reactors. Using electricity for oxidation and reduction processes allows for simple process control and can replace aeration or dosage of chemicals. In fuel cells, energy can be recovered from organic substances, ammonia and urea, but the amount of recoverable energy is small. Electrolytic reactors require energy, which can be substantial, if the electric conductivity of the solution is low and high overpotentials are needed. Nevertheless, simple reactor operation makes electrochemical systems an interesting option for nitrogen removal and phosphorus recovery (through precipitation). Furthermore, electrochemical processes can be a suitable post-treatment step for the removal of micropollutants and pathogens. Critical aspects of high-voltage electrochemical processes are the production of halogenated by-products, chlorate and perchlorate and the costs of electrodes for specific processes, such as ammonia oxidation. To make bio-electrochemical processes efficient and cost-effective, better design and operational strategies (e.g., pH buffering, avoidance of precipitation/clogging) and improved scale-up, reactor materials and concepts will be required. Future studies must also consider how the degradation processes in urine and brownwater influence the performance of electrochemical processes.

REFERENCES

Ahn Y. and Logan B. E. (2010). Effectiveness of domestic wastewater treatment using microbial fuel cells at ambient and mesophilic temperatures. *Bioresour. Technol.*, **101**, 469–475.

Amstutz V., Katsaounis A., Kapałka A., Comninellis C. and Udert K. M. (2012). Effects of carbonate on the electrolytic removal of ammonia and urea from urine with thermally repared IrO_2 electrodes. *J. Appl. Electrochem*, **42**(9), 787–795.

Anglada A., Urtiaga A. and Ortiz I. (2009). Contributions of electrochemical oxidation to waste-water treatment: Fundamentals and review of applications. *J. Chem. Technol. Biot.*, **84**(12), 1747–1755.

Barr K. G., Solley D. O., Starrenburg D. J. and Lewis R. G. (2008). Evaluation, selection and initial performance of a large scale centralised biosolids facility at Oxley Creek water reclamation plant, Brisbane. *Wat. Sci. Technol.*, **57**(10), 1579–1586.

Boggs B. K., King R. L. and Botte G. G. (2009). Urea electrolysis: Direct hydrogen production from urine. *Chem. Commun.*, **32**, 4859–4861.

Boggs B. K. and Botte G. G. (2009). On-board hydrogen storage and production: An application of ammonia electrolysis. *J. Power Sources*, **192**(2), 573–581.

Bunce N. J. and Bejan D. (2011). Mechanism of electrochemical oxidation of ammonia. *Electrochim. Acta*, **56**(24), 8085–8093.

Chen G. (2004). Electrochemical technologies in wastewater treatment. *Sep. Purif. Technol.*, **38**(1), 11–41.

Clauwaert P., Tolêdo R., van der Ha D., Crab R., Verstraete W., Hu H., Udert K. M. and Rabaey K. (2008). Combining biocatalyzed electrolysis with anaerobic digestion. *Wat. Sci. Technol.*, **57**(4), 575–579.

Comninellis C. and Chen G. (2010). *Electrochemistry for the Environment.* Springer, New York, USA.

De Gusseme B., Hennebel T., Vanhaecke L., Soetaert M., Desloover J., Wille K., Verbeken K., Verstraete W. and Boon N. (2011). Biogenic palladium enhances diatrizoate removal from hospital wastewater in a microbial electrolysis cell. *Environ. Sci. Technol.*, **45**(13), 5737–5745.

Dutta P. K., Rabaey K., Yuan Z. and Keller J. (2008). Spontaneous electrochemical removal of aqueous sulfide. *Wat. Res.*, **42**(20), 4965–4975.

El-Naggar M. Y., Wanger G., Leung K. M., Yuzvinsky T. D., Southam G., Yang J., Lau W. M., Nealson K. H. and Gorby Y. A. (2010). Electrical transport along bacterial nanowires from shewanella oneidensis MR-1. *P. Natl. Acad. Sci. USA*, **107**(42), 18127–18131.

Fangzhou D., Zhenglong L., Shaoqiang Y., Beizhen X. and Hong L. (2010). Electricity generation directly using human feces wastewater for life support system. *Acta Astronaut.*, **68**(9–10), 1537–1547.

Friedler E., Butler D. and Alfiya Y. (2013). Wastewater composition. In: *Source Separation and Decentralization for Wastewater Treatment*, T. A. Larsen, K. M. Udert and J. Lienert (eds.), IWA Publishing, London, UK, pp. 241–257.

Freguia S., Rabaey K., Yuan Z. and Keller J. (2007). Non-catalyzed cathodic oxygen reduction at graphite granules in microbial fuel cells. *Electrochim. Acta.*, **53**(2), 598–603.

Freguia S., Teh E. H., Boon N., Leung K. M., Keller J. and Rabaey K. (2010). Microbial fuel cells operating on mixed fatty acids. *Bioresour. Technol.*, **101**(4), 1233–1238.

Hinden J. M. and Beer H. B. (1982). Dimensionally stable coated electrode for electrolytic process, comprising a protective oxide interface on valve metal base. European Patent EP0046449.

Holt P. K., Barton G. W. and Mitchell C. A. (2005). The future for electrocoagulation as a localised water treatment technology. *Chemosphere*, **59**(3), 355–367.

Hug A. and Udert K. M. (2012). Struvite precipitation from urine with electrochemical magnesium dosage. *Wat. Res.* In Press, Corrected Proof. doi: 10.1016/j.watres.2012.09.036

Ikematsu M., Kaneda K., Iseki M., Matsuura H. and Yasuda M. (2006). Electrolytic treatment of human urine to remove nitrogen and phosphorus. *Chem. Lett.*, **35**(6), 576–577.

Kapałka A., Cally A., Neodo S., Comninellis C., Wächter M. and Udert K. M. (2010a). Electrochemical behavior of ammonia at $Ni/Ni(OH)_2$ electrode. *El. Commun.*, **12**(1), 18–21.

Kapałka A., Fierro S., Frontistis Z., Katsaounis A., Frey O., Koudelka M., Comninellis C. and Udert K. M. (2009). Electrochemical behaviour of ammonia (NH_4^+/NH_3) on electrochemically grown anodic iridium oxide film (AIROF) electrode. *Electrochem. Commun.*, **11**(8), 1590–1592.

Kapałka A., Joss L., Anglada A., Comninellis C. and Udert K. M. (2010c). Direct and mediated electrochemical oxidation of ammonia on boron-doped diamond electrode. *Electrochem. Commun.*, **12**(12), 1714–1717.

Kapałka A., Katsaounis A., Michels N. L., Leonidova A., Souentie S., Comninellis C. and Udert K. M. (2010b). Ammonia oxidation to nitrogen mediated by electrogenerated active chlorine on $Ti/PtOx\text{-}IrO_2$. *El. Commun.*, **12**(9), 1203–1205.

Kumar P. R., Chaudhari S., Khilar K. C. and Mahajan S. P. (2004). Removal of arsenic from water by electrocoagulation. *Chemosphere*, **55**(9), 1245–1252.

Kuntke P., Śmiech K. M., Bruning H., Zeeman G., Saakes M., Sleutels T. H. J. A., Hamelers H. V. M. and Buisman C. J. N. (2012). Ammonium recovery and energy production from urine by a microbial fuel cell. *Wat. Res.*, **46**(8), 2627–2636.

Lan R. and Tao S. (2011). Preparation of nano-sized nickel as anode catalyst for direct urea and urine fuel cells. *J. Power Sources*, **196**(11), 5021–5026.

Lan R., Tao S. and Irvine T. S. (2010). A direct urea fuel cell – power from fertiliser and waste. *Energ. Environ. Sci.*, **3**, 438–441.

Logan B. (2008). *Microbial fuel cells.* Wiley-Interscience, Hoboken NJ, USA.

Marinčić L. and Leitz F. B. (1978). Electro-oxidation of ammonia in waste water. *J. Appl. Electrochem.*, **8**(4), 333–345.

Marti B. (2010). *Bioelectrochemical Removal of Sulfur from Source-Separated Urine.* Master's thesis. Institute of Environmental Engineering (IfU), Swiss Federal Institute of Technology (ETH) Zurich, Switzerland.

Morgenroth E. (2008). Biofilm reactors. In: *Biological Wastewater Treatment. Principles, Modelling and Design*, M. Henze, M. C. M. van Loosdrecht, G. A. Ekama and D. Brdjanovic (ed.), IWA Publishing, London, UK.

Mu Y., Rabaey K., Rozendal R. A., Yuan Z. and Keller J. (2009). Decolourization of azo dyes in bio-electrochemical systems. *Environ. Sci. Technol.*, **43**(13), 5137–5143.

Mu Y., Radjenovic J., Shen J. Y., Rozendal R. A., Rabaey K. and Keller J. (2011). Dehalogenation of iodinated X-ray contrast media in a bioelectrochemical system. *Environ. Sci. Technol.*, **45**(2), 782–788.

Mu Y., Rozendal R. A., Rabaey K. and Keller J. (2009a). Nitrobenzene removal in bio-electrochemical systems. *Environ. Sci. Technol.*, **43**(22), 8690–8695.

Picioreanu C., Katuri K. P., Head I. M., van Loosdrecht M. C. M. and Scott K. (2008). Mathematical model for microbial fuel cells with anodic biofilms and anaerobic digestion. *Wat. Sci. Tech.*, **57**(7), 965–971.

Radjenovic J., Bagastyo A., Rozendal R., Mu Y., Keller J. and Rabaey K. (2011). Electrochemical oxidation of trace organic contaminants in reverse osmosis concentrate using RuO_2/IrO_2-coated titanium anodes. *Wat. Res.*, **45**(4), 1579–1586.

Radjenovic J., Farré M. J., Mu Y., Gernjak W. and Keller J. (2012). Reductive electrochemical remediation of emerging and regulated disinfection byproducts. *Wat. Res.*, **46**(6), 1705–1714.

Ronteltap M., Maurer M., Hausherr R. and Gujer W. (2010). Struvite precipitation from urine – Influencing factors on particle size. *Wat. Res.*, **44**(6), 2038–2046.

Rozendal R. A., Jeremiasse A. W., Hamelers H. V. M. and Buisman C. J. N. (2008). Hydrogen production with a microbial biocathode. *Environ. Sci. Technol.*, **42**(2), 629–634.

Rozendal R. A., Leone E., Rabaey K. and Keller J. (2009). Efficient hydrogen peroxide generation from organic matter in a bioelectrochemical system. *Electrochem. Commun.*, **11**(9), 1752–1755.

Siegrist H., Laureni M. and Udert K. M. (2013). Transfer into the gas phase: ammonia stripping. In: *Source Separation and Decentralization for Wastewater Treatment*, T. A. Larsen, K. M. Udert and J. Lienert (eds.), IWA Publishing, London, UK, pp. 337–350.

Simka W., Piotrowski J. and Nawrat G. (2007). Influence of anode material on electrochemical decomposition of urea. *Electrochim. Acta.*, **52**(18), 5696–5703.

Udert K. M. and Jenni S. (2013). Biological nitrogen conversion processes. In: *Source Separation and Decentralization for Wastewater Treatment*, T. A. Larsen, K. M. Udert and J. Lienert (eds.), IWA Publishing, London, UK, pp. 291–305.

Udert K. M., Larsen T. A. and Gujer W. (2006). Fate of major compounds in source-separated urine. *Wat. Sci. Technol.*, **54**(11–12), 413–420.

Udert K. M., Larsen T. A., Biebow M. and Gujer W. (2003). Urea hydrolysis and precipitation dynamics in a urine-collecting system. *Wat. Res.*, **37**(11), 2571–2582.

Virdis B., Rabaey K., Rozendal R. A., Yuan Z. and Keller J. (2010). Simultaneous nitrification, denitrification and carbon removal in microbial fuel cells. *Wat. Res.*, **44**(9), 2970–2980.

Wang Q., Noguchi C., Hara Y., Sharon C., Kakimoto K. and Kato Y. (1997). Studies on anaerobic digestion mechanism: influence of pretreatment temperature on biodegradation of waste activated sludge. *Environ. Technol.*, **18**(10), 999–1008.

Winker R. (2009). *Urine Treatment with Microbial Fuel Cells. Inhibition Effects of Ammonia Compounds*. Master's thesis. Institute of Environmental Engineering (IfU), Swiss Federal Institute of Technology (ETH) Zurich, Switzerland.

Yan W., Wang D. and Botte G. G. (2012). Nickel and cobalt bimetallic hydroxide catalysts for urea electro-oxidation. *Electrochim. Acta*, **61**(1), 25–30.

Zang G. L., Sheng G. P., Li W. W., Tong Z. H., Zeng R. J., Shi C. and Yu H. Q. (2012). Nutrient removal and energy production in a urine treatment process using magnesium ammonium phosphate precipitation and a microbial fuel cell technique. *Phys. Chem. Chem. Phys.*, **14**(6), 1978–1984.

Zheng X.-Y., Kong H.-N., Wu D.-Y., Wang C., Li Y. and Ye H.-R. (2009). Phosphate removal from source separated urine by electrocoagulation using iron plate electrodes. *Wat. Sci. Technol.*, **60**(11), 2929–2938.

Zheng X. Y., Shen Y. H., Ye H. R., Yan L., Zhang Y. J., Wang C. and Kong H. N. (2010). Phosphate removal from source-separated urine by electrocoagulation using aluminium plate electrodes. *Fresen. Environ. Bull.*, **19**(4 A), 635–640.

Zöllig H. (2008). *Urine Treatment in Microbial Fuel Cells*. Master's thesis. Institute of Environmental Engineering (IfU), Swiss Federal Institute of Technology (ETH) Zurich, Switzerland.

Chapter 23

Transfer into the gas phase: ammonia stripping

Hansruedi Siegrist, Michele Laureni and Kai M. Udert

23.1 INTRODUCTION

Stripping is the most common process for the selective recovery of ammonia from wastewater. Two stripping technologies have been used for source-separated urine: first, air stripping of ammonia with subsequent ammonia adsorption in acid, and second, steam stripping with ammonia recovery in the condensate. In this chapter, we discuss the basic concepts of air stripping, because it is probably more suitable for decentralized reactors than steam stripping. We will also present results of air stripping/acid adsorption and steam stripping experiments with urine. Furthermore, we will discuss the conditions that can lead to passive ammonia stripping in urine-collecting systems. In the future, passive stripping could be a low-energy option for ammonia recovery in on-site systems.

The focus of this chapter is on the treatment of source-separated urine. However, we will also refer to the treatment of digester supernatant during the discussion of the basic concepts of air stripping. Since our literature review did not reveal any report about ammonia stripping from blackwater, we will not discuss this application.

Throughout the text, we will use the term ammonia for the sum of free ammonia (NH_3) and ammonium (NH_4^+). To specifically address the two forms of ammonia, we will use the chemical formula NH_3 and NH_4^+. The terms for carbonate and phosphate will be used accordingly.

23.2 WASTEWATERS WITH HIGH AMMONIA CONTENT

Urine and blackwater are the two source-separated waste streams with high ammonia concentrations (Table 23.1). The main nitrogen compound in fresh urine is urea. During urine collection in toilets and subsequent storage and transport, urea is degraded by the bacteria-produced enzyme urease (Udert *et al.*

2003). The hydrolysis of one mole of urea releases two moles of free ammonia (NH_3) and one mole of carbon dioxide (CO_2). Due to the release of NH_3 the alkalinity and the pH increase from approximately 6.2 to values around 9.0 (Udert *et al.* 2003):

$$NH_2 - CO - NH_2 + H_2O \leftrightarrow 2NH_3 + CO_2 \tag{23.1}$$

We use the term "stored urine" for urine after urea hydrolysis. In theory, stored urine can contain ammonia concentrations of $8000\,mgN{\cdot}L^{-1}$. In existing urine-collection systems or urine diverting dry toilets, the ammonia content of the stored urine is often substantially lower. As an example, we show measurements in the urine collected in the NoMix system of Forum Chriesbach (Table 23.1). Here, ammonia volatilizes due to passive ventilation or is diluted with flushing water. We will discuss the mechanisms of ammonia volatilization in detail in section 23.6.3. Urea hydrolysis is also the main source of ammonia in blackwater.

Table 23.1 Typical concentrations in high-strength ammonia solutions.

		Stored urine w/o losses (1)	**Stored urine with losses and dilution**(2)	**Blackwater**(3)	**Digester supernatant**(4)
Ammonia	$[mgN{\cdot}L^{-1}]$	8000	2400	1100	790
NH_3	$[mgN{\cdot}L^{-1}]$	1700	370	36	11
Carbonate	$[mgC{\cdot}L^{-1}]$	3200	1200	800	920
Alkalinity (5)	$[mmol{\cdot}L^{-1}]$	580	240	70	77
Phosphate	$[mgP{\cdot}L^{-1}]$	540	210	43	7.3
Ionic strength (5)	$[mol{\cdot}L^{-1}]$	0.6	0.3	0.09	0.08
pH	[-]	8.9	8.7	7.9	7.5

(1) Simulated concentrations based on medical data (Udert 2002) assuming that all magnesium and calcium precipitated as a consequence of the pH increase due to urea hydrolysis. (2) Urine collection tank at Eawag, Dübendorf (Udert and Wächter 2012), (3) Pilot project in Sneek, The Netherlands (Vlaeminck *et al.* 2009), (4) WWTP Werdhölzli, Zürich, Switzerland (Gur 2012). (5) Simulated values, except for alkalinity of digester supernatant. All simulations were done with PHREEQC (Parkhurst and Appelo 1999, Davies approach). The ionic strength is defined with equation 23.7.

The ammonia of digester supernatant is released by the degradation of proteins and other nitrogen containing compounds during anaerobic digestion. In contrast to stored urine, the pH value is lower and the alkalinity consists nearly completely of bicarbonate. Table 23.1 shows that urine has the highest pH value and the highest ammonia concentration of all three solutions. This is an advantage for ammonia stripping, nevertheless, the pH value is still too low for efficient ammonia stripping. Although stored urine already has a high pH value, the pH increase in stored urine requires more base than in blackwater or digester supernatant, since stored urine is a strongly buffered solution.

23.3 CHEMICAL EQUILIBRIA

For efficient stripping most of the ammonia has to be present as NH_3. The ratio NH_3/NH_4^+ depends on the pH value and the temperature. It can be increased by adding a base, for example sodium hydroxide (NaOH), by removing carbon dioxide (CO_2) or by raising the temperature. In this section, we will present basic concepts and equations that can be used to calculate the NH_3 concentration.

23.3.1 Acid-base equilibrium

The acidity constant K_a is a product of activities {i}. The activity of a compound i can be calculated by multiplying its concentration C_i with the corresponding activity coefficients $f_{a,i}$. The following equations describe the acid-base equilibria for NH_4^+/NH_3 and CO_2/HCO_3^- (Stumm and Morgan 1996):

$$NH_4^+ \leftrightarrow NH_3 + H^+ \tag{23.2}$$

$$K_{a,NH_4} = \frac{f_{a,NH_3} \cdot C_{NH_3} \cdot 10^{-pH}}{f_{a,NH_4} \cdot C_{NH_4}} = 5.0 \cdot 10^{-10} \cdot e^{0.07 \cdot (T-298)} \tag{23.3}$$

$$H_2O + CO_2 \leftrightarrow H^+ + HCO_3^- \tag{23.4}$$

$$K_{a,CO_2} = \frac{f_{a,HCO_3} \cdot C_{HCO_3} \cdot 10^{-pH}}{f_{a,CO_2} \cdot C_{CO_2}} = 4.5 \cdot 10^{-7} \cdot e^{0.008 \cdot (T-298)} \tag{23.5}$$

The unit of K_{a,NH_4} and K_{a,CO_2} is [mol·L^{-1}] and the temperature T is given in [K]. The activity of protons is 10^{-pH}, because pH electrodes measure directly the activity of the protons. Since the K_a values are very small, they are usually given as negative logarithm

$$pK_a = -\log_{10}(K_a) \tag{23.6}$$

Activity coefficients depend on the charge of the compound. Ions have activity coefficients smaller than 1. At ionic strengths typical for digester supernatant, blackwater and stored urine, the activity coefficients decrease when the ionic strength rises. The ion activity coefficient for solutions with an ionic strength $I < 0.5$ mol·L^{-1} can be calculated with the equation by Davies (Stumm and Morgan 1996)

$$\log_{10}(f_{a,i}) = -0.5 \cdot z_i^2 \cdot \left(\frac{\sqrt{I}}{1+\sqrt{I}} - 0.2 \cdot I\right) \tag{23.7}$$

$$I = 0.5 \cdot \sum C_i \cdot z_i^2 \tag{23.8}$$

where I is the ionic strength in [mol·L^{-1}] and z_i [-] the charge of the ion i.

Table 23.2 Acidity constants based on activities (pK_a) and concentrations (pK'_a, conditional acidity constant) for NH_4^+ and carbon dioxide at 298 K.

	pK_a	pK'_a $I = 0.1\ mol \cdot L^{-1}$	pK'_a $I = 0.4\ mol \cdot L^{-1}$
$NH_4^+ \leftrightarrow NH_3 + H^+$	9.3	9.4	9.45
$CO_2 + H_2O \leftrightarrow HCO_3^- + H^+$	6.35	6.25	6.2

The activity coefficients of uncharged volatile compounds such as NH_3 and carbon dioxide (CO_2) are larger than 1 and increase with rising ionic strength. However, the influence of the ionic strength is relatively low, thus an activity coefficient of 1 is usually assumed for uncharged compounds in wastewater.

More accurate activity calculations, especially for ionic strengths above $I = 0.5\ mol \cdot L^{-1}$ require more complex models. An often used approach is the Pitzer theory, which is implemented in a series of chemical speciation programs, such as PHREEQC (Parkhurst and Appelo 1999).

The ionic strength dependencies influence the chemical equilibria differently (see Table 23.2). To simplify speciation calculations conditional equilibrium constants are defined for solutions with a given ionic strength. The conditional equilibrium constant K'_a (and its negative logarithm pK'_a) is the quotient of the concentrations and not the activities. One exception are the protons: for practical reasons, 10^{-pH} is used, which is the proton activity. The following example shows how K'_a can be derived from the K_a value when the activity coefficients are known:

$$K'_{a,NH_4} = K_{a,NH_4} \cdot \frac{f_{a,NH_4}}{f_{a,NH_3}} \tag{23.9}$$

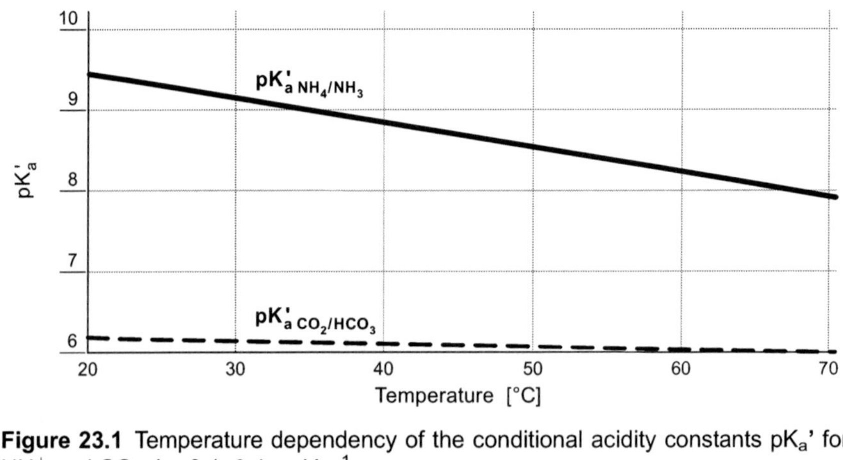

Figure 23.1 Temperature dependency of the conditional acidity constants pK_a' for NH_4^+ and CO_2. $I = 0.1–0.4\ mol \cdot L^{-1}$.

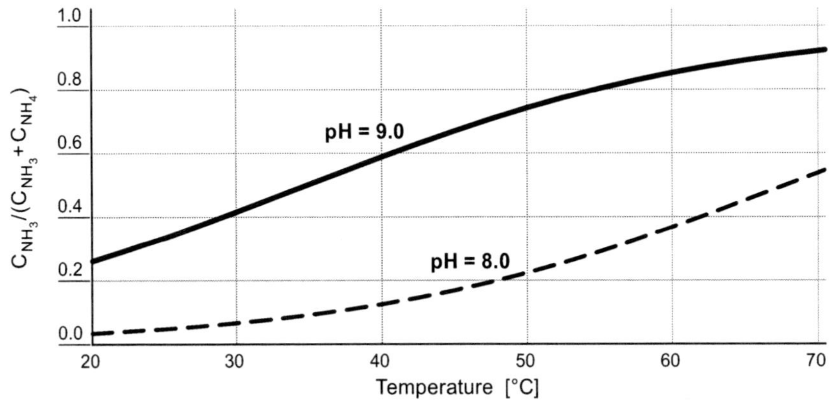

Figure 23.2 Fraction of NH_3 depending on temperature at pH 8 and 9.

Figure 23.1 demonstrates that the equilibrium of NH_4^+/NH_3 has a much stronger temperature dependency than the CO_2/HCO_3^- equilibrium. Just by raising the temperature the fraction of NH_3 increases substantially (Figure 23.2). As a consequence, ammonia can be stripped efficiently at a lower pH value and less base has to be dosed.

23.3.2 Gas exchange equilibrium

The Henry's law constant H describes the distribution of a volatile compound between gas phase and water phase at thermodynamic equilibrium. If the partial pressure of the compound is known, the equilibrium concentration in the water phase can be calculated according to the following relationship:

$$H_i^C = \frac{p_i}{f_{a,i} \cdot C_i} \approx \frac{p_i}{C_i} \tag{23.10}$$

p_i [bar] is the partial pressure of compound i, C_i the concentration of compound i in the water phase and H_i^C [bar·L·mol^{-1}] is the Henry constant. The dimensionless version of the Henry's law constant H_i is useful when performing mass balances in air stripping processes:

$$H_i = \frac{p_i}{R \cdot T} \cdot \frac{1}{C_i} = \frac{H_i^C}{R \cdot T} \approx \frac{C_{i,\text{gas}}}{C_i} \tag{23.11}$$

As mentioned before, the activity coefficient for gases is assumed to be 1 for low strength wastewaters, but more accurate values can be calculated with a chemical speciation program.

Surfactants can decrease the gas transfer from water to the air. The dominant mechanism is the accumulation of the surfactants at the air/water interface, thereby decreasing the mole fraction of the volatile compound at the interface.

The effect of temperature can be estimated with the van't Hoff equation:

$$H_{i,T1} = H_{i,T2} \cdot e^{\frac{\Delta H^0_{diss}}{R} \cdot \left(\frac{1}{T2} - \frac{1}{T1}\right)} \tag{23.12}$$

ΔH^0_{diss} [J·mol^{-1}] is the standard change of enthalpy for the dissolution of compound i in water, R (8.314 J·K^{-1}·mol^{-1}) the ideal gas constant and T is given in [K]. Values for the dimensionless Henry's law constants for ammonia and carbon dioxide can be calculated for common temperatures in wastewater treatment according to the following equations (Crittenden *et al.* 2005)

$$H_{NH_3,T} = 0.0006 \cdot e^{4340 \cdot \left(\frac{1}{293} - \frac{1}{T}\right)} \tag{23.13}$$

$$H_{CO_2,T} = 1.1 \cdot e^{2400 \cdot \left(\frac{1}{293} - \frac{1}{T}\right)} \tag{23.14}$$

Since carbon dioxide is about one thousand times more volatile than NH_3 (Figure 23.3) it could be stripped in a first stripper column without losing a high amount of NH_3 with the off-gas. CO_2 stripping raises the pH value in the water phase. This helps to reduce the amount of base required to shift the NH_3/NH_4^+ ratio towards NH_3.

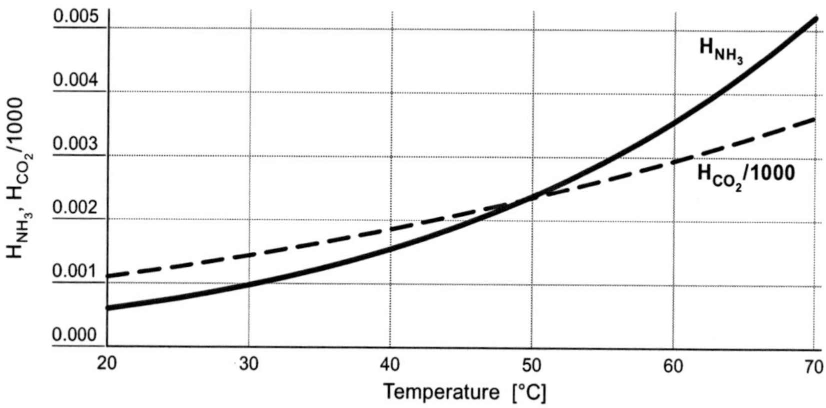

Figure 23.3 Dimensionless Henry's law constant ($L_{water} \cdot L_{air}^{-1}$) for NH_3 and CO_2. NH_3 is a strongly soluble gas whereas CO_2 is rather volatile.

23.4 AMMONIA STRIPPING WITH AIR

Some municipal wastewater treatment plants are equipped with NH_3 stripping to treat the ammonia rich digester supernatant. The most common process is ammonia stripping with air and the later adsorption of ammonia in acid. Packed columns are used to increase the water/air interface. In the stripper, air and digester supernatant flow in opposite directions in order to maximize ammonia stripping (Figure 23.4). Furthermore, the digester liquid is heated and NaOH is dosed to move the acid/base equilibrium towards NH_3. The ammonia-rich air is transferred to the sorber column. This column is also operated in counter-current mode. Highly concentrated H_2SO_4 is used for ammonia adsorption. Due to the low pH value, all adsorbed ammonia is directly converted to NH_4^+. The effluent is an ammonium sulfate solution ($(NH_4)_2SO_4$) with about 10% ammonia and a pH value of approximately 5. In order to minimize the heat loss, the off-gas of the sorber column is recycled back to the stripper column. Fresh air can be introduced to the stripper in order to remove carbon dioxide thereby increasing the pH value. However, fresh air (and the associated heat loss) is not necessary, if sufficient base is added.

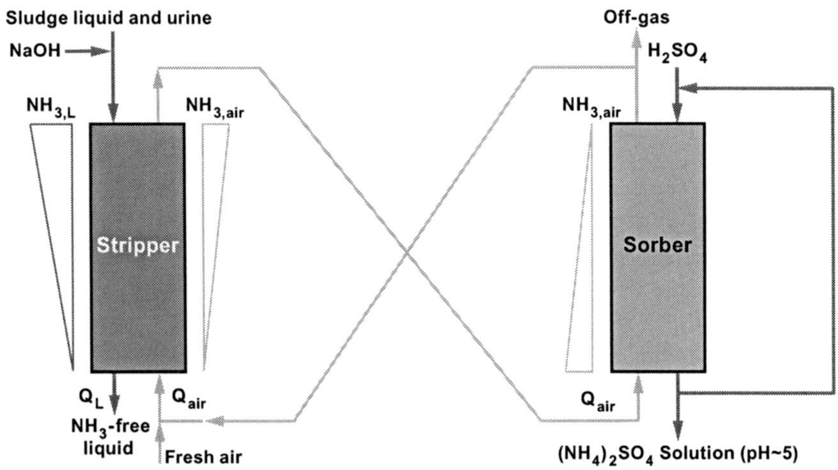

Figure 23.4 Flow scheme of ammonia stripping with air. NaOH is added and the influent is heated to increase the NH_3 content. H_2SO_4 is added to the sorber protonating NH_3 and producing a concentrated $(NH_4)_2SO_4$ solution.

Source-separated urine can be treated together with the sludge liquid. However, mixing digester supernatant with stored urine can result in precipitation of struvite and calcium phosphate: urine has a high pH value and high concentrations of phosphate, while the digester supernatant delivers the calcium and magnesium ions necessary for precipitation (Udert *et al.* 2003). To prevent scaling of the sorber material, the phosphate in the urine has to be precipitated in an upstream reactor by adding magnesium.

The required ratio of air to liquid flow Q_{air}/Q_L can be calculated with Equation 23.15 assuming that

- the inlet NH_4^+ is completely converted to NH_3
- the dissolved inlet NH_3 is in equilibrium with the off-gas NH_3 and

- the inlet air as well as the treated liquid is free of ammonia.

$$Q_L \cdot C_{NH_3,L} = Q_{\text{air}} \cdot C_{NH_3,\text{air}} \Longrightarrow \frac{Q_{\text{air}}}{Q_L} = \frac{C_{NH_3,L}}{C_{NH_3,\text{air}}} = \frac{1}{H_{NH_3}} \tag{23.15}$$

In practice, the dissolved NH_3 is hardly in equilibrium with the off-gas ammonia and Q_{air}/Q_L has to be larger than $1/H_{NH_3}$. In this case, the stripping factor S is used to calculate the required air flow:

$$S = H_{NH_3} \cdot \frac{Q_{\text{air}}}{Q_L} \tag{23.16}$$

S depends on the design of the column. Typical values range from 1.5 to 5. The stripping column height *Z* can be determined with the following equation (Tchobanoglous *et al.* 2003):

$$\begin{aligned} Z &= HTU \cdot NTU \\ HTU &= \frac{Q_L}{A \cdot K_L a} \\ NTU &= \frac{S}{S-1} \cdot ln\left[\left(\frac{C_{NH_3,\text{in}}}{C_{NH_3,\text{out}}} \cdot (S-1) + 1\right) \cdot \frac{1}{S}\right] \end{aligned} \tag{23.17}$$

where HTU [m] is the height of one transfer unit, NTU [-] is the number of transfer units, Q_L [$m^3 \cdot h^{-1}$] is the flow rate of the liquid, A [m^2] the cross-sectional area of the column, K_La [h^{-1}] the volumetric mass transfer coefficient and $C_{NH3,in}$ and $C_{NH3,out}$ the aqueous concentrations of NH_3 is the inlet and outlet, respectively.

The volumetric mass transfer coefficient K_La for NH_3 depends on the packing material, specific packing surface *a* [$m^2 \cdot m^{-3}$], temperature and the liquid composition. Typical K_La values for NH_3 are in the range of 2–10 h^{-1} for 30–60°C (Crittenden *et al.* 2005, Mackowiak 1990).

23.5 AMMONIA STRIPPING WITH AIR AND CARBON DIOXIDE PRE-STRIPPING

NaOH addition can be significantly reduced with a CO_2 pre-stripper (Figure 23.5). Since CO_2 is about one thousand times more volatile than NH_3 (Figure 23.3), it can be stripped in a pre-stripper, which is operated with a significantly lower airflow (less than 5%) than the ammonia stripper. Ammonia and energy losses are minimal in the CO_2 pre-stripper. Introducing fresh air to the air that is circulated between the stripper and sorber and introducing the off-gas to the pre-stripper additionally reduces the need for base addition.

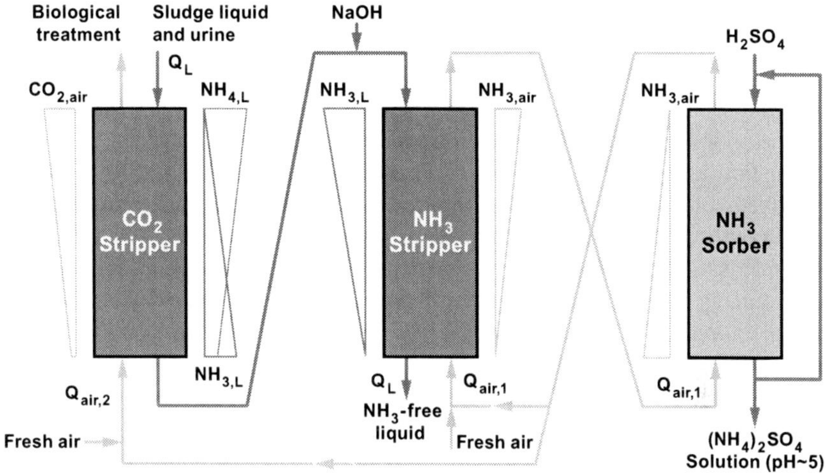

Figure 23.5 Flow scheme of ammonia stripping with a carbon dioxide pre-stripping column. This additional stripper reduces significantly the necessary base addition to the ammonia stripper.

23.6 AMMONIA STRIPPING FROM STORED URINE

The two main processes that have been intensively tested for ammonia recovery from urine are air stripping with consecutive ammonia adsorption in H_2SO_4 (as described above) and steam stripping. Kuntke *et al.* (2012) suggested using microbial fuel cells to enhance ammonia stripping. This process is described in more detail in Udert and Jenni (2013). A special case of ammonia stripping is the passive stripping observed in urine-collecting systems: the experiences from the NoMix system at Eawag will also be presented in this section.

23.6.1 Stripping reactors with adsorption in acid

Ammonia recovery from stored urine via air stripping and adsorption in acid has been studied by several research groups. Basakcilardan-Kabakci *et al.* (2007) investigated the basic influence factors with a laboratory set-up, while Paris *et al.* (2007) and Antonini *et al.* (2011) reported experiments with a full-scale reactor. The basic reactor set-up corresponds to Figure 23.4. Antonini *et al.* (2011) did several experiments in batch mode. The input was stored urine (50 L, ammonia 4,500 $mgN{\cdot}L^{-1}$, phosphate 310 $mgP{\cdot}L^{-1}$, pH 9). In order to prevent clogging of the stripper, phosphate had been precipitated as struvite in an upstream reactor by adding magnesium oxide at a ratio of 1.5 $mol\ Mg{\cdot}mol\ P^{-1}$. The NH_3 content was increased by heating the urine to 40°C and by adding 1 L of a 50% NaOH solution to 50 L of urine. The resulting pH value was 10. The flow rate of urine to the stripper was either 10 or 80 $L \cdot h^{-1}$, the flow rate of H_2SO_4 (1:10 vol. H_2SO_4) to the sorber was 55 $L \cdot h^{-1}$ and the airflow rate was 130 $m^3 \cdot h^{-1}$. The energy consumption was measured with a conventional electricity meter.

Maximum ammonia recovery at relatively low energy consumption was achieved when the urine was recirculated to the stripper. The experiment lasted for 3.5 h and the urine flow rate was 80 $L \cdot h^{-1}$. 94% of the ammonia was stripped and 100% of the stripped ammonia was later adsorbed in the acid. Based on the experimental results Antonini *et al.* calculated that the stripping and adsorption process required an energy input of 18.8 to 28.2 $kWh \cdot kgN^{-1}$ as electricity, which is 60.6 to 91.0 $kWh \cdot kgN^{-1}$ primary energy assuming a conversion efficiency of 31% for electricity production (average European electricity mix, UCPTE 1994). This is considerably higher than the calculated 13 $kWh \cdot kgN^{-1}$ (primary energy) required by an air stripping/acid adsorption reactor used for ammonia recovery from the digester supernatant of a wastewater treatment plant with 100,000 person equivalent (WWTP Kloten Opfikon/Switzerland, Böhler *et al.* 2012). Smaller reactors are often less energy-efficient, but it should also be kept in mind that the reactor of Antonini *et al.* was a prototype and optimization of the energy demand is possible.

23.6.2 Steam stripping

Tettenborn *et al.* (2007) used steam stripping to transfer ammonia from stored urine into the condensate formed from the steam. They conducted experiments with laboratory reactors and with a pilot plant designed for 800 person equivalent. The pilot reactor consisted of a 4.8 m high stripping column, which was operated in counter-current. The residence time of the urine was 15 min. The steam had a pressure of 6 bar and a temperature of 160°C. Various process combinations were tested: urine flow rates between 70 and 110 $L \cdot h^{-1}$, steam flows between 15 and 35 $kg \cdot h^{-1}$ and pH values between 8.5 and 11. The pH was increased by the addition of NaOH or potassium hydroxide (KOH). 91 to 100% of the ammonia could be removed. A silicone-based anti-foaming agent was used to prevent strong foaming. Depending on the amount of steam, the ammonia concentrations in the condensate were between 15 to 34 times higher than in the initial stored urine solution (2 to 7.4 $gNH_3 \cdot L^{-1}$). About 25 kg steam was used for 100 L of urine. This corresponds to 188 $kWh \cdot m^{-3}$ or 30.8 $kWh \cdot kgN^{-1}$, if the initial ammonia concentration in stored urine was 7.4 $gNH_3 \cdot L^{-1}$. This calculated energy demand accounts only for steam production, but it shows that steam stripping has a similar energy demand than air stripping and adsorption in H_2SO_4. Tettenborn *et al.* (2007) mentioned that insulation and energy recovery could lower the energy demand by more than a factor of three (50 to 55 $kWh \cdot m^{-3}$). Nevertheless, even if the energy demand for steam stripping were lower than for air stripping/acid adsorption, the complexity of steam production makes this technology less suitable for small decentralized reactors.

23.6.3 Passive ammonia stripping in urine-collecting systems

Conventional sanitary installations have rising pipes, which are open to the atmosphere to prevent low pressure when flushing water flows down the pipe. The open rising pipes also help to remove odors. Some urine-collecting systems are built in a similar way. One example is the main building of Eawag, Forum Chriesbach (Goosse *et al.* 2009). This building has two urine-collecting systems with open rising pipes directly connected to the urine collection tanks (Figure 23.6). The two openings to the atmosphere allow air to circulate. In analogy to naturally ventilated buildings, air flows through the pipes and the storage tank, driven by buoyancy forces (Linden 1999). The heat of the building, the evaporation of water and, to a lower extent, the volatilization of ammonia reduce the density of the air inside the pipe. The resulting density ρ [$kg \cdot m^{-3}$] can be expressed with the following equation derived from the ideal gas law:

$$\rho = \frac{1}{R \cdot T} \cdot \left[\left(P - p_w - p_{NH_3} \right) \cdot M_{air} + p_w \cdot M_w + p_{NH_3} \cdot M_{NH_3} \right] \tag{23.18}$$

R (8.314 J K^{-1} mol^{-1}) is the ideal gas constant, T [K] is the temperature, P [P_a] is the reference pressure (usually the pressure of the ambient air), p_w [Pa] and p_{NH_3} [Pa] are the partial pressures of water vapor and ammonia, respectively. M_{air} (0.02895 $kg \cdot mol^{-1}$), M_w (0.01802 $kg \cdot mol^{-1}$) and M_{NH_3} (0.017 $kg \cdot mol^{-1}$) are the molecular weights of dry air, water vapor and ammonia, respectively.

The density gradients, induced by the progressive air saturation with water, play a primary role in driving the airflow especially at high atmospheric temperatures (e.g., summer). The airflow velocity v [$m \cdot s^{-1}$] induced by buoyancy forces (stack effect without wind) can be approximated, based on Bernoulli's principle, as follows (Laureni *et al.* 2012):

$$v = \vartheta \cdot \sqrt{\frac{\rho_{in}}{\rho_{out}} - 1} \tag{23.19}$$

ρ_{in} [$kg \cdot m^{-3}$] is the density of the air just before entering the urine-collecting system, ρ_{out} [$kg \cdot m^{-3}$] is the density of the air just before leaving the urine-collecting system (in general saturated with water) and ϑ [$m \cdot s^{-1}$] an empirical coefficient, to be determined experimentally, which accounts for a variety of pressure losses along the flow path.

In the two urine-collecting systems of Forum Chriesbach, nearly all urea is hydrolyzed before the urine reaches the collection tanks (Laureni *et al.* 2012). As a consequence, the rising pipes act as counter-current stripping columns, in which the ammonia-rich urine flows down to the collection tanks and the humid and warm air rises up to the roof (Figure 23.6). Laureni *et al.* (2012) found that up to 47% of the ammonia was lost by volatilization. Most of the ammonia loss occurred during working hours, when urine was present in the pipes (Figure 23.6). The airflow through the pipes was between 0.3–0.6 m·s^{-1}. The volumetric air to liquid ratios were in the operational range of industrial stripping columns: 2000–6000 m^3·m^{-3} (Tchobanoglous *et al.* 2003). Peak ammonia concentrations in the off-gas were as high as 1200 mgN·m^{-3}.

The high ammonia volatilization in urine-collecting systems is problematic, because the nitrogen is lost for recycling and the ammonia degassing can lead to environmental pollution and odor nuisance. Interrupting the airflow in the pipes, for example by installing a one-way valve, is a simple solution to solve this problem. Anyway, in new energy-efficient buildings, the airflow through rising pipes is restricted by a valve on top of the rising pipe to prevent energy losses. A completely different approach would be to enforce the airflow and thereby increase the ammonia removal with the aim of recovering ammonia at the top of the rising pipe, for example in an acid trap.

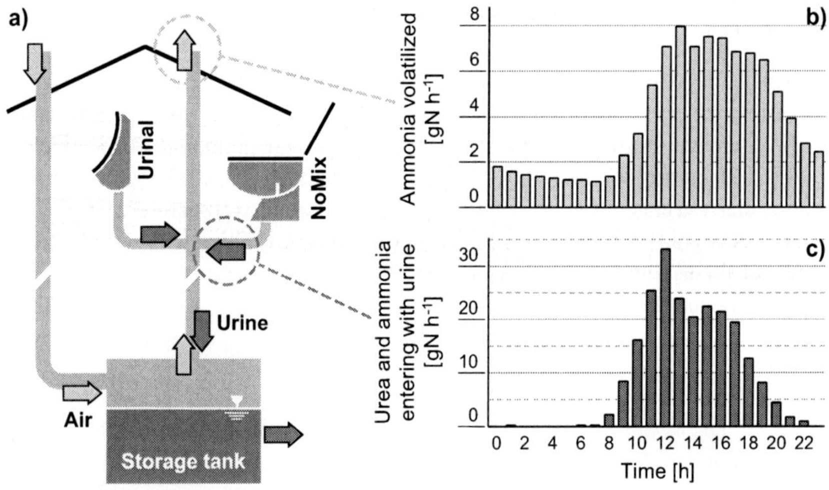

Figure 23.6 Simplified scheme (a) of a urine-collecting system at Forum Chriesbach, Eawag. The arrows represent air (light) and urine (dark). The typical daily profile of urea and ammonia entering the system with fresh urine (c) and NH_3 leaving the system with the air (b) are plotted for a typical working day in this office building.

23.7 PRODUCTS OF AMMONIA STRIPPING

Ammonia stripping produces solutions, which can be used in agriculture or industry. When ammonia is adsorbed in H_2SO_4, $(NH_4)_2SO_4$ is formed, which can be applied as fertilizer for example as annual ammonia depot fertilization in spring (Spiess *et al.* 2006). The concentrated ammonia solution from the steam stripping process can be used as raw product for fertilizer production or for the reduction of NOx of flue gas to molecular nitrogen in the Denox process.

23.8 CONCLUSIONS

Ammonia stripping can be considered to be a proven technology for nitrogen recovery from source-separated urine. Nearly complete recovery of ammonia is possible. However, the need for strong bases and acids for the air stripping/acid adsorption method and the need of steam under high pressure and high temperature for steam stripping are challenges for small decentralized reactors. Furthermore, small on-site reactors will probably not achieve the high energy efficiency of large-scale reactors. Based on the current stage of knowledge, we conclude that ammonia stripping is a suitable process for medium-sized reactors that treat the urine of several hundred people. Combining ammonia stripping with an upstream struvite precipitation reactor is an interesting process combination for the recovery of the two main nutrients from urine, nitrogen and phosphorus.

REFERENCES

Antonini S., Paris S., Eichert T. and Clemens J. (2011). Nitrogen and phosphorus recovery from human urine by struvite precipitation and air stripping in Vietnam. *CLEAN– Soil, Air, Water*, **39**(12), 1099–1104.

Başakçılardan-Kabakcı S., İpekoğlu A. N. and Talınlı İ. (2007). Recovery of ammonia from human urine by stripping and absorption. *Environmental. Engineering and Science*, **24** (5), 615–624.

Böhler M., Büttner S., Liebi C. and Siegrist H. (2012). Ammoniakstrippung mittels Luft zur Behandlung von Faulwasser und Urin auf der Kläranlage Kloten/Opfikon (Air stripping of ammonia for the treatment of digester supernatant and urine at the WWTP Kloten/ Opfikon). *Aqua & Gas*, **1**, 26–31.

Crittenden J., Trussell R., Hand D., Howe K. and Tchobanoglous G. (2005). Water treatment: Principles and design. 2nd edn. John Wiley & Sons, New York, USA.

Goosse P., Steiner M., Udert K. M. and Neuenschwander W. (2009). NoMix-Anlage. Erste Monitoringergebnisse im Forum Chriesbach (NoMix Toilet System. First monitoring results in Forum Chriesbach). *Gas Wasser Abwasser*, 7/2009, 567–574.

Gur E. (2012). *Mineral precipitation in anaerobic digester supernatant: A case study on WWTP Werdhölzli.* Master's thesis. Institute of Environmental Engineering (IfU), Swiss Federal Institute of Technology (ETH), Zurich, Switzerland.

Kuntke P., Śmiech K. M., Bruning H., Zeeman G., Saakes M., Sleutels T. H. J. A., Hamelers H. V. M. and Buisman C. J. N. (2012). Ammonium recovery and energy production from urine by a microbial fuel cell. *Water Research*, **46**(8), 2627–2636.

Laureni M., Wächter M., Durciova B., Mohn J. and Udert K. M. (2012). Transformations and losses of major urine compounds in a urine-collecting systems. *Submitted to Water Research.*

Linden P. F. (1999). The fluid mechanics of natural ventilation. *Annual Review of Fluid Mechanics*, **31**(1), 201–238.

Mackowiak J. (1990). Einsatz von modernen Füllkörpern zur Reduzierung von Schadstoffen aus Abluft und Abwasser (Use of modern filling material for the reduction of pollutants from offgas and wastewater). *Staub - Reinhaltung der Luft*, **50**, 221–227.

Paris S., Schlapp C. and Netter T. (2007). A contribution to sustainable growth by research and development. Proceedings of the International Symposium on 'Water Supply and Sanitation for All', Berching (Germany), 27–28 Sep 2007.

Parkhurst D. L. and Appelo C. A. J. (1999). *User's Guide to PHREEQC (Version 2)*, US Geological Survey, Denver, Colorado, USA.

Spiess E., Irla E. and Heusser J. (2006). *Depot-Injektion von ammoniumhaltigen Düngern nach dem CULTAN-System* (Depot injection of ammonium-containing fertilizers with the CULTAN system). ART Report No 657, Forschngsanstalt Agroscope Reckenholz–Tänikon (ART), Switzerland.

Stumm W. and Morgan J. J. (1996). *Aquatic Chemistry*, 3rd Edition. Wiley, New York, USA.

Tchobanoglous G., Burton F. L. and Stensel H. D. (2003). *Metcalf &Eddy Wastewater Engineering*. Treatment and Reuse, 4th (edn.), McGraw–Hill, Boston, USA.

Tettenborn F., Behrendt J. and Otterpohl R. (2007). *Resource recovery and removal of pharmaceutical residues. Treatment of separate collected urine*. Institute of Wastewater Management and Water Protection, Technical University of Hamburg, Germany.

UCPTE (1994). *Yearly Report 1993*. Union pour la coordination de la production et du transport de l'électricité. Vienna, Austria.

Udert K. M. (2002). *The fate of phosphorus and nitrogen in source-separated urine. PhD thesis*. Swiss Federal Institute of Technology (ETH), Zurich, Switzerland.

Udert K. M., Larsen T. A., Biebow M. and Gujer W. (2003). Urea hydrolysis and precipitation dynamics in a urine-collecting system. *Water Research*, **37**(11), 2571–2582.

Udert K. M. and Wächter M. (2012). Complete nutrient recovery from source-separated urine by nitrification and distillation. *Water Research*, **46**(2), 453–464.

Udert K. M and Jenni S. (2013). Biological nitrogen conversion processes. In: *Source Separation and Decentralization for Wastewater Management*, T. A. Larsen, K. M. Udert and J. Lienert (eds.), IWA Publishing, London, UK, pp. 291–305.

Vlaeminck S. E., Terada A., Smets B. F., van der Linden D., Boon N., Verstraete W. and Carballa M. (2009). Nitrogen removal from digested black water by one-stage partial nitritation and anammox. *Environmental Science and Technology*, **43**(13), 5035–5041.

Chapter 24

Transfer into the solid phase

Işık Kabdaşlı, Olcay Tünay and Kai M. Udert

24.1 INTRODUCTION

Several treatment methods can be used to transfer dissolved compounds from source-separated wastewater streams onto or into a solid phase. A critical review of these methods will be given here. The emphasis will be placed on the recovery of nutrients, especially from human urine; the treatment of other waste streams will be described in brief.

24.2 STRUVITE PRECIPITATION

Struvite ($MgNH_4PO_4{\cdot}6H_2O$) precipitation has been proposed as a technique for recovering both ammonia and phosphate from human urine (Lind *et al.* 2000, Kabdaşlı *et al.* 2006a, Wilsenach *et al.* 2007, Tilley *et al.* 2008) and blackwater (de Graaf *et al.* 2011). Most of the research so far has focused on the treatment of source-separated urine. Struvite precipitation can be used to recover phosphate or ammonia from fresh or ureolyzed human urine. For the recovery of phosphate, only magnesium, sometimes along with a base to reach the optimum pH value for struvite precipitation, has to be added to urine or blackwater.

Struvite precipitation can also be applied to human urine in order to recover potassium (Kuşçuoğlu 2008, Wilsenach *et al.* 2007). In this approach, a special kind of struvite is formed: magnesium potassium phosphate (also called K-struvite; $MgKPO_4{\cdot}6H_2O$). To ensure that potassium instead of ammonia precipitates, nearly all ammonia has to be removed beforehand. In this chapter, we will focus on phosphorus and ammonia recoveries and will therefore not discuss the formation of magnesium potassium phosphate in more detail.

24.2.1 Conditions for struvite formation

Struvite ($MgNH_4PO_4{\cdot}6H_2O$; MAP) is a white crystalline mineral consisting of magnesium, ammonia and phosphate in equal molar amounts. Ronteltap *et al.* (2007a) determined the solubility product K_{so} to be $10^{-13.26\pm0.057}$, a value which is in accordance with previous studies. The recommended operating pH range is 8.0 to 9.5 and the process efficiency increases as the pH rises within this interval (Kabdaşlı *et al.* 2006a, 2006b, 2006c; Tünay *et al.* 2009). Working at pH values greater than 9.0 can cause ammonia loss if the solution is strongly stirred. It may also lead to the formation of magnesium carbonate compounds such as nesquehonite ($MgCO_3{\cdot}3H_2O$, Sakthivel *et al.* 2012).

The solution matrix influences struvite precipitation in several ways. High ionic strength increases the solubility of struvite (Tatlı 2006, Tünay *et al.* 2009). Some specific ions such as calcium, carbonate, sulfate and sodium prolong the induction time for struvite crystallization, while carbonate has only a minor inhibiting effect (Kabdaşlı *et al.* 2006d). The presence of organic complexing agents such as citrate and phosphocitrate results in a dramatic increase in induction time (Kofina *et al.* 2007). Nevertheless, struvite crystallization is a fast process compared to the formation of other phosphate minerals such as hydroxylapatite, and precipitation already starts at low supersaturation (Udert *et al.* 2003a).

Ronteltap *et al.* (2010) showed that strong mixing increases the particle size, probably because it prevents high local supersaturation, a factor known to be responsible for small crystals (Wilsenach *et al.* 2007). Ronteltap *et al.* (2010) showed that larger particles form at higher temperatures (up to 30°C) and at lower pH values (down to 8).

24.2.2 Magnesium sources

The salts $MgCl_2$, MgO and $MgSO_4$ are common magnesium sources for struvite precipitation (Zeng and Li 2006). $MgCl_2$ and $MgSO_4$ dissolve very quickly, but when MgO is used, appropriate mixing conditions must be provided together with prolonged reaction times to ensure its complete dissolution (Tatlı 2006, Tünay *et al.* 2009). Using pure chemicals such as $MgCl_2$, $MgSO_4$ or MgO can add considerably to the financial costs of struvite production. Alternative low-cost sources are bittern, a waste product of salt production from seawater, calcined magnesite rock (Etter *et al.* 2011) or wood ash (Sakthivel *et al.* 2012). Their transport determines the costs of these compounds. Etter *et al.* (2011) estimated that calcined magnesite would be the cheapest magnesium source in Kathmandu, Nepal, because minable magnesite is available close to the city. Wood ash can be utilized as a magnesium source to precipitate struvite, but the final product contains many minerals besides struvite, including some heavy metals, which is a considerable drawback for the reuse of the precipitated phosphate (Sakthivel *et al.* 2012).

24.2.3 Struvite precipitation in urine

Fresh urine has a pH value of 6.2 $\pm$ 0.5 and low ammonia concentrations (480 $\pm$ 140 $mgN \cdot L^{-1}$) because most of the nitrogen is bound in the urea (7600 $\pm$ 1400 $mgN \cdot L^{-1}$) (Udert *et al.* 2003a). A base and a magnesium source would have to be added to recover the phosphorus as struvite from fresh urine. However, if urine is collected in NoMix toilets or urinals, bacterial urease will hydrolyze urea completely to ammonia and carbon dioxide, causing the pH to rise to values around 9 (Udert *et al.* 2003b). During urea hydrolysis (also called ureolysis), all magnesium present in fresh urine precipitates as struvite, while calcium precipitates as hydroxylapatite ($Ca_{10}(PO_4)_6(OH)_2$) (Udert *et al.* 2003b). At high dilution with tap water, calcite ($CaCO_3$) also precipitates (Udert *et al.* 2003a). In undiluted urine, close to 30% of the phosphate precipitates with the magnesium and calcium available in urine. Some dilution with tap water will even increase this fraction further (Udert *et al.* 2003a).

After urea hydrolysis, the conditions are very favourable for struvite precipitation. Only a magnesium source has to be added to recover nearly all the phosphate. More than 90% of the phosphate can be recovered with a 1.1:1 mol $Mg \cdot mol^{-1}P$ magnesium dose. In most applications, the phosphate recovery is determined by solid recovery and not struvite crystallization (Wilsenach *et al.* 2007, Etter *et al.* 2011). The pH decrease during phosphate recovery from undiluted urine is small due to the high buffer capacity of ureolyzed urine. Liu *et al.* (2008) reported a decrease of 9.23 to 9.14. If no phosphate is added, only about 5% of the ammonia is precipitated as struvite (Liu *et al.* 2008).

More than 95% of the nitrogen can be recovered from ureolyzed urine if magnesium and phosphate are dosed at molar ratios of 1.0 mol $Mg \cdot mol^{-1}NH_4$ and 1.0 mol $PO_4 \cdot mol^{-1}NH_4$ and the pH is fixed at values between 8.0 and 9.5 respectively (Kabdaşlı *et al.* 2006a, 2006b; Tünay *et al.* 2009). High pH values not only increase the ammonia removal efficiency, but also prevent foaming. When magnesium chloride and phosphoric acid or phosphate salts such as sodium phosphates are used to remove nitrogen from source-separated urine, the pH decreases and strong foaming can occur during stirring, probably due to carbon dioxide loss (Liu *et al.* 2008, Kabdaşlı *et al.* 2006a). However, the amount of base required is substantial: to keep the pH value at 8.5 Liu *et al.* (2008) had to add 10 g of sodium hydroxide per litre of undiluted ureolyzed urine.

24.2.4 Struvite precipitation in blackwater

de Graaff *et al.* (2011) investigated struvite precipitation as a technique to recover phosphorus from the effluent of an Upflow Anaerobic Sludge Blanket (UASB) reactor treating blackwater (see Zeeman and Kujawa-Roeleveld 2013). At least 90% of the phosphate in the effluent could be removed either by adding a surplus of magnesium (1.5 mol $Mg \cdot mol^{-1}PO_4$) at the original pH value of 8.0 or by increasing the pH value to 9.0 with sodium hydroxide and applying a minimum magnesium dosage ratio of 1.3 mol Mg·mol^{-1}PO_4. The authors calculated using their experimental data that 0.15 kg PO_4-P·p^{-1}·y^{-1} can be recovered as struvite. Since suspended solids in the UASB effluent would be removed together with the freshly formed struvite, the overall recovery of particulate phosphate could be 0.22 kg PO_4-P·p^{-1}·y^{-1}.

24.2.5 Micropollutants and pathogens

Ronteltap and co-workers (2007b) performed a lab-scale experimental study in order to assess the risk of incorporating pharmaceuticals (carbamazepine, diclofenac, ibuprofen, and propranolol), hormones (estradiol, estron, and ethinylestradiol), and heavy metals (As, Co, Cd, Cu, Cr, Ni, and Pb) into struvite recovered from source-separated urine. They demonstrated that (i) over 98% of hormones and pharmaceuticals remained in the filtrate after struvite precipitation; and (ii) only a small fraction of the metals in urine was attached to struvite. The content of heavy metals in struvite was much lower than in commercially available fertilizers. This is mainly due to the very low heavy metal content of urine and the use of analytical grade magnesium chloride as a precipitant. The heavy metal content in the final precipitate can exceed the permitted levels in phosphate fertilizers if alternative magnesium sources such as wood ash are used (Sakthivel *et al.* 2012). When struvite is recovered from human waste streams, they may contain pathogens. Decrey *et al.* (2011) reported that air-drying of struvite is a possible method to inactivate viruses (human virus surrogate phage ΦX174) and helminth eggs (*Ascaris suum*) in struvite recovered from source-separated urine. However, some viable *Ascaris* eggs and infective phages were still detected after several days of drying. Heating is not a good way to kill pathogens in struvite, because struvite decomposes and releases gaseous ammonia at temperatures above 40 to 55°C (Bhuiyan *et al.* 2008).

24.2.6 Use of struvite

Struvite can be used as a slow-release phosphorus fertilizer. Johnston and Richards (2003) compared dry matter and phosphate off-take from ryegrass fertilized with different kinds of phosphate compounds. Struvite obtained from wastewater and synthetically prepared struvite had similar dry matter yields and phosphate off-takes from grass to monocalcium phosphate, which is considered to be fully plant-available. The struvite compounds were substantially better than recovered calcium phosphate and synthetically prepared iron phosphate. Struvite can also be used in the production of fire-resistant panels and cement, and as a low-price phosphate source in the phosphate industry (de-Bashan and Bashan 2004).

24.3 OTHER PRECIPITATION PROCESSES

Chemical phosphate precipitation is a well-established process in centralized wastewater treatment plants. The same processes can be applied to source-separated waste streams such as urine, blackwater or greywater. This section gives a short overview of the processes and materials involved. In-depth information about removal mechanisms, application bases, design criteria and costs of phosphorus treatment by chemical precipitation is accessible in textbooks and manuals (Tchobanoglous *et al.* 2004, US EPA 2009).

Aluminium and iron salts, such as alum ($Al_2(SO_4)_3{\cdot}18H_2O$) or ferric chloride ($FeCl_3{\cdot}6H_2O$) are the most common precipitants. Lime ($Ca(OH)_2$) is a cheap alternative, but it is less common because the added calcium ions react not only with phosphate but also with carbonate, which results in high sludge production together with large amount of lime consumption (Sertaç 2012). Since source-separated urine and blackwater have high carbonate concentrations, lime is not recommended as a precipitant.

With the exception of studies on the electrochemical dissolution of aluminium and iron for chemical precipitation (see Udert *et al.* 2013), only a few studies report on the use of conventional chemical precipitation for the treatment of source-separated waste streams: Pidou *et al.* (2008) report about these processes for greywater treatment and Sertaç (2012) investigated them for source-separated urine. These studies report almost complete phosphate removal together with up to 75% removal of organic matter.

The use of iron and aluminium salts makes the precipitate unrecoverable for possible industrial processing into fertilizer (de-Bashan and Bashan 2004). Additionally, owing to the high adsorption capacity of metal hydroxide flocs, the precipitate may also contain considerable amounts of micropollutants such as pharmaceuticals or heavy metals.

24.4 PHOSPHATE ADSORPTION

24.4.1 Phosphate removal with adsorption

Adsorption is a process in which dissolved substances are adsorbed or accumulated onto a solid phase and thus removed from the liquid (Crittenden *et al.* 2005). Phosphate adsorption is a well understood and widely used mechanism for phosphorus removal (Rittmann *et al.* 2011).

24.4.2 Applications

Phosphate adsorption is one of the mechanisms behind phosphate removal by precipitation with iron salts, aluminium salts, and lime (see Section 24.3, Rittmann *et al.* 2011). Filters for phosphate removal are common in on-site wastewater treatment (Johansson Westholm 2006), especially for constructed wetlands (Drizo *et al.* 2002). They have long standing times and do not require process control. However, the materials used for on-site phosphate removal often have only low phosphorus adsorption capacities and their adsorption properties are not well defined (Cucarella and Renman 2009). Recently, materials with high phosphate adsorption capacities have been proposed to polish the effluent of membrane bioreactors (MBR, Ernst *et al.* 2007). Despite the possible benefits of phosphate adsorption, we were unable to find any published studies on phosphate adsorption from source-separated waste streams such as urine or blackwater.

24.4.3 Adsorbents

Johansson Westholm (2006) gives an overview of the wide range of materials used as filter substrates in on-site wastewater treatment. They comprise minerals and rocks, soils, marine sediments, industrial by-products and designed adsorbents (*e.g.,* filtralite®, LECA). Most adsorbents contain iron, aluminium, or calcium. Cucarella and Renman (2009) suggested a classification of filter materials for on-site treatment based on their phosphorus sorption capacity: soils and gravels have very low sorption capacities (<0.1 gP·kg^{-1}), whereas those of blast furnace slags and fly ashes can be very high (>10 gP·kg^{-1}).

Further high-capacity adsorbents are granulated ferric hydroxide (GFH) and activated alumina (AA). Genz *et al.* (2004) reported phosphorus adsorption capacities of 23.3 gP·kg^{-1} and 13.8 gP·kg^{-1} for GFH and AA respectively. These capacities were determined at pH 5.5 with synthetically prepared solutions containing 4 mgP·L^{-1} and particle sizes $<$ 63 μm. Ernst *et al.* (2007) determined an adsorption capacity of 65 gP·kg^{-1} at pH 4 for GFH using synthetically prepared solutions containing phosphorus of 100 mgP·L^{-1} (particle size $<$ 63 μm). Recently, a wide range of further high-capacity adsorbents have been developed. An overview is given by Rittmann *et al.* (2011).

24.4.4 Mechanisms

In adsorption filters, phosphate is not only removed by physical adsorption but also by chemical precipitation. Cucarella and Renman (2009) therefore suggested that the process of phosphate removal in filtration should be described as sorption rather than adsorption.

The pH value determines the charge of the adsorbent. At pH values below the point of zero charge (pH_{PZC}), the adsorbent is positively charged, which enhances the adsorption of phosphate anions. GFH has a pH_{PZC} between 7.5 and 8.0 (Ernst *et al.* 2007). Phosphate adsorption is therefore higher under neutral to acidic conditions. Inorganic anions such as carbonate or sulfate compete with phosphate (Dzombak and Morel 1990). Genz *et al.* (2004) showed that organic compounds can be even stronger competitors than inorganic anions.

Cucarella and Renman (2009) pointed out that the phosphorus adsorption capacities given in the literature are hardly comparable because they are determined under different conditions. The main parameters affecting the adsorption are the form and amount of the material, the adsorbent-to-solution ratio, the pH, the initial phosphate concentration, the solution matrix as well as agitation, temperature and contact time. Especially when the initial concentrations are high, it can take several days for the maximum adsorption capacity to be reached.

24.4.5 Phosphorus recovery

Currently, phosphate adsorption in on-site wastewater treatment aims to prevent eutrophication. The regeneration of the adsorbent and the recovery of phosphorus are of minor importance. Regeneration and phosphorus recovery are important for synthetically prepared high-capacity adsorbents such as GFH and AA due to the high material costs. Sodium hydroxide solutions are usually used for desorbing phosphate. Donnert and Sackeler (1999) proposed to regenerate AA with NaOH and then to precipitate calcium phosphate by adding $Ca(OH)_2$. When using $Ca(OH)_2$ instead of $CaCl_2$ or another calcium salt, no anions such as chloride are added and the pH is not decreased. This simplifies reuse of the solution after the phosphate has been precipitated.

The regeneration efficiency depends on the wastewater treated. Ernst *et al.* (2007) could desorb 90% of the phosphate from GFH with a 0.1 mol·L^{-1} NaOH solution when the columns had been fed with a synthetically prepared solution containing 100 mgP·L^{-1}. The regeneration efficiency was only 50% in experiments with GFH, which was loaded with MBR effluents containing between 2.5 and 4.0 mgP·L^{-1}.

24.5 NUTRIENT REMOVAL BY ION EXCHANGE

Ion exchange is a type of adsorption process. In wastewater treatment, it describes the exchange of ions in the aqueous phase with ions in the solid phase (Crittenden *et al.* 2005). This section focuses on zeolites, a group of natural minerals with ion exchange capacities that are often used for the removal of cations such as ammonium or potassium.

24.5.1 Operational conditions

Ammonia removal by zeolites has been well defined in terms of its thermodynamics and kinetics; fixed bed models for column operation are also available (Kithome *et al.* 1999, Cincotti *et al.* 2001, Saltalı *et al.* 2007, Wang and Peng 2010). Temperature is not an important operating parameter at moderate values. The maximum ammonia removal was shown to occur between pH 6 and 8 (Hedström 2001, Saltalı *et al.* 2007). Ammonium adsorption is a fast process and is completed within 10 minutes. Typical grain sizes are between 0.125 and 2 mm (Lind *et al.* 2000, Beler-Baykal *et al.* 2009).

The exchange capacity and selectivity of zeolites can be modified by pre-treatment such as washing with water, acid treatment, treatment with a base and heating. The most common pre-treatment for ammonia removal is saturation of zeolites with sodium ions (Booker *et al.* 1996, Hedström 2001).

The exhausted zeolite can be chemically or biologically regenerated after ammonium adsorption. NaCl solution is commonly used for chemical regeneration. NaCl concentrations vary from 0.1 to 0.6 mol·L^{-1} (Hedström 2001). Ødegaard (1992) recommended a mixture of NaCl and NaOH to decrease the need for brine.

Ion exchange treatment is easy to apply and suitable for decentralized treatment. This is not an energy-intensive technology. The energy is mainly required for the pumps and mixers used for solution preparation.

24.5.2 Ion exchange in source-separated urine

The treatment of source-separated urine with zeolite ion exchangers requires the urea to be hydrolyzed. Urine storage is usually sufficient to achieve the release of ammonia, since ureolyzing bacteria are commonly present in urine collection systems (Udert *et al.* 2003b). However, the pH in ureolyzed urine can be too high for effective ammonia adsorption on zeolites. Due to the high buffer capacity of ureolyzed urine, substantial amounts of acids are required to lower the pH value (Ek *et al.* 2006).

A combination of ammonia removal with zeolites and phosphate removal by struvite precipitation was first proposed and investigated by Lind *et al.* (2000). Other studies then followed (Bán and Dave 2004, Ganrot *et al.* 2007). Lind *et al.* (2000) dosed 2.4 g MgO and 20 g non-pretreated clinoptilolite to remove phosphate and ammonia from 1 litre of a synthetic solution representing fresh urine. Interestingly, the authors observed urea hydrolysis after the addition of the MgO salts. This effect helped to remove the phosphate as struvite. When they added the clinoptilolite after struvite precipitation, they could achieve 75% ammonia removal, while this figure rose to 80% when they added the zeolite simultaneously with the MgO. Ganrot *et al.* (2007) conducted similar experiments, however with real ureolyzed urine. Surprisingly, they reported that the clinoptilolite dose did not lead to any significant ammonia removal, although they did observe substantially higher phosphate removal.

Beler-Baykal *et al.* (2009) studied the effect of the initial loading on the removal of ammonia and potassium from ureolyzed human urine with clinoptilolite. The initial loading is defined as the amount of adsorbable ions (in this case ammonium and potassium) at the beginning of the experiment divided by the amount of adsorbent added. Experiments were conducted at pH 6.5–7.5 with packed columns and continuous urine recycling (batch mode). The clinoptilolite had been pre-treated with sodium chloride to increase its ammonium exchange capacity. The initial urine loadings varied between 5.2–33.6 mg NH_4^+/g of clinoptilolite. The results indicated that up to an initial loading of 10 mg NH_4^+/g clinoptilolite, removal efficiencies of 94 and 99% were obtained for ammonia and potassium respectively. In an earlier study (Beler-Baykal *et al.* 2004), the authors showed that ammonia desorption became more effective as the pH in the desorbing water rose from 7 to 13.

24.5.3 Amount of zeolite

Large amounts of zeolite are required to remove ammonia from urine. If a capacity of 10 mg NH_4^+/g of zeolite is assumed and daily ammonia production per capita is 12 $g{\cdot}p^{-1}{\cdot}d^{-1}$ (ureolyzed urine), the amount of zeolite needed to treat one person's ammonia load is about 1.2 $kg{\cdot}p^{-1}{\cdot}d^{-1}$ or 440 $kg{\cdot}p^{-1}{\cdot}y^{-1}$. For comparison, if the ammonia was to be recovered by struvite precipitation, the resulting amount of solid would be 5.5 times less (approximately 80 $kg{\cdot}p^{-1}{\cdot}y^{-1}$). The loaded zeolite can be used directly as fertilizer, or the ammonia can be recovered by regenerating zeolite with NaCl and NaOH. In the latter mode of operation, treatment with zeolite is a method used to separate and possibly concentrate ammonia and potassium.

24.5.4 Blackwater and greywater

Li *et al.* (2009) reviewed various technological approaches to greywater treatment and reuse, while Widiastuti *et al.* (2008) looked specifically at potential applications of zeolite for greywater treatment. Studies in the literature on the use of ion exchange material during or after aerobic and anaerobic treatment of wastewater suggest that ion exchange with zeolites is a potential approach for ammonia removal from blackwater (Milan *et al.* 1997, Tada *et al.* 2005, Thornton *et al.* 2007).

24.5.5 Use of zeolites as fertilizer carrier

Due to their high ion-exchange capacity, zeolites are used as soil conditioners as well as carriers of fertilizer, pesticides and fungicides. Their advantage as a carrier is that the fertilizer is released gradually. As the zeolite carrying the fertilizer is applied at the beginning of vegetation growth, it supplies fertilizers almost evenly during the whole plant growing season (Reháková *et al.* 2004).

Zeolites not only adsorb macronutrients such as ammonium or potassium, but can also adsorb heavy metals such as copper, cadmium, lead, zinc and chromium (Cincotti *et al.* 2001, Inglezakis *et al.* 2004). They may also transfer heavy metals from source-separated streams, especially blackwater or greywater, to agricultural soils. Heavy metal transfer should be less of a concern for source-separated urine, since its heavy metal content is very low (Ronteltap *et al.* 2007b).

Another concern besides heavy metal uptake is the adsorption of organic micropollutants. Studies on the removal of organics such as chlorophenols, nitrophenols, benzene and ethylbenzene by the use of zeolites indicated that the latter have complex interactions with organic matter (Cincotti *et al.* 2001, Sismanoglu and Pura 2001, Apreutesei *et al.* 2009, Jorgensen and Weatherley 2003). More research is needed to evaluate the possible take up and release of micropollutants from source-separated waste streams.

24.6 CONCENTRATION PROCESSES

Water removal is probably the only process that allows the recovery of all nutrients in a single solid product (Udert and Wächter 2012). Water removal processes, which have been tested with source-separated urine, are presented in the following subsections.

24.6.1 Freeze and thaw

The freeze and thaw process is based on the fact that ice crystals grow by incorporating only water molecules; solid and dissolved impurities are forced to the boundaries as freezing takes place. Lind *et al.* (2001) studied the concentrations of human urine and synthetically prepared solutions. Their results indicated that 80% of the nitrogen and phosphorus can be concentrated in 25% of the initial urine volume. In similar studies, Gulyas *et al.* (2004) and Ganrot *et al.* (2007) reported that the frozen ice still contained large amounts of impurities. However, Schmidt and Alleman (2006) achieved more than 99% reduction in contaminant concentrations and water recoveries of 85–95% with sequential steps of conventional eutectic freeze crystallization. The study was designed to provide a water recovery technology for spacecraft.

24.6.2 Electrodialysis

In the electrodialysis process, the liquid is exposed to an electric field, thus forcing the anions and cations to move in opposite directions. The ion flow is interrupted by several alternating anion and cation exchange membranes. As a result, the ions are accumulated. In treating source-separated urine by this process, Pronk *et al.* (2007) achieved concentration factors of 3.5 and 4.1 in continuous and in batch mode respectively. The nutrient concentration was limited due to the water transport by osmosis and electro-osmosis (Pronk *et al.* 2006). About 15% of the ions remained in the diluate in continuous-mode experiments (Pronk *et al.* 2006). A key advantage of electrodialysis is the removal of micropollutants by membranes (Escher *et al.* 2006).

24.6.3 Reverse osmosis

Reverse osmosis is a very common process for producing drinking water from seawater with a typical water recovery of 35 to 45% (Greenlee *et al.* 2009). This process was also tested with untreated and acidified stored urine (Ek *et al.* 2006). Nitrogen recovery was substantially higher with acidified urine. When 80% of the water was removed, thus concentrating the solution by a factor of five, only 79% of the nitrogen was recovered with urine at a pH value of 9.2, while up to 98% could be recovered at a pH value of 6.0. Some phosphorus loss occurred, probably by particle removal in the pre-filtration step.

24.6.4 Distillation

Distillation is a process in which high temperatures or low pressures are used to remove water or other liquids from a solution. Energy can be recovered by vapour compression. To prevent the volatilization of ammonia (NH_3) from stored urine, the pH has to be greatly decreased: by adding 13 g H_2SO_4 to one litre of stored urine, Ek *et al.* (2006) recovered 95% of the nitrogen during distillation. The pH value after acid additions was 4.5. Udert and Wächter (2012) used a biological process, namely nitrification, to lower the pH. The product of nitrification was an ammonia nitrate solution with a molar ratio of 1:1 and pH values of between 6.2 and 7.0. They captured more than 97% of the total nitrogen in the final solid product.

24.6.5 Energy demand

Udert and Wächter (2012) estimated the primary power required to remove the water from one person's urine. Lyophilization (freeze drying) had by far the highest power demand (510 $W{\cdot}p^{-1}$), followed by freeze and thaw (49 $W{\cdot}p^{-1}$), distillation without energy recovery (44 $W{\cdot}p^{-1}$), distillation with 85% energy recovery (21 $W{\cdot}p^{-1}$), electrodialysis (6.1 $W{\cdot}p^{-1}$) and reverse osmosis (1.9 $W{\cdot}p^{-1}$). However, only lyophilization and distillation allow complete water removal. The authors pointed out that the power demand for distillation could be reduced to 6.0 $W{\cdot}p^{-1}$ if 80% of the water was removed by reverse osmosis before distillation with vapour compression.

24.7 CONCLUDING REMARKS

Struvite precipitation from source-separated urine and blackwater is a proven technology. The process is simple and its energy demand low. The only chemicals to be added are magnesium salts, plus a base in the case of blackwater. The main disadvantage of this process is its partial recovery of nutrients: if only magnesium is dosed, phosphate is the only compound to be recovered to any extent. If ammonia is to be recovered, additional phosphate has to be dosed. Iron, aluminium or calcium salts are less favourable for phosphate precipitation: iron and aluminium phosphates are not well suited for recycling, and $Ca(OH)_2$ addition produces high amounts of the co-precipitate $CaCO_3$. Adsorption processes, including ion-exchange, can be very interesting for nutrient removal in on-site reactors. However, they are less suitable for direct nutrient recycling to agriculture: the nutrient load is low, some adsorbents are too expensive, or the

Kofina A. N., Demadis K. D. and Koutsoukos P. G. (2007). The effect of citrate and phosphocitrate on struvite precipitation. *Crys. Growth & Des.*, **7**(12), 2705–2712.

Kuşçuoğlu S. (2008). Determination of the Factors Affecting on K-struvite Precipitation. MSc Thesis, Institute of Sciences, İstanbul Technical University, İstanbul (in Turkish).

Li F., Wichmann K. and Otterpohl R. (2009). Review of the technological approaches for grey water treatment and reuses. *Sci. Total Environ.*, **407**, 3439–3449.

Lind B-B., Ban Z. and Byden S. (2000). Nutrient recovery from human urine by struvite crystallization with ammonia adsorption on zeolite and wollastonite. *Bioresource Technol.*, **73**, 169–174.

Lind B-B., Ban Z. and Byden S. (2001). Volume reduction and concentration of nutrients in human urine. *Ecol. Eng.*, **16**, 561–566.

Liu Z. G., Zhao Q. L., Wang K., Qui W., Li W. and Wang J. F. (2008). Comparison between complete and partial recovery of N and P from stale human urine with MAP crystallization. *J. Environ. Eng. Sci.*, **7**, 223–228.

Milan Z., Sànchez E., Weiland P., de Las Pozas C., Borja R., Mayari R. and Rovirosa N. (1997). Ammonia ramoval from anaerobically treated piggery manure by ion exchange in columns packed with homoionic zeolite. *Chem. Eng. J.*, **66**, 65–71.

Ødegaard H. (1992). *Rensing av avøpsvann*, Tapir forlag, Trondheim, (in Norwegian).

Pidou M., Avery L., Stephenson T., Jefrey P, Parsons S. A., Liu S., Memon F. A. and Jeferson B. (2008). Chemical solutions for greywater recycling. *Chemosphere*, **71**, 147–155.

Pronk W., Biebow M. and Boller M. (2006). Electrodialysis for recovering salts from a urine solution containing micropollutants. *Environ. Sci. Technol.*, **40**(7), 2414–2420.

Pronk W., Zuleeg S., Lienert J., Escher B., Koller M., Berner A., Koch G. and Boller M. (2007). Pilot experiments with electrodialysis and ozonation for the production of a fertiliser from urine. *Water Sci. Technol.*, **56**(5), 219–227.

Reháková M., Cuvanová S., Dzivák M., Rimár J. and Gaval'Ová Z. (2004). Agricultural and agrochemical uses of natural zeolite of clinoptilolite type. *Curr. Opin. Solid St. M.*, **8**, 397–404.

Rittmann B. E., Mayer B., Westerhoff P. and Edwards M. (2011). Capturing the lost phosphorus. *Chemosphere*, **84**(6), 846–853.

Ronteltap M., Maurer M. and Gujer W. (2007a). Struvite precipitation thermodynamics in source–separated urine. *Water Res.*, **41**, 977–984.

Ronteltap M., Maurer M. and Gujer W. (2007b). The behaviour of pharmaceuticals and heavy metals during struvite precipitation in urine. *Water Res.*, **41**, 1859–1868.

Ronteltap M., Maurer M., Hausherr R. and Gujer W. (2010). Struvite precipitation from urine – influencing factors on particle size. *Water Res.*, **44**, 2038–2046.

Sakthivel S. R., Tilley E. and Udert K. M. (2012). Wood ash as a magnesium source for phosphorus recovery from source-separated urine. *Sci. Total Environ.*, **419**, 68–75.

Saltalı K., Sarı A. and Aydın M. (2007). Removal of ammonium from aqueous solution by natural Turkish (Yıldızeli) zeolite for environmental quality. *J. Hazard. Mater.*, **141**, 258–263.

Schmidt J. M. and Alleman J. E. (2006). Macro- and micro-scale contaminant removal and resource recovery from urine via the freeze concentration process. Earth&Space 2006: Engineering, Construction, and Operations in Challenging Environments Proceedings of the 10th Biennial ASCE.

Sertaç B. (2012). Recovery of Nitrogen and Phosphorus Removal from Source-Separated Urine by Chemical Precipitation. MSc Thesis, Institute of Sciences, İstanbul Technical University, İstanbul, Turkey (in Turkish).

Sismanoglu T. and Pura S. (2001). Adsorption of aqueous nitrophenols on clinoptilolite. *Colloid. Surface A*, **180**, 1–6.

Tada C., Yang Y., Hanaoka T., Sonoda A., Ooi K. and Sawayama S. (2005). Effect of natural zeolite on methane production for anaerobic digestion of ammonium rich organic sludge. *Bioresource Technol.*, **96**, 459–464.

Tatlı M. B. (2006). Determination of Conditions Influencing in Nitrogen Recycling from Human Urine by Struvite Precipitation. MSc Thesis, Institute of Sciences, İstanbul Technical University, İstanbul, Turkey (in Turkish).

Tchobanoglous G., Burton F. L. and Stensel H. D. (2004). Metcalf &Eddy Wastewater Engineering. Treatment and Reuse, fourth ed. McGraw–Hill, Boston.

Tilley E., Atwater J. and Mavinic D. (2008). Recovery of struvite from stored human urine. *Environ. Technol.*, **29**(7), 797–806.

Thornton A., Pearce P. and Parsons S. A. (2007). Ammonium removal of digested sludge liquors using ion exchange. *Water Res.*, **41**, 433–439.

Tünay O., Kabdaşlı I. and Tatlı M. B. (2009). Nitrogen removal and recovery from human urine by struvite precipitation. *Int. J. Environ. Waste Man.*, **3**(3/4), 382–392.

Udert K. M., Larsen T. A. and Gujer W. (2003a). Estimating the precipitation potential in urine –collection system. *Water Res.*, **37**, 2667–2677.

Udert K. M., Larsen T. A., Biebow M. and Gujer W. (2003b). Urea hydrolysis and precipitation dynamics in a urine collection system. *Water Res.*, **37**, 2571–2582.

Udert K. M. and Wächter M. (2012). Complete nutrient recovery from source-separated urine by nitrification and distillation. *Water Res.*, **46**(2), 453–464.

Udert K. M., Brown-Malker S. and Keller J. (2013). Electrochemical systems. In: *Source Separation and Decentralization for Wastewater Management*, T. A. Larsen, K. M. Udert and J. Lienert (eds.), IWA Publishing, London, UK, pp. 321–335.

US EPA (2009). Nutrient Control Design Manual, State of Technology Review Report. EPA/600/012, United States Environmental Protection Agency.

Wang S. and Peng Y. (2010). Natural zeolites as effective adsorbents in water and wastewater treatment. *Chem. Eng. J.*, **156**, 11–24.

Widiastuti N., Wu H., Ang M. and Zhang D. (2008). The potential application of natural zeolite for greywater treatment. *Desalination*, **218**, 271–280.

Wilsenach J. A., Schuurbiers C. A. H. and van Loosdrecht M. C. M. (2007). Phosphate and potassium recovery from source separated urine through struvite precipitation. *Water Res.*, **41**, 458–466.

Zeeman G. and Kujawa-Roeleveld K. (2013). Anaerobic treatment of source-separated domestic wastewater. In: *Source Separation and Decentralization for Wastewater Management*, T. A. Larsen, K. M. Udert and J. Lienert (eds.), IWA Publishing, London, UK, pp. 307–320.

Zeng L. and Li X. (2006). Nutrient removal from anaerobically digested cattle manure by struvite precipitation. *J. Environ. Eng. Sci.*, **5**, 285–294.

Chapter 25

Membrane processes

Gregory Leslie and Zenah Bradford-Hartke

25.1 INTRODUCTION

The full spectrum of liquid phase, pressure driven, membrane separation processes may be used in decentralized wastewater treatment applications. Micro-porous membranes such as microfiltration (MF) and ultrafiltration (UF) membranes may be used in place of a clarifier to remove micron-sized particles such as microorganisms and suspended solids; reducing effluent turbidity and providing partial or full disinfection. These membranes separate components by size exclusion in the same way a sieve or filter does and can be used to remove suspended particles and macromolecules. Semi-permeable membranes such as reverse osmosis (RO) and nanofiltration (NF) membranes can be deployed for removing monovalent, divalent and trivalent ions, residual carbonaceous, nitrogenous and phosphorous nutrients. In addition, NF and RO membranes may also be used for the removal of some trace organics and micropollutants.

Membranes offer many advantages over conventional solid-liquid separation techniques. The technology is very compact, does not rely on gravity separation and provides consistent product quality over a range of pollutant loading rates. Membranes are modular and readily scalable for decentralized applications ranging from a single domicile, through to apartment/office buildings or entire sub-divisions. Unlike conventional separation processes, the challenge with membrane processes in wastewater applications is maintaining consistent throughput capacity; filtrate or product quality is generally very consistent. Consequently, the focus is on developing a design that is appropriate for the feedwater and an operating regime to manage surface fouling which reduces treatment throughput. This includes selection of appropriate pre-treatment of the feed, membrane loading rate, and backwash/cleaning strategy.

The following chapter contains information on the use of membranes to treat grey- and blackwater streams in decentralized applications as well as source-separated urine treatment. Emphasis is placed on key design parameters and reported performance.

25.2 BASIC FEATURES OF MEMBRANE SYSTEMS

There are several excellent texts on the features of membrane systems (Mulder 1991). Specifics on the use of membranes in wastewater treatment for reuse can be found in Chapters 7 to 9 of the Metcalf and Eddy book written by Asano *et al.* (2007) while the definitive text on membrane bioreactors (MBRs) in its second edition is provided by Judd and Judd (2006).

The modular nature of membrane systems is ideally suited to decentralized applications. The modules commonly used in decentralized applications include variants of flatsheet and hollowfibre/tubular configurations. Sheet membranes can be configured as plate and frame or spiral wound modules; while fibres or tubes can be housed in pressure vessels or immersed in a process tank. Modules can be operated in either dead end or cross flow mode.

Microfiltration and ultrafiltration systems usually consist of pressurized vessels or submerged modules. In submerged systems, permeate is collected by applying a vacuum pressure to the permeate rather than pressurizing the feed. The filtration process is interrupted every 10 to 30 min and the direction of the flow across the membrane is reversed for 1 to 2 min to backwash retained solids off the membrane. The waste stream generated by the backwash is typically 5 to 10% of the total flow treated in a day. MBRs typically use submerged hollow fibre or flat plate configurations which are either gravity driven by maintaining a set water height or pressure driven (vacuum).

Nanofiltration and reverse osmosis modules are packaged in spiral wound configurations. Separation is achieved across a thin film deposited on a porous support. Consequently, flow across the membrane only occurs in one direction as reverse flow, or backwashing, would detach the film from the support. As a result NF and RO are operated in continuous, cross-flow mode to avoid the accumulation of material on the surface. The concentrate stream that discharges on a continuous basis accounts for 15 to 25% of the feed flow depending on the feedwater. NF and RO rely on extensive pre-treatment to prevent clogging of the narrow feed channels of the spiral wound modules.

The treatment capacity of membrane systems in decentralized applications is determined by the membrane flux. Flux is defined as volume treated per unit area per unit time. Non-SI units of $L{\cdot}m^{-2}{\cdot}h^{-1}$ (Lmh) or $gal{\cdot}ft^{-2}{\cdot}d^{-1}$ are used in preference to the SI units of $m^3{\cdot}m^{-2}{\cdot}s^{-1}$. Use of non-SI units results in a flux for wastewater applications that ranges from 10–100 Lmh for MF and UF systems and 15–25 Lmh for NF and RO systems. The flux is directly proportional to the applied pressure and indirectly proportional to the viscosity and the resistance to flow across the membrane.

25.2.1 Precautions for decentralized systems

Membranes are very susceptible to damage by gross pollutants, tissues, sanitary items, kitchen scraps, plastics, hair and lint. Mechanical screening using self-cleaning metal screens is recommended by most membrane manufacturers. Many decentralized systems use a maximum screen size of 3 mm (Melin *et al.* 2006), however smaller apertures of 1 mm have been used in some applications (Paris and Schlapp 2010, Merz *et al.* 2007, Friedler *et al.* 2006).

Membrane processes used in decentralized wastewater treatment applications must be adapted to the wide variations in diurnal flow profiles. Intermittent operation is possible provided the system can be run in idle mode without fouling of the membrane surface (Ravazzini *et al.* 2005). In extreme cases, such as decentralized seasonal treatment of waste from a holiday resort, it is necessary to discontinue operation and preserve the membranes. As a rule of thumb it is possible for the membranes to idle for two to three weeks without preservation, provided the potential for biological growth is low.

25.3 DECENTRALIZED APPLICATIONS

25.3.1 Introduction

Membranes can be used in decentralized systems to treat both source-separated and combined streams. Source-separated applications include treatment of greywater, blackwater and urine. The key design issues such as hydraulic loading, separation efficiency, fouling and cleaning are specific to each application and are summarized in the following sections.

25.3.2 Greywater

Greywater can be treated by membrane processes for the purposes of on-site reuse via toilet flushing or other non-potable applications. MF and UF membranes are used to remove suspended solids and microorganisms and thereby reduce the potential for human contact with microbial pathogens at the point of reuse. Notwithstanding this, it is not uncommon to deploy a dedicated disinfection system, such as UV as a final barrier for disinfection, after the membrane process. In some cases, such as the London Millennium Dome, the membranes were incorporated into a bioreactor to treat greywater, however, in such applications, the variability of nutrients in the influent limits the performance of the biological system (Smith *et al.* 2000). Greywater effluent after treatment in an MBR is expected to be odourless (Merz *et al.* 2007) and colourless (Kim *et al.* 2009), which makes reuse less offensive to users. Energy use in greywater MBRs can range from 0.5 to 1.5 $kWh{\cdot}m^{-3}$ (Friedler and Hadari 2006).

25.3.2.1 Performance

Operating fluxes on greywater can range from 8 to 38 Lmh (Kraume *et al.* 2010) and achieve a net water recovery of greater than 80%, however, the recovery will vary depending on the suspended solids content of the greywater (Taniguchi 1994). Removal of turbidity and suspended solids is consistently high for greywater applications and can range from 98–100%.

Greywater is the only separated stream with a high soap content and surfactant removal in MBRs treating greywater are reported to range from 97% (Merz *et al.* 2007) to more than 99% (Huelgas and Funamizu 2010).

Microbial removal efficiencies are normally reported as log removal (LR) of feed concentration (C_{feed}) to permeate concentration (C_{permeate}):

$$LR = \log_{10}\frac{C_{\text{feed}}}{C_{\text{permeate}}}$$

LR vary greatly. This is not due to the properties of the membranes used but rather the variable microbial loads in lightly contaminated versus highly contaminated greywater. Reported faecal coliform removals for MBRs range from 99% (Merz *et al.* 2007) to 100% (Kraume *et al.* 2010) while total coliform removal of more than 91% is expected (Winward *et al.* 2008). Winward *et al.* (2008) also reported successful Clostridia, E. coli and enterococci removal and predicted that no pathogenic protozoa would pass through the membrane based on their size. Kim *et al.* (2009) showed that a MF membrane with a pore size of 0.4 μm achieved 100% E. coli removal and 30% Salmonella and Staphylococcus removal.

Removal of organic material and nutrients from greywater using MBRs is dependent on the biological processes in the MBR rather than the membranes. Studies of MBRs treating greywater consistently report removals in the range of 85–99% though Friedler and Gilboa (2010) report a lower COD removal efficiency of 70% with a SRT of 15–20 d (Table 25.1). Winward *et al.* (2008) reported very low COD removal of 46% when treating low strength and 89% for high strength greywater in an MBR. The difference in removal efficiencies is a result of the higher influent concentration; the permeate quality does not vary considerably under the increased organic load suggesting that the biological degradation rates were similar in both set-ups (Table 25.1). Details of the set-ups are given in Table 25.2.

25.3.2.2 Challenges and limitations

Membrane fouling and hence cleaning will depend on the feedwater quality and examples cited are to be used as a rough guide only. Kraume *et al.* (2010) recommends cleaning every 3 to 4 months. Contamination of the filtrate was observed by Merz *et al.* (2007) when treating greywater in a MBR. The proximity of the treated water tank to the bioreactor was suggested as a possible cause as the tank was open. Contamination was also reported by Friedler *et al.* (2006) who suspected sample contamination or bacterial transfer as aerosols.

Table 25.1 COD and BOD removal from greywater in membrane bioreactors (MBRs).

COD Removal [%]	BOD Removal [%]	Reference
99		Kim *et al.* (2009)
96		Huelgas and Funamizu (2010)
91		Kraume *et al.* (2010)
92.2		Paris and Schlapp (2010)
46	96	Winward *et al.* (2008), low strength
89	99	Winward *et al.* (2008), high strength
70	>96	Friedler and Gilboa (2010)
85	94	Merz *et al.* (2007)
	100	Taniguchi (1994)

25.3.3 Blackwater

MF and UF membranes used to treat blackwater in decentralized applications will form part of an integrated suspended growth bioreactor. The biological systems are generally operated aerobically, however, there are several examples of anaerobic systems followed by membrane filtration or a fully integrated anaerobic MBRs.

25.3.3.1 Performance

Typical fluxes for MBRs operating on blackwater range from 8 and 10 Lmh (van Voorthuizen *et al.* 2008) to 16 Lmh (Boehler *et al.* 2007). Removal of organic material and nutrients from blackwater using MBRs is dependent on the biological processes in the MBR rather than the membranes. For aerobic MBRs, COD removal is around 90% (van Voorthuizen *et al.* 2008). Removal of organic material in the anaerobic MBR studied by van Voorthuizen *et al.* (2008) was slightly lower at 86%.

Complete nitrogen removal is possible in MBRs (Boehler *et al.* 2007). Without enhanced biological phosphorus removal, the amount of phosphorus removed is limited to the amount used for biomass growth and therefore linked to sludge production. Boehler *et al.* (2007) reported that with intermittent aeration up to 65% of the incoming phosphorus was removed. In this study, intermittent

Table 25.2 Decentralised membrane applications for treatment of greywater.

	Lesjean and Gnirss (2006)	**Merz *et al.* (2007)**	**Friedler and Gilboa (2010)**	**Taniguchi (1994)**	**Kim *et al.* (2009)**	**Paris and Schlapp (2010)**
Type	MBR	MBR (3L lab-scale reactor)	Pilot plant MBR (UF)	Biological treatment, UF tubular module	Anaerobic, anoxic, aerobic submerged MF	MBR with submerged UF modules
Location	Berlin-Stahnsdorf WWTP	Sports and leisure club, Morocco	Israel	Department store Tokyo	Yonsei Univ. Seoul, Korea	Dormitory, Vietnam
Water source	GW from 10 private apartments and one office building (50 p.e.)	GW from showers	GW (from baths, showers and wash-basins) of 14 apartments	GW from kitchen wastes in department store	GW from 170 apartments	GW from personal hygiene, laundry, washing and cooking
Flux/flow	21 $L \cdot h^{-1}$	8 Lmh	>6.25 $L \cdot h^{-1}$	Outlet 24,000 $L \cdot h^{-1}$		Q_{max} of 229 $L \cdot h^{-1}$ av. 125–188 $L \cdot h^{-1}$
Pore size		0.1 µm	100,000 Da	Cut-off 20,000 Da	0.4 µm	38 nm
Detail	SRT 4d, HRT 2h	Hollow fibre UF (Zenon)	Berghoff tubular UF, Crossflow	80% recovery	Hollow fibre HRT 6h	
Pressure		TMP 0.025 bar		8.3–9.3 bar		

aeration fostered the growth of phosphorus accumulating bacteria. Nutrients should pass through the MF or UF membranes with only the particulate bound fraction removed as demonstrated by van Voorthuizen *et al.* (2008) who report 95 to 100% transmission of phosphorus and 84–97% transmission of nitrogen through the membrane. Details of selected case studies are given in Table 25.3.

Table 25.3 Decentralized membrane applications for treatment of blackwater.

	Boehler *et al.* (2007)	**van Voorthuizen *et al.* (2010)**	**van Voorthuizen *et al.* (2010)**	**van Voorthuizen *et al.* (2010)**
Redox conditions	Aerobic	Anaerobic	Aerobic	Anaerobic
Reactor type	MBR 3–3.4 m^3	4 L MBR, submerged tubular UF	MBR 4 L, submerged tubular UF	UASB, submerged UF membranes
Location	Zermatt, ski resort	The Netherlands	The Netherlands	The Netherlands
Water source	Cable car station blackwater	School toilet blackwater using ~5 L water per flush	School toilet blackwater using ~5 L water per flush	School toilet blackwater using ~5 L water per flush
Application	Flushwater and sinks	Nutrient recovery from MBR effluent	Nutrient recovery from MBR effluent	Nutrient recovery from MBR effluent
Flux/flow	16 Lmh/ 300 $L \cdot h^{-1}$	8 Lmh	9.7 Lmh	10 Lmh
Pore size	0.04 µm	Cut-off of 250 kDa	Cut-off of 250 kDa	Cut-off of 250 kDa
Detail	Flat sheet Martin system AG	MEMOS VFU 250-a	MEMOS VFU 250-a	MEMOS VFU 250-a
Membrane material		PVDF	PVDF	PVDF

25.3.3.2 Challenges and limitations

The permeate from MBRs can have a yellowish colour (Abegglen and Siegrist 2006). Colour is an aesthetic problem rather than a safety concern but will still hinder the implementation of decentralized treatment and reuse due to user acceptance for indoor uses such as laundry or toilet flushing.

The energy requirements of decentralized MBRs can range from 8.1 $kWh{\cdot}m^{-3}$ for a decentralized MBR to 11.5 $kWh{\cdot}m^{-3}$ for combined MBR UV systems (Boehler *et al.* 2007). In contrast an MBR system with UV in a centralized application benefits from economies of scale and requires typically only 3–5 $kWh{\cdot}m^{-3}$.

Noise may be an issue for household or apartment block decentralized systems, preventing continuous operation particularly during the night (Abegglen and Siegrist 2006). Intermittent operation is possible provided measures are taken to prevent anaerobic conditions and odour generation (Boehler *et al.* 2007). However, in one case the installation of a ventilation system and the use of an airtight MBR plant did not prevent noticeable odours during periods of intermittent operation (Abegglen and Siegrist 2006).

Van Voorthuizen *et al.* (2008) conducted trials on aerobic and anaerobic MBRs treating blackwater over a period of 165 days. The permeability of the membranes was maintained by a combination of relaxation, backwash and chemical cleaning. Interestingly, the membrane permeability steadily declined for the anaerobic system, while the permeability in the aerobic MBR stabilized after an initial period of decline. Notwithstanding this, anaerobic MBR systems generally require less than one third of the energy needed by aerobic MBRs. Consequently, anaerobic MBRs are the subject of a number of research efforts, to identify and manage the cause of fouling, to assess expected membrane life and refine treatment costs (Fane and Fane 2005).

25.3.4 Source-separated urine

Membranes are used to treat urine for the purposes of nutrient recovery. The objective of the treatment step may be to either concentrate phosphate and nitrogenous components using a membrane with high salt rejection, such as reverse osmosis, prior to further treatment. Alternatively, the objective may be to separate the trivalent phosphates and micropollutants from urea via a nanofiltration membrane to produce a pollutant free nitrogen source. In either case the use of flat sheet, spiral wound NF or RO membranes, necessitates the use of physical pre-treatment to prevent blocking of the membrane module and chemical pre-treatment to prevent the accumulation of salts and precipitates on the membrane surface.

25.3.4.1 Performance

Ek *et al.* (2006) investigated the possibility of using polyamide (PA) reverse osmosis membranes to concentrate source-separated urine in which the urea was hydrolyzed during storage. The tested pre-treatment methods were a 0.5 mm sieve, 5 μm cartridge filters and ultrafiltration membranes with a cut-off of 100,000 Daltons. The membrane flux increased with the improved feed quality after UF though the difference was not enough to account for the higher cost of the ultrafiltration membranes.

Pronk *et al.* (2006) applied nanofiltration membranes to separate pharmaceutical and estrogenic compounds from source-separated urine and to produce a nitrogen enriched permeate as liquid fertilizer. Filtration was carried out at 20 bar with fluxes up to ~125 Lmh. The nanofiltration membranes only removed 30–40% of the COD which indicates a high fraction of low molecular weight organics present in the urine (Pronk *et al.* 2006). Nitrogen transmission through NF was initially 80% for fresh urine at pH 7 and declined as the uncharged, polar urea hydrolyzed into the charged ammonium ions which can only be partially rejected by the NF membrane. In contrast, rejection of the trivalent phosphate was consistently more than 90% which could facilitate nutrient recovery from the concentrate stream.

Unlike nanofiltration, reverse osmosis membranes reject both nitrogen and phosphorus which are transferred to the concentrate stream. Trials at 80% recovery by Ek *et al.* (2006) produced a concentrate containing 95% of the influent total N and more than 99% of the total phosphorus. The pH of the hydrolyzed urine was adjusted to pH 7. At higher pH, the phosphorus separation remained unchanged but the nitrogen removal was less (80–90%).

25.3.4.2 Challenges and limitations

In decentralized applications, the flow of urine from a single household would be low and the cost of installing RO membranes would likely inhibit applications. Membrane treatment of urine may be more feasible if the scale were increased to an apartment block or complex, but this would require urine storage. The spontaneous hydrolysis of urea in stored urine is unfavourable for micropollutant removal using nanofiltration as the rejection decreases with the increase in pH (Pronk *et al.* 2006). Furthermore, the precipitates which are formed as a consequence of urea hydrolysis (Udert *et al.* 2003) can lead to scaling on the membranes. Based on pilot scale tests, Ek *et al.* (2006) reported the specific energy demand for RO treatment as 8 $kWh{\cdot}m^{-3}$ when operated at 30°C with an additional 4$kWh{\cdot}m^{-3}$ to reach this temperature.

25.3.5 Combined wastewater

In set-ups without source separation, membranes are used to treat combined wastewater with the objective of either avoiding the cost of connecting a new development to a centralized collection system, producing recycled water for local reuse or reducing the costs associated with flow and load to a centralized treatment plant. Depending on the number of connections to the membrane system, or the availability of space, the membrane processes may be used as the final filtration step in a conventional wastewater system (Ravazzini *et al.* 2005), or as an integrated MBR (Zanetti *et al.* 2010, Gnirss *et al.* 2003). Ravazzini *et al.* (2005) conducted trials on raw sewage (RS) and clarifier effluent, while Zanetti *et al.* (2010) and Gnirss *et al.* (2003) operated on screened raw sewage.

25.3.5.1 Performance

Fluxes in MBRs operating on combined wastewater are similar to fluxes for blackwater feed: Gnirss *et al.* (2003) reported fluxes between 10.8–12.2 Lmh, and Annaka *et al.* (2006) observed a maximum flux of 31 Lmh. In either application, particulate removal efficiency based on the reduction in turbidity and suspended solids was greater than 99%.

Similarly, both configurations achieved very high levels of pathogen removal. The Ravazzini system operating on primary screened effluent using pressurized microfiltration membranes achieved 5 and 7 LR of total and faecal coliforms, respectively, while the MBR process achieved 6.02 LR for total coliform and 6.98 LR for faecal coliforms (Asano *et al.* 2007, Zanetti *et al.* 2010). Coliphage removal ranges from 2 LR (Asano *et al.* 2007) to 4.44 LR (Zanetti *et al.* 2010). High LR by MF is also reported for E. coli (6.77), enterococci (5.77), and thermo-tolerant coliforms (6.72) (Zanetti *et al.* 2010). Notwithstanding these high LR values, regulations and guidelines including the US EPA Filtration Guidance Manual limit the LR for a single unit process to 4 (USEPA 2005). Consequently, an additional disinfection process such as chlorination or ultraviolet irradiation is typically applied to MBR filtrate (WJP Solutions 2011).

Organic removal in MBRs is high (>95%) while reported removals by direct filtration are much lower at 37–42% (Table 25.4). Microfiltration and ultrafiltration will remove only the particulate organic fraction which means the removal efficiency is directly related to the feedwater characteristics.

Table 25.4 Organic and particulate removal from combined wastewater.

COD Removal [%]	Particulate removal [%]	Reference
37		Ravazzini *et al.* 2005, raw sludge
42		Ravazzini *et al.* 2005, primary effl.
96		Gnirss *et al.* (2003)
95		Lesjean *et al.* (2003)
>97	Susp. solids ~100	Zanetti *et al.* (2010)
95		Abegglen and Siegrist (2006)

Dissolved nutrients will not be removed by the membranes, but nutrient removal can be achieved when coupled with biological treatment in a MBR. Nitrogen and phosphorus transmission through the membrane range from 90 to 100% and

81 to 92%, respectively, but again this is related to the particulate bound fraction in the influent, the biological processes (denitrification, biological phosphorus removal) and the chemical processes (phosphate precipitation) in the reactor (Table 25.5). High levels of nutrient removal can be achieved in MBRs with reported values of 99% TP and 94% TN (Lesjean *et al.* 2003) with enhanced biological phosphorus removal. It should be noted that salts will pass through the treatment process and may accumulate in the reuse system. Further details of selected cases are given in Table 25.6.

Table 25.5 Nutrient transmission for membranes treating combined wastewater.

Phosphorus [%]		Nitrogen [%]		Reference
Total P	81	Total N	90	Ravazzini *et al.* (2005), raw sludge
Phosphate	98	Ammonia	100	
Total P	92	Total N	100	Ravazzini *et al.* (2005), primary effl.
Phosphate	94	Ammonia	100	
Total P	<1	Total N	18	Gnirss *et al.* 2003
Phosphate	1	Ammonia	<2	
Total P	1	Total N	6	Lesjean *et al.* 2003
		Ammonia	<2	

25.3.5.2 Challenges and limitations

Challenges for decentralized applications include maintenance, cleaning and monitoring of membrane integrity (Annaka *et al.* 2006). Other issues such as noise, odour and permeate reuse are equally important but must be evaluated on a case by case basis.

25.4 INDUSTRY TRENDS

Greater emphasis on the need to reuse water and reduce the load on existing wastewater treatment plants will drive the use of membranes in decentralized applications. In addition, broader changes in the water industry such as carbon accounting, the persistence of trace contaminants and the application of risk management techniques will impact the use, design and performance of membrane processes. The following section considers the effect of these industry trends on the use of membranes in decentralized applications.

25.4.1 Accounting for the cost of carbon

The amount of carbon emitted per cubic meter of water (kg $CO_2 \cdot m^{-3}$) will depend on the energy source for electricity generation, the amount of chemicals used in the

Table 25.6 Decentralized membrane applications for treatment of combined wastewater.

	de Koning *et al.* (2008)	**de Koning *et al.* (2008)**	**Zanetti *et al.* (2010)**	**WJP solutions (2011)**	**Gnirss *et al.* (2003)**	**Lesjean *et al.* (2003)**	**Ravazzini *et al.* (2005)**	**Ravazzini *et al.* (2005)**	**Abegglen and Siegrist (2006)**
Type	MF, RO, chlorination	Screening, chlorination, MF, CF, RO, UV, infiltr. pond	MBR	MBR,UV, RO, chlorination	Bio-P, MBR	Bio-P, MBR	UF	UF	Aerobic MBR, activated carbon
Location	Sydney Olympic park	Torreele, Belgium	University of Bologna	Melbourne University	Berlin	Berlin	Delft University	Delft University	Swiss private home
Water source	Secondary effluent and stormwater	Secondary effluent	WW from 8,000 p.e.	Blackwater and greywater	Degritted raw sewage from 10,000 p.e	Degritted raw sewage from 10,000 p.e.	Raw sewage	Primary clarifier effluent	Greywater and brown-water
Effluent reuse	On-site	Groundwater recharge	None	On-site	Groundwater recharge	Groundwater recharge	Pilot	Pilot	Flushwater, gardening
Flux / flow	310,000 $L \cdot h^{-1}$	290,000 $L \cdot h^{-1}$	21 Lmh	1,300 $L \cdot h^{-1}$	10.8–12.2 Lmh	10.8 Lmh	120 Lmh	160 Lmh	15 Lmh, max 25 Lmh
Pore size	0.2 μm MF		0.4 μm		0.2 μm	0.2 μm			0.04 μm

process and life of consumable items such as the membrane. Using estimates from the Australian Greenhouse Gas office (2010) the carbon emission can range from 0.5 kg $CO_2 \cdot m^{-3}$ for microfiltration and ultrafiltration to 1.5 to 2.5 kg $CO_2 \cdot m^{-3}$ for reverse osmosis systems. In these applications approximately 60% of the emissions are associated with energy consumption. Reducing the carbon footprint of decentralized systems, particularly on combined and blackwater applications using MBR processes will require a move from aerobic to anaerobic operation.

25.4.2 Increased monitoring and regulation of trace contaminants

The persistence of trace organic contaminants through biological treatment processes poses a challenge for membrane processes in decentralized wastewater treatment and recycling applications. In decentralized applications an increased focus on the persistence of trace organics could result in extra measures for the on-site handling of residuals, particularly biosolids and backwash streams containing higher concentrations of these molecules. The additional regulations for monitoring and compliance will result in additional costs. It will be a challenging task for the local and national authorities to set commensurate regulations, which account for the relative importance of the different sources of trace organics in the urban environment.

25.4.3 Application of risk management procedures

Recent changes in water recycling guidelines and regulations have included the use of risk management approaches to ensure process reliability and final water quality through the continuous monitoring of performance along the treatment process. This is achieved through the use of Hazard Analysis and Critical Control Point (HACCP) techniques. HACCP relies on continuous monitoring of a critical performance parameter, such as particle counts, that indicates whether the membrane is performing as intended. The implication of these changes for decentralized systems will be an increase in the complexity and cost of on-site instrumentation and control. In decentralized facilities, the individual user may not be present to monitor or capable of maintaining these systems. Ideally, centralized monitoring would be required with technical staff that could be dispatched to sites as required. This would likely be too costly to implement at small facilities. This problem could be solved with low cost, reliable testing systems (Fane and Fane 2005)

25.5 CONCLUSIONS

Membrane processes can be used to deliver consistently high quality water in decentralized applications for the purpose of water reuse. The literature provides information on typical membrane loading rates and water recovery from pilot and some full-scale applications. A variety of systems are available for greywater, blackwater, urine and combined streams, however, the critical issues are associated with managing the variations in wastewater flow and providing adequate pre-treatment to protect the membrane from damage. Adopting appropriate cleaning strategies to manage system permeability is always challenging, however, based on information on energy consumption, a key research need is to reduce energy consumption through the greater use of anaerobic systems. More research is also required for low-cost reliable process control systems, which ensure stable operation of small on-site membrane systems without the need for frequent maintenance.

REFERENCES

Abegglen C. and Siegrist H. (2006). Domestic wastewater treatment with a small-scale membrane bioreactor. *Water Sci. Technol.*, **53**(3), 69–78.

Annaka Y., Hamamoto Y., Akatsu M., Maruyama K., Oota S. and Murakami T. (2006). Development of MBR with reduced operational and maintenance costs. *Water Sci. Technol.*, **53**(3), 53–60.

Asano T., Burton F., Leverenz H. L., Tsuchihashi R. and Tchobanoglous G. (2007). Water Reuse : Issues, Technologies, and Applications, Metcalf and Eddy: McGraw–Hill, New York, USA.

Boehler M., Joss S., Buetzer D., Holzapfel M., Mooser H. and Siegrist H. (2007). Treatment of toilet wastewater for reuse in a membrane bioreactor. *Water Sci. Technol.*, **56**(5), 63–70.

de Koning J., Bixio D., Karabelas A., Salgot M. and Schäfer A. (2008). Characterisation and assessment of water treatment technologies for reuse. *Desalination*, **218**(1–3), 92–104.

Ek M., Bergström R., Bjurhem J. E., Bjorlenius B. and Hellström D. (2006). Concentration of nutrients from urine and reject water from anaerobically digested sludge. *Water Sci. Technol.*, **54**(11–12), 437–44.

Fane A. G. and Fane S. A. (2005). The role of membrane technology in sustainable decentralized wastewater systems. *Water Sci. Technol.*, **51**(10), 317–325.

Friedler E. and Hadari M. (2006). Economic feasibility of on-site greywater reuse in multi-storey buildings. *Desalination*, **190**(1–3), 221–234.

Friedler E. and Gilboa Y. (2010). Performance of UV disinfection and the microbial quality of greywater effluent along a reuse system for toilet flushing. *Sci. Total Environ.*, **408**(9), 2109–2117.

Friedler E., Kovalio R. and Ben-Zvi A. (2006). Comparative study of the microbial quality of greywater treated by three on-site treatment systems. *Environ. Technol.*, **27**(6), 653–663.

Gnirss R., Lesjean B., Adam C. and Buisson H. (2003). Cost effective and advanced phosphorus removal in membrane bioreactors for a decentralised wastewater technology. *Water Sci. Technol.*, **47**(12), 133–139.

Huelgas A. and Funamizu N. (2010). Flat-plate submerged membrane bioreactor for the treatment of higher-load graywater. *Desalination*, **250**(1), 162–166.

Judd S. and Judd C. (2006). The MBR Book : Principles and Applications of Membrane Bioreactors in Water and Wastewater Treatment. 2nd edn, Elsevier, Amsterdam, The Netherlands.

Kim J., Song I., Oh H., Jong J., Park J. and Choung Y. (2009). A laboratory-scale graywater treatment system based on a membrane filtration and oxidation process – characteristics of graywater from a residential complex. *Desalination*, **238**(1–3), 347–357.

Kraume M., Scheumann R., Baban A. and El Hamouri B. (2010). Performance of a compact submerged membrane sequencing batch reactor (SM-SBR) for greywater treatment. *Desalination*, **250**(3), 1011–1013.

Lesjean B. and Gnirss R. (2006). Grey water treatment with a membrane bioreactor operated at low SRT and low HRT. *Desalination*, **199**(1–3), 432–434.

Lesjean B., Gnirss R., Adam C., Kraume M. and Luck F. (2003). Enhanced biological phosphorus removal process implemented in membrane bioreactors to improve phosphorous recovery and recycling. *Water Sci. Technol.*, **48**(1), 87–94.

Melin T., Jefferson B., Bixio D., Thoeye C., de Wilde W., de Koning J., van der Graaf J. and Wintgens T. (2006). Membrane bioreactor technology for wastewater treatment and reuse. *Desalination*, **187**(1–3), 271–282.

Merz C., Scheumann R., El Hamouri B. and Kraume M. (2007). Membrane bioreactor technology for the treatment of greywater from a sports and leisure club. *Desalination*, **215**(1–3), 37–43.

Mulder M. (1991). Basic Principles of Membrane Technology. Kluwer Academic, Dordrecht, The Netherlands.

Paris S. and Schlapp C. (2010). Greywater recycling in Vietnam – Application of the HUBER MBR process. *Desalination*, **250**(3), 1027–1030.

Pronk W., Palmquist H., Biebow M. and Boller M. (2006). Nanofiltration for the separation of pharmaceuticals from nutrients in source-separated urine. *Water Res.*, **40**(7), 1405–1412.

Ravazzini A. M., van Nieuwenhuijzen A. F. and van der Graaf J. H. M. J. (2005). Direct ultrafiltration of municipal wastewater: comparison between filtration of raw sewage and primary clarifier effluent. *Desalination*, **178**(1–3), 51–62.

Smith A., Khow J., Hills S. and Donn A. (2000). Water reuse at the UK's Millennium Dome. *Membrane Technol.*, **118**, 5–8.

Taniguchi Y. (1994). State of The Art Microfiltration Technology for Drinking Water and Waste Water Treatment in Japan. Conference Proceedings, Microfiltration I: Microfiltration for Water Treatment Symposium, National Water Research Institute (NWRI), Fountain Valley, CA, USA.

Udert K. M., Larsen T. A., Biebow M. and Gujer W. (2003). Urea hydrolysis and precipitation dynamics in a urine-collecting system. *Water Res.*, **37**(11), 2571–2582.

van Voorthuizen E., Zwijnenburg A., van der Meer W. and Temmink H. (2008). Biological black water treatment combined with membrane separation. *Water Res.*, **42**(16), 4334–4340.

Winward G. P., Avery L. M., Frazer-Williams R., Pidou M., Jeffrey P., Stephenson T. and Jefferson B. (2008). A study of the microbial quality of grey water and an evaluation of treatment technologies for reuse. *Ecol. Eng.*, **32**(2), 187–197.

WJP Solutions (2011). University of Melbourne Water Treatment Solution Profile. http://www.wjpsolutions.com.au/Projects (accessed on 28 April 2012)

Zanetti F., De Luca G. and Sacchetti R. (2010). Performance of a full-scale membrane bioreactor system in treating municipal wastewater for reuse purposes. *Bioresour. Technol.*, **101**(10), 3768–3771.

Chapter 26

Chemical oxidation processes

Urs von Gunten

26.1 INTRODUCTION

Various chemical oxidants have been widely used for many decades in water treatment, mainly for disinfection, but also to remove colour, taste and odour, as well as iron and manganese. In the last two decades, conventional oxidation with chemical oxidants and advanced oxidation processes (AOPs: use of $^{\bullet}OH$ radicals as the main oxidants) have also come into more widespread use for removing micropollutants first from drinking water and in recent years also from wastewater effluents. The knowledge gained in these applications can be used to assess the feasibility of oxidative treatment of source-separated wastewater (greywater, urine) and the effluent from decentralized reactors.

26.1.1 Common chemical oxidants

An overview of the main chemical oxidants used in water treatment is given in Table 26.1. Ozone is the best disinfectant and can even inactivate protozoan cysts and oocysts such as *Giardia lamblia* cysts (*G. lamblia*) and *Cryptosporidium parvum* oocysts (*C. parvum*) with reasonable exposures (integral of ozone concentration over reaction time). In contrast, these microorganisms (especially *C. parvum*) require prohibitive exposures to another oxidant, namely chlorine, which may lead to high concentrations of disinfection by-products. In any case, the use of chlorine dioxide is not advised for the inactivation of protozoa. Although the novel oxidant ferrate(VI) is a good disinfectant for bacteria (Sharma 2007), it is not expected to efficiently inactivate protozoa in view of its general reactivity, which is similar to or lower than chlorine (Lee and von Gunten 2010).

It can be seen that the DOC decreases dramatically from hydrolyzed urine to wastewater effluent. This is caused partially by dilution and partially by the DOC removal during activated sludge treatment. The DOC concentration in surface waters and groundwaters is typically dominated by other natural processes (soil weathering, algal growth, etc.) and is even lower. Because the oxidant demand is closely related to the DOM concentration, it is evident that waste streams with high concentrations of particulate organic substances, such as faeces, brownwater or blackwater, are not well suited for chemical oxidation treatment. However, it has to be considered that human urine represents $<1\%$ of the total flow of municipal wastewater and contains a disproportionately high fraction of many biologically active compounds (Larsen and Gujer 1996, Lienert *et al.* 2007). Therefore, it might still be a feasible option to eliminate the micropollutants from source-separated urine.

Chlorine is a special case in terms of its reactivity, because unlike the other chemical oxidants it reacts quickly with ammonia to form chloramines (Wolfe *et al.* 1984). Since these are significantly less efficient disinfectants (about 500 times less than HOCl), waters with high ammonia concentrations (e.g., hydrolyzed urine, non-nitrified wastewater) are only properly disinfected at high chlorine doses above the breakpoint (Wolfe *et al.* 1984). However, very high chlorine doses may be required, so that upstream ammonia removal (biological or physical, see Udert and Jenni and Siegrist *et al.* 2013) before chlorination may be a better solution to this problem. In the following, the effect of the DOC concentration on the oxidation efficiency is illustrated for ozonation because sufficient data are available to make such a comparison for this oxidant. Similar considerations apply to other oxidants.

26.2.3 Efficiency of oxidation/disinfection with ozone: the role of DOC concentration

Table 26.3 shows the required ozone doses for 90% elimination of two selected micropollutants during ozonation of hydrolyzed urine, electrodialized urine diluate, wastewater effluent and two surface waters. Data are given for 17α-ethinylestradiol (EE2), a synthetic steroid estrogen, and ibuprofen (IP), an antiphlogistic. This exercise can also be done for other micropollutants as discussed by Kümmerer (2013). While EE2 reacts quickly with ozone and $^{\cdot}OH$ (pH 7: $k_{O3} \approx 3 \times 10^6\ mol^{-1} \cdot L \cdot s^{-1}$, $k_{OH} = 9.8 \times 10^9\ mol^{-1} \cdot L \cdot s^{-1}$), IP reacts mostly with $^{\cdot}OH$, which are formed as secondary reactions from ozone decomposition (pH 7: $k_{O3} \approx 9.6\ mol^{-1} \cdot L \cdot s^{-1}$, $k_{OH} = 7.4 \times 10^9\ mol^{-1} \cdot L \cdot s^{-1}$) (Huber *et al.* 2003).

The efficiency of this process increases with decreasing DOC concentrations. This is mainly due to the competition between DOM and micropollutants for the chemical oxidant. For 90% elimination of EE2, the ozone dose varies over more than three orders of magnitude, reflecting the difference in DOC concentration between hydrolyzed urine and the pre-treated river water used for drinking water. While this difference seems quite large, it has to be considered that the volumes of urine and wastewater differ by at least two orders of magnitude. This is reflected by the ozone dose normalized to the DOC concentration, which results in similar values for hydrolyzed urine and wastewater (within a factor of 2) for compounds with high ozone reactivity and ozone resistance. Table 26.3 also shows the absolute amount of ozone in $g{\cdot}pers^{-1}{\cdot}d^{-1}$ for the case of ibuprofen (IP). These data show that the treatment of wastewater effluent is about three times more efficient than for hydrolyzed urine, although the values become more comparable when the urine is pre-treated (e.g., by electrodialysis). Urine treatment at household level consequently seems feasible with pre-treatment, but as mentioned above, only part of the micropollutant load in the wastewater will be treated at this level. Furthermore, Table 26.3 illustrates that for compounds reacting quickly with ozone, the ozone doses involved in drinking water treatment are again significantly lower. A similar approach can be taken for disinfection processes. The ozone doses would probably lie between the values for EE2 and IP elimination.

Table 26.3 Ozone doses required ($mg{\cdot}L^{-1}$) for 90% elimination of 17α-ethinylestradiol (EE2) and ibuprofen (IP) in various water.

Water type	EE2		IP		
	O_3 dose [$mg{\cdot}L^{-1}$]	O_3/DOC [$g{\cdot}g^{-1}$]	O_3 dose [$mg{\cdot}L^{-1}$]	O_3/DOC [$g{\cdot}g^{-1}$]	O_3 per pers. [$g{\cdot}p^{-1}{\cdot}d^{-1}$]
Hydrolyzed urine	~500	0.25	1000	0.5	1.3
Urine diluate (electrodialysis)(2)	~150	0.375	~600	1.5	0.78
Wastewater 1 (7.7 $mgDOC{\cdot}L^{-1}$)	>1	>0.13	n.d.(1)	n.d.(1)	n.d.(1)
Wastewater 2 (5.0 $mgDOC{\cdot}L^{-1}$)	0.5	0.1	~4	0.8	0.4
Lake water (3.0 $mgDOC{\cdot}L^{-1}$)	0.1	0.03	n.d.(1)	n.d.(1)	n.d.(1)
River water (1.0 $mgDOC{\cdot}L^{-1}$)	<0.1	<0.08	>2	>2	0.28

(1)n.d.: not determined;
(2)contained some methanol from dosing of micropollutants.
Sources: Huber *et al.* (2003), Huber *et al.* (2005), Dodd *et al.* (2008), Lee and von Gunten (2010).

26.2.4 Application of oxidation/disinfection processes to source-separated urine

On the basis of the data shown in Table 26.3 and further discussions (Dodd *et al.* 2008), it can be concluded that ozonation of source-separated urine to remove micropollutants reacting quickly with ozone is significantly more energy-intensive than the same process for wastewater, whereas the difference is smaller for ozone-resistant compounds. If the treatment of source-separated urine is combined with nutrient recovery (N, P), the overall energy requirement even becomes favourable for urine treatment (Dodd *et al.* 2008). However, micropollutants from sources other than urine cannot be treated by this approach. Disinfection of source-separated urine may be achieved for ozone doses similar to those required for IP oxidation. Significant improvement can be achieved by pre-treatment of urine, for example, by electrodialysis. For urine disinfection by chlorine, the ammonia level has to be reduced significantly before this oxidant can be considered (see above). No data is available for urine disinfection with chlorine dioxide or ferrate(VI).

26.2.5 Application of oxidation/disinfection processes to greywater

There is not much if any information on applying oxidation processes to greywater. The main purpose of these processes is disinfection and decoloration. Oxidation of undesired micropollutants can also be an issue. On the basis of the water matrix composition of greywater shown in Table 26.2 and the ozone doses shown in Table 26.3, ozone doses of the order of 3.5–35 $mg{\cdot}L^{-1}$ can be expected for the inactivation of bacteria (Zimmermann *et al.* 2011). Decoloration is an important factor with regard to aesthetics, especially if greywater is reused for toilet flushing, for instance. An 80% removal of true colour (455 nm) was achieved for two wastewaters for an O_3/DOC (w/w) ratio of 0.6–0.8 (Wert *et al.* 2009). This is in a similar range of ozone doses as for disinfection and transformation of ozone-resistant micropollutants. So if greywater is optimized for colour removal, efficient inactivation of its bacteria can be expected. Since greywater has a significantly lower DOC concentration than urine, other disinfectants such as chlorine or chlorine dioxide may also be feasible for inactivating microorganisms. Disinfection may require pre-treatment for particle removal to avoid reduced efficiency due to the association of microorganisms with particles.

26.2.6 Application of oxidation/disinfection processes to effluents of biological reactors

There is a wealth of information on ozone application to secondary wastewater effluents (von Sonntag and von Gunten 2012). These results are well transferrable to biological reactors. It has been shown in laboratory tests (ferrate, ozone; Huber *et al.* 2003, Lee *et al.* 2009, Wert *et al.* 2009, Lee and von Gunten 2010), at pilot-scale (ozone; Ternes *et al.* 2003, Huber *et al.* 2005) and in full-scale operation (ozone; Hollender *et al.* 2009, Reungoat *et al.* 2010, Zimmermann *et al.* 2011) that micropollutants can be efficiently removed from secondary wastewater effluents by oxidative processes with ozone (ferrate). Furthermore, a high degree of inactivation of microorganisms can be achieved with the same ozone doses (Zimmermann *et al.* 2011). Compared to household treatment, this also allows micropollutants derived from sources other than households to be oxidized (e.g., hospitals, industries, etc.).

26.3 CONCLUSIONS

Several chemical oxidants are available for oxidative water treatment in the urban water cycle. Depending on the location of the oxidation process, micropollutants from different sources can be treated. Typically, pre-treatment steps are necessary (particle removal, removal of DOC, ammonia and nitrite) to increase the lifetime of the oxidants and the overall efficiency of the processes. Ozone, $^{\bullet}OH$ radicals and possibly ferrate are the oxidants with the best overall performance for the elimination of micropollutants on the basis of their reaction kinetics, oxidant stability and by-product formation. It has been shown that ozonation can be used for disinfection and removal of micropollutants from source-separated urine, greywater, and for wastewater effluent polishing. Chlorine is not well suited for micropollutant removal, but can be used for disinfection. Pre-treatment for ammonia removal is then necessary to avoid the formation of less efficient chloramines.

Micropollutants are typically not mineralized in oxidative water treatment and the toxicity of the transformation products is a major concern. *In vivo* tests have shown that ozonation leads to reduced toxicity, and the same applies to *in vitro* test systems in combination with biological filtration or other separation steps. Only a few cases of the formation of more toxic products are known so far. More information is needed to fully assess this issue.

Oxidation/disinfection by-products are formed from the reaction of chemical oxidants with water matrix components. Bromide plays an important role here, since it can be oxidized to bromine by chlorine and ozone. This results in the formation of bromo-organic compounds during chlorination, whereas both bromo-organic compounds and bromate can be formed during ozonation. Higher concentrations of these toxicologically relevant compounds can be expected to occur with increasing bromide levels.

REFERENCES

Cao N., Yang M., Zhang Y., Hu J., Ike M., Hirotsuji J., Matsui H., Inoue D. and Sei K. (2009). Evaluation of wastewater reclamation technologies based on in vitro and in vivo bioassays. *Science of The Total Environment*, **407**(5), 1588–1597.

Dodd M. C., Kohler H.-P.E. and von Gunten U. (2009). Oxidation of Antibacterial Compounds by Ozone and Hydroxyl Radical: Elimination of Biological Activity during Aqueous Ozonation Processes. *Environmental Science and Technology*, **43**(7), 2498–2504.

Dodd M. C., Zuleeg S., von Gunten U. and Pronk W. (2008). Ozonation of Source-Separated Urine for Resource Recovery and Waste Minimization: Process Modeling, Reaction Chemistry, and Operational Considerations. *Environmental Science and Technology*, **42**(24), 9329–9337.

Escher B. I., Bramaz N. and Ort C. (2009). JEM Spotlight: Monitoring the treatment efficiency of a full scale ozonation on a sewage treatment plant with a mode-of-action based test battery. *Journal of Environmental Monitoring*, **11**(10), 1836–1846.

Gordon G., Slootmaekers B., Tachiyashiki S. and Wood D. W. (1990). Minimizing Chlorite Ion and Chlorate Ion in Water Treated with Chlorine Dioxide. *Journal American Water Works Association*, **82**(4), 160–165.

Hammes F., Salhi E., Koster O., Kaiser H. P., Egli T. and von Gunten U. (2006). Mechanistic and kinetic evaluation of organic disinfection by-product and assimilable organic carbon (AOC) formation during the ozonation of drinking water. *Water Research*, **40**(12), 2275–2286.

Hoigné J. and Bader H. (1994). Kinetics of reactions of chlorine dioxide (OClO) in water – I. Rate constants for inorganic and organic compounds. *Water Resesarch*, **28**(1), 45–55.

Hollender J., Zimmermann S. G., Koepke S., Krauss M., McArdell C. S., Ort C., Singer H., von Gunten U. and Siegrist H. (2009). Elimination of Organic Micropollutants in a Municipal Wastewater Treatment Plant Upgraded with a Full-Scale Post-Ozonation Followed by Sand Filtration. *Environmental Science and Technology*, **43**(20), 7862–7869.

Huber M. M., Canonica S., Park G. Y. and von Gunten U. (2003). Oxidation of pharmaceuticals during ozonation and advanced oxidation processes. *Environmental Science and Technology*, **37**(5), 1016–1024.

Huber M. M., Gobel A., Joss A., Hermann N., Loffler D., McArdell C. S., Ried A., Siegrist H., Ternes T. A. and von Gunten U. (2005). Oxidation of pharmaceuticals during ozonation of municipal wastewater effluents: A pilot study. *Environmental Science and Technology*, **39**(11), 4290–4299.

Huber M. M., Ternes T. A. and von Gunten U. (2004). Removal of estrogenic activity and formation of oxidation products during ozonation of 17 alpha-ethinylestradiol. *Environmental Science and Technology*, **38**(19), 5177–5186.

Katsoyiannis I. A., Canonica S. and von Gunten U. (2011). Efficiency and energy requirements for the transformation of organic micropollutants by ozone, O_3/H_2O_2 and UV/H_2O_2. *Water Research*, **45**(13): 3811–3822.

Krasner S.W. (2009). The formation and control of emerging disinfection by-products of health concern. *Philosophical Transactions of the Royal Society A-Mathematical Physical and Engineering Sciences*, **367**(1904), 4077–4095.

Krasner S. W., Weinberg H. S., Richardson S. D., Pastor S. J., Chinn R., Sclimenti M. J., Onstad G. D. and Thruston A. D. (2006). Occurrence of a new generation of disinfection byproducts. *Environmental Science and Technology*, **40**(23), 7175–7185.

Lange F., Cornelissen S., Kubac D., Sein M. M., von Sonntag J., Hannich C. B., Golloch A., Heipieper H. J., Moder M. and von Sonntag C. (2006). Degradation of macrolide antibiotics by ozone: A mechanistic case study with clarithromycin. *Chemosphere*, **65** (1), 17–23.

Larsen T. A. and Gujer W. (1996). Separate management of anthropogenic nutrient solutions (human urine). *Water Science and Technology*, **34**(3–4): 87–94.

Lee Y., Escher B. I. and von Gunten U. (2008). Efficient removal of estrogenic activity during oxidative treatment of waters containing steroid estrogens. *Environmental Science and Technology*, **42**(17), 6333–6339.

Lee Y. and von Gunten U. (2010). Oxidative transformation of micropollutants during municipal wastewater treatment: Comparison of kinetic aspects of selective (chlorine, chlorine dioxide, ferrate(VI), and ozone) and non-selective oxidants (hydroxyl radical). *Water Research*, **44**(2), 555–566.

Lee Y., Zimmermann S. G., Kieu A. T. and von Gunten U. (2009). Ferrate (Fe(VI)) Application for Municipal Wastewater Treatment: A Novel Process for Simultaneous Micropollutant Oxidation and Phosphate Removal. *Environmental Science and Technology*, **43**(10), 3831–3838.

Lienert J., Güdel K. and Escher B. I. (2007). Screening Method for Ecotoxicological Hazard Assessment of 42 Pharmaceuticals Considering Human Metabolism and Excretory Routes. *Environmental Science and Technology*, **41**(12), 4471–4478.

Kümmerer K. (2013). The issue of micropollutants in urban water management. In: *Source Separation and Decentralization for Wastewater Management*, T. A. Larsen, K. M. Udert and J. Lienert (eds.), IWA Publishing, London, UK, pp. 71–84.

Macova M., Escher B. I., Reungoat J., Carswell S., Chue K. L., Keller J. and Mueller J. F. (2010). Monitoring the biological activity of micropollutants during advanced wastewater treatment with ozonation and activated carbon filtration. *Water Research*, **44**(2), 477–492.

Petala M., Kokokiris L., Samaras P., Papadopoulos A. and Zouboulis A. (2009). Toxicological and ecotoxic impact of secondary and tertiary treated sewage effluents. *Water Research*, **43**(20), 5063–5074.

Pinkernell U. and von Gunten U. (2001). Bromate minimization during ozonation: mechanistic considerations. *Environmental Science and Technology*, **35**(12), 2525–2531.

Pronk W., Biebow M. and Boller M. (2006). Electrodialysis for Recovering Salts from a Urine Solution Containing Micropollutants. *Environmental Science and Technology*, **40**(7), 2414–2420.

Ramseier M. K., Peter A., Traber J. and von Gunten U. (2011). Formation of assimilable organic carbon during oxidation of natural waters with ozone, chlorine dioxide, chlorine, permanganate, and ferrate. *Water Research*, **45**(5), 2002–2010.

Reungoat J., Macova M., Escher B. I., Carswell S., Mueller J. F. and Keller J. (2010). Removal of micropollutants and reduction of biological activity in a full scale reclamation plant using ozonation and activated carbon filtration. *Water Research*, **44** (2), 625–637.

Richardson S. D., Fasano F., Ellington J. J., Crumley F. G., Buettner K. M., Evans J. J., Blount B. C., Silva L. K., Waite T. J., Luther G. W., McKague A. B., Miltner R. J., Wagner E. D. and Plewa M. J. (2008). Occurrence and Mammalian Cell Toxicity of Iodinated Disinfection Byproducts in Drinking Water. *Environmental Science and Technology*, **42**(22), 8330–8338.

Richardson S. D., Simmons J. E. and Rice G. (2002). Disinfection byproducts: The next generation. *Environmental Science and Technology*, **36**(9), 198A–205A.

Rook J. J. (1974). Formation of haloforms during chlorination of natural waters. *Water Treatment and Examination*, **23**, 234–243.

Sarathy S. and Mohseni M. (2009). The fate of natural organic matter during UV/H_2O_2 advanced oxidation of drinking water. *Canadian Journal of Civil Engineering*, **36**(1), 160–169.

Schmidt C. K. and Brauch H.-J. (2008). N,N-Dimethylsulfamide as Precursor for N-Nitrosodimethylamine (NDMA) Formation upon Ozonation and its Fate During Drinking Water Treatment. *Environmental Science and Technology*, **42**(17), 6340–6346.

Sharma V. K. (2007). Disinfection performance of Fe(VI) in water and wastewater: a review. *Water Science and Technology*, **55**(1–2). 225–232.

Siegrist H., Laureni M. and Udert K. M. (2013). Transfer into the gas phase: ammonia stripping. In: *Source Separation and Decentralization for Wastewater Management*, T. A. Larsen, K. M. Udert and J. Lienert (eds.), IWA Publishing, London, UK, pp. 337–350.

Stalter D., Magdeburg A. and Oehlmann J. (2010). Comparative toxicity assessment of ozone and activated carbon treated sewage effluents using an in vivo test battery. *Water Research*, **44**(8): 2610–2620.

Stalter D., Magdeburg A., Weil M., Knacker T. and Oehlmann J. (2010). Toxication or detoxication? In vivo toxicity assessment of ozonation as advanced wastewater treatment with the rainbow trout. *Water Research*, **44**(2), 439–448.

Suarez S., Dodd M. C., Omil F. and von Gunten U. (2007). Kinetics of triclosan oxidation by aqueous ozone and consequent loss of antibacterial activity: Relevance to municipal wastewater ozonation. *Water Research*, **41**(12), 2481–2490.

Ternes T. A., Stuber J., Herrmann N., McDowell D., Ried A., Kampmann M. and Teiser B. (2003). Ozonation: a tool for removal of pharmaceuticals, contrast media and musk fragrances from wastewater? *Water Research*, **37**(8), 1976–1982.

Udert K. M., Larsen T. A., Biebow M. and Gujer W. (2003). Urea hydrolysis and precipitation dynamics in a urine-collecting system. *Water Research*, **37**(11), 2571–2582.

Udert K. M. and Jenni S. (2013). Biological nitrogen conversion processes. In: *Source Separation and Decentralization for Wastewater Management*, T. A. Larsen, K. M. Udert and J. Lienert (eds.), IWA Publishing, London, UK, pp. 291–305.

von Gunten U. and Hoigné J. (1992). Factors controlling the formation of bromate during ozonation of bromide-containing waters. *Journal of Water Supply Research and Technology-Aqua*, **41**(5), 299–304.

von Gunten U. and Hoigné J. (1994). Bromate formation during ozonation of bromide-containing waters: Interaction of ozone and hydroxyl radical reactions. *Environmental Science and Technology*, **28**(7), 1234–1242.

von Gunten U. and Oliveras Y. (1998). Advanced Oxidation of bromide-containing waters: bromate formation mechanisms. *Environmental Science and Technology*, **32**(1), 63–70.

von Gunten U., Salhi E., Schmidt C. K. and Arnold W. A. (2010). Kinetics and Mechanisms of N-Nitrosodimethylamine Formation upon Ozonation of N,N-Dimethylsulfamide-Containing Waters: Bromide Catalysis. *Environmental Science and Technology*, **44** (15), 5762–5768.

von Sonntag C. (2008). Advanced oxidation processes: mechanistic aspects. *Water Science and Technology*, **58**(5), 1015–1021.

von Sonntag C. and von Gunten U. (2012). Chemistry of Ozone in Water and Wastewater Treatment; from Basic Principles to Applications. IWA publishing, London.

Werdehoff K. S. and Singer P. C. (1987). Chlorine Dioxide Effects on THMFP, TOXFP, and the Formation of Inorganic By-Products. *Journal American Water Works Association*, **79**(9), 107–113.

Wert E. C., Rosario-Ortiz F. L. and Snyder S. A. (2009). Effect of ozone exposure on the oxidation of trace organic contaminants in wastewater. *Water Research*, **43**(4), 1005–1014.

Wert E. C., Rosario-Ortiz F. L. and Snyder S. A. (2009). Using Ultraviolet Absorbance and Color To Assess Pharmaceutical Oxidation during Ozonation of Wastewater. *Environmental Science and Technology*, **43**(13), 4858–4863.

WHO (2004). *Guidelines for drinking water quality*. World Health Organisation, Geneva, Switzerland.

Wolfe R. L., Ward J. F. and Olson G. R. (1984). Inorganic Chloramines as Drinking Water Disinfectants: A Review. *Journal American Water Works Association*, **77**(5), 74–88.

Zhang H., Yamada H. and Tsuno H. (2008). Removal of Endocrine-Disrupting Chemicals during Ozonation of Municipal Sewage with Brominated Byproducts Control. *Environmental Science and Technology*, **42**(9), 3375–3380.

Zimmermann S. G., Wittenwiler M., Hollender J., Krauss M., Ort C., Siegrist H. and von Gunten U. (2011). Kinetic assessment and modeling of an ozonation step for full-scale municipal wastewater treatment: Micropollutant oxidation, by-product formation and disinfection. *Water Research*, **45**(2), 605–617.

Chapter 27

Enhanced fractionation of mixed wastewater as an alternative to separation at the source

Willy Verstraete, Vasileios Diamantis and Bert Bundervoet

27.1 INTRODUCTION

The efficiency and sustainability of urban wastewater treatment can be improved by separating the solid and the liquid fraction of wastewater to a maximum extent (Diamantis *et al.* 2011). Fractionation and pre-concentration of sewage is an alternative to the separate collection of liquid and solid phases at the source. Fractionation requires fewer changes of an existing wastewater system than source separation. Fractionation and pre-concentration can help to minimize the carbon footprint, reduce capital expenditures and operational costs, and maximize energy, water and nutrient recovery.

The main goal of fractionation is to concentrate as much as possible of the wastewater organics in a separate stream, which can later be used for energy recovery. Various technologies are available for fractionation, among them chemically enhanced sedimentation, dissolved air flotation, bioflocculation and direct sewage filtration (Verstraete and Vlaeminck 2011). The concentrate stream consists of ~5–10% of the incoming wastewater flow and contains at least 70% of the wastewater organics.

Due to its high organic load, the concentrate can easily be subjected to anaerobic digestion to permit energy and nutrient recovery. It can be treated on site, but is preferably collected by truck or returned to the sewer: this allows it to be treated more efficiently at a centralized or satellite digester which acts as an energy generator.

The water exiting the concentrator consists of 90–95% of the incoming flow and contains 20–30% of the organics as well as 60–70% of the nitrogen and phosphorus content. Depending on the pre-concentration technology, the phosphorus content can be decreased significantly, for example by dosing iron or aluminium salts during the concentration step.

In case of reuse or effluent disposal to sensitive water bodies, a post-treatment step is necessary. Advanced treatment processes such as membrane filtration, activated carbon or reverse osmosis can be used to obtain the desired quality (Diamantis *et al.* 2010a). For nitrogen removal, a simple trickling filter or a rotating biological contactor (based on the oxygen-limited autotrophic nitrification/denitrification process, OLAND) can reduce the ammonia-nitrogen load with minimum energy input (de Clippeleir *et al.* 2011). The water can be made suitable for disposal to a watercourse or a soil-based infiltration system. Figure 27.1 shows the concept of pre-concentration treatment for mixed municipal wastewater.

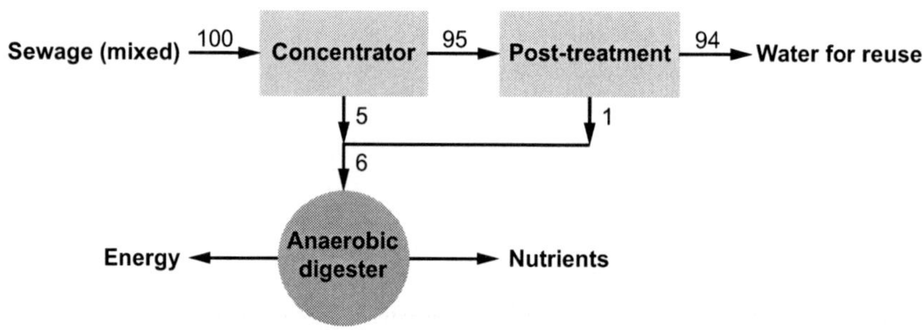

Figure 27.1 The concept of sewage pre-concentration, including enhanced separation of wastewater solids coupled with anaerobic digestion and treatment for effluent reuse. The numbers indicate the relative flow rates. The quality of the water for reuse is brought to the desired level by advanced treatment processes (not shown in figure).

Both large and small wastewater treatment plants can benefit from the early fractionation of solids and liquids. The solid fraction is rich in organic matter and may be used for energy recovery by anaerobic digestion. In contrast, the liquid fraction, which contains dissolved nutrients, can be reclaimed as irrigation, fertilization or even drinking water in combination with some post-treatment steps. This can be of importance in arid environments (e.g., the Greek islands) (Diamantis *et al.* 2010b). This chapter explores and discusses the potential of pre-concentration technologies at a small-scale level ranging from small towns to individual households.

27.2 MOTIVES FOR PRE-CONCENTRATION

Conventional activated sludge (CAS) treatment plants, especially with extensive aeration, do not represent optimal treatment for energy recovery from wastewater

(Verstraete and Vlaeminck 2011). The process requires significant energy input for COD (35–40 $kWh{\cdot}IE^{-1}{\cdot}y^{-1}$, IE stands for inhabitant equivalent) and ammonia oxidation (extra 10–15 $kWh{\cdot}IE^{-1}{\cdot}y^{-1}$), only a small part of which (20% according to Müller and Kobel 2004) is recovered by anaerobic sludge digestion. In treatment plants, the intrinsic energy content of the wastewater organics is largely lost.

The pre-concentration concept, on the other hand, produces an organically rich stream that is suitable for anaerobic digestion. Thus the sewage energy content can be harvested to a maximum extent, resulting in a zero carbon footprint (Wett *et al.* 2007, Verstraete and Vlaeminck 2011). These technologies can be implemented on a large scale, and a few wastewater treatment plants with a zero carbon footprint already exist. Nevertheless, the central question at present remains to what extent an approach focused on fractionation and recovery of resources from sewage can be implemented on a medium and small-scale level.

Even if the technologies are applicable on a smaller scale, the central question relates to their economic feasibility, since the techniques of reusing the recovered products in useful amounts are quite complex and at the very forefront of current technological and societal developments.

Decentralized or satellite wastewater treatment plants have several advantages. First of all, they avoid sewerage, which is currently responsible for a dilution factor of 3 to 4 due to storm and infiltration water (Brombach *et al.* 2005). This results in higher strength wastewater, which can be treated more efficiently by pre-concentration. It also enhances reuse at the local community level, for example as irrigation water, which is considered to be more beneficial since the transport of nutrients and water is no longer needed, thus narrowing the water and nutrient cycle.

27.3 APPROACHES FOR SMALL COMMUNITIES

27.3.1 Fractionation of mixed wastewater

A variety of techniques can be used to separate solids from wastewater. Many of them are already used in large-scale units but not necessarily for pre-concentration. Some of them are related to the techniques used to treat sewer overflows such as parallel plate decanters, dissolved air flotation (DAF) devices as well as various types of fine sieves (van Dijk *et al.* 2011). Chemically enhanced primary treatment and physical pre-concentration by flotation are of particular interest. Direct membrane filtration of sewage with a UF membrane is another interesting process because it already combines an pre-concentration step and a post-treatment step. However, it suffers from severe membrane fouling in practice, which results in fluxes as low as 10 $L{\cdot}m^{-2}{\cdot}h^{-1}$, even with high trans-membrane pressures (Diamantis *et al.* 2009). This makes it difficult to apply, certainly for decentralized or small communities. So it will not be discussed further. In contrast, bioflocculation looks most promising, a technique which uses the strong adsorptive properties of fresh activated sludge as a way to enhance sedimentation.

27.3.1.1 Chemically enhanced primary treatment (CEPT) and sedimentation

A well-known and widely used technique for enhancing particle separation is the addition of metal salts to raw sewage. These salts destabilize the colloids and promote coagulation. The coagulant is mostly an iron salt such as $FeCl_3$ and $FeSO_4$, but aluminium salts are also used. The small aggregates of primary destabilized particles can then be flocculated to facilitate the sedimentation process. Besides removing organic solids, coagulants also allow the precipitation of phosphorus, heavy metals, bacteria and micropollutants (Suarez *et al.* 2009, Carballa *et al.* 2005).

Operational data from 35 chemically enhanced primary treatment (CEPT) plants with a capacity of less than 2,000 IE in Norway, demonstrate the applicability of CEPT as a fractionation technique with a COD removal efficiency of approximately 75% (Odegaard 2001). But CEPT also eliminates an average of 90% of the phosphorus and suspended solids with effluent COD, SS and P concentrations of 121 ± 72, 22 ± 16 and 0.5 ± 0.4 mg·L^{-1}, respectively.

A small treatment plant of 1,000 m^3·d^{-1} capacity (designed for 4,000 IE) requires a slow mixing tank (for coagulation) of 20 m^3 volume (hydraulic retention time, HRT = 30 min) and a final clarifier of 60 m^3. Compared to a treatment plant for biological COD removal (500–1,000 m^3 reactor volume), a significant decrease in overall footprint and consequently in capital costs is achieved.

Overall, the costs of CEPT on a small to medium scale are of the order of 0.10–0.25 EUR·per m^3 water treated (Culp *et al.* 1978).

27.3.1.2 Dissolved air flotation

The plant footprint can be kept quite small by using a dissolved air flotation (DAF) module instead of a clarifier to concentrate the coagulated organics. A DAF offers significant advantages over conventional sedimentation processes. It is usually designed for surface loading rates of 1 to 15 m·h^{-1} and provides excellent separation efficiency, since smaller particles can be removed. Indeed, coupled with coagulation/flocculation, it achieved a COD removal in the order of 70–90%, with effluent COD concentrations of less than 100 mg·L^{-1} (Odegaard 2001). DAF using poly-aluminium chloride as the coagulant also reduced the number of enteric microbes by two log values and the total phosphorus by 90% (Koivunen and Heinonen-Tanski 2008). The area required by a DAF module for a 1,000 m^3·d^{-1} facility (designed for 4,000 IE) is only 5 m^2 (at 10 m·h^{-1} surface loading rate), compared to 50 m^2 for a typical sedimentation tank (at 1 m·h^{-1} surface loading rate). The corresponding area footprint can therefore be reduced by a factor of 5 to 10 relative to conventional sedimentation. This process is unsuitable for individual households but may be used for small community applications, since it needs a minimum operational flow rate of about 2 $m^3 \cdot h^{-1}$.

An interesting feature of the application of the DAF process is its ability to produce sludges of relatively high solids concentration of up to 4–5% (Mels *et al.* 2001). This is particularly attractive because it produces small volumes of relatively concentrated sludge which require small-scale storage facilities and are easier to dry or require less digester volume.

Overall, the costs for DAF on a small to medium scale are in the order of 0.10–0.25 $EUR \cdot m^{-3}$ water treated, similar to the CEPT process (Culp *et al.* 1978).

27.3.1.3 Rapidly activated sludge bioflocculation and sedimentation

In conventional activated sludge processes, the excess sludge production is a function of the biological activity, which depends on the sludge retention time (SRT) and the temperature. The overall biomass yield decreases as the SRT increases due to biomass loss by endogenous respiration. The yield is higher when no primary treatment is used, as more non-biodegradable solids remain in the influent wastewater. At short SRT (less than 5 days), as used in the high-rate bioflocculation processes, most of the organics from the wastewater are transferred to the sludge, mainly by adsorption and not by assimilation (Bohnke and Diering 1986). Accordingly, a larger share of the energy content of the sewage is conserved and transferred to the sludge, which can easily be concentrated to 2–4%. When operating on a full scale, the "Strass" wastewater treatment plant (Austria), which implemented a high-rate biosorption process at a 0.5 h HRT and 0.5 d SRT, achieved energy self-sufficiency because more biogas could be produced due to the higher sludge production (Wett *et al.* 2007). The plant could even become a net energy producer by implementing more efficient post-treatment to remove the nitrogen (e.g., anammox).

A 1,000 $m^3 \cdot d^{-1}$ treatment plant (handling about 4,000 IE) requires a bioreactor volume of 40 m^3 (HRT = 1 h) and a clarification tank of 60 m^3 for sludge separation and recycling. This is 12–25 times smaller than for conventional COD removal with the aerated activated sludge process. If 5–10 mg of iron or aluminium are added per L of sewage, not only is the COD significantly removed but the phosphorus concentration is also reduced by 80%: the discharge limits for organic matter and P can thus be attained.

Other biological pre-treatment methods integrating the aspect of a selector-sorption or high-rate biosorptive activated sludge can be explored to facilitate the overall upfront harvesting of the organics (Akanyeti *et al.* 2010). In this respect, a high rate Biological Aerated Filter (BAF) can also be considered as an up-concentration technology (Mendoza-Espinosa and Stephenson 1999).

Overall, the costs for rapid sludge biosorption and sedimentation are of the order of 0.15–0.30 EUR·per m^3 water treated (Haruvy 1997, van Haandel and van der Lubbe 2007).

27.3.2 Post-treatment options for water reuse

In many cases, the effluent will not yet be suitable for further reuse, especially for discharge into sensitive bodies of water. Further post-treatment will therefore be necessary after the pre-concentration stage to meet the discharge limits.

One option would be to introduce conventional aerobic biological treatment after the pre-concentration stage, as this enables the transformation of organic matter oxidation to CO_2 and excess biomass, nitrogen conversion to nitrogen gas in denitrifying systems and some phosphorus incorporation into the biomass. The effluent discharged by the conventional activated sludge process is low in COD but contains nitrate due to limited nitrogen removal. The total nitrogen (TN) concentrations are between 30–50 $mgN{\cdot}L^{-1}$ by most small-scale technologies (Rich 2008). The phosphorus and pathogens in the effluent are usually not sufficiently reduced by the conventional activated sludge process. Finally, several pharmaceuticals and trace organic chemicals can be degraded, while others are unaffected (Onesios *et al.* 2009, Conn *et al.* 2006). Consequently, many of the advantages of pre-concentration are lost if the effluent is treated with a conventional aerobic activated sludge process. In contrast, low-rate trickling filters constitute an elegant technology for water upgrading. Their energy requirements are acceptable (less than 0.2 $kWh{\cdot}m^{-3}$ water treated), as are the maintenance conditions (no sludge recirculation and mechanical aeration). This technology enables both COD removal and nitrification (Tchobanoglous *et al.* 2003) sufficient to meet the discharge limits if phosphorus is removed during pre-concentration, for instance by bioflocculation combined with chemical P-removal.

Oxygen-limited nitrification and denitrification (OLAND) is an alternative process to remove nitrogen (de Clippeleir *et al.* 2011). It can be operated with rotating biological contactors (RBC) or suspended biomass reactors. The energy input is minimal (below 0.1 $kWh{\cdot}m^{-3}$). The required reactor volume is also small compared to conventional systems, with HRTs of the order of 2–3 h.

Natural treatment systems are a possible option for the effluent, especially in temperate and warm climates (Sundaravadivel and Vingeswaran 2001). Their main advantages are low energy input, lower operation and maintenance, the provision of natural habitats and their aesthetic value. Their major disadvantages include their large footprint, which makes process selection for densely populated areas difficult. Indeed, a total area of 20 m^2 (50 $L{\cdot}m^{-2}{\cdot}d^{-1}$ design) is required for a single family (designed for 1 $m^3{\cdot}d^{-1}$).

The quality of the final effluent can be further improved by using advanced treatment processes (e.g., membrane filtration, activated carbon, reverse osmosis) (Antakyali *et al.* 2008). If the water is to be reused for irrigation and nutrient removal is not desired, simple UF-post treatment can be enough to reduce the pathogen levels. The filtration of effluent water coming from an pre-concentration stage such as CEPT or bioflocculation is generally considered to be 2–3 times more efficient in terms of flux than direct-membrane filtration of sewage (Even-Ezra *et al.* 2010).

27.3.3 The potential for energy recovery

The main advantage of working with pre-concentration instead of a conventional aerobic sludge system is the larger share of energy recovered from the sewage. Moreover, the resulting concentrates can be more efficiently digested than secondary sludge, which only breaks down by 25–40%. Thus bioflocculation sludge is digestible up to 70–80%. The methane yield from the digestion of CEPT and DAF sludges can become lower in case of high concentration of Fe or Al (Smith and Carliell-Marquet 2008).

The choice of concentrative (instead of dissipative) treatment of municipal wastewater for small communities is largely determined by the availability of an anaerobic digester in the vicinity of the point of separation. At present, anaerobic digesters are considered to be economically justifiable from a scale of 1,000 m^3 or 100,000 IE onwards. Recently, new approaches such as pocket digesters with a 10 kW electric output and an overall investment in the order of 100,000 EUR indicate that small-scale installations in the order of 10,000 IE are becoming feasible (Pointon and Langan 2002).

Since the installed power of micro-CHP modules (combined heat power) is currently at least 10 kW, electricity production from anaerobic digesters is feasible if a minimum conversion of 240–280 $kgCOD{\cdot}d^{-1}$ to biogas is ensured. As indicated, this situation corresponds to more than 3,000 IE when 70–80% of the organics are concentrated and 70% of the concentrate is degraded to biogas. Accordingly, in terms of economics, it is rather difficult to apply anaerobic digesters to small communities of less than 2,000 IE. This problem can be overcome if anaerobic co-digestion with organically rich materials is considered, for instance using residues from food treatment facilities, rendering operations, olive mills, cheese production (whey and lactose), biodiesel production (glycerin) and concentrated manure. These co-substrates may contain up to 100–200 kg COD per m^3 and can therefore contribute to enhanced biogas and electricity production without requiring significant investment costs.

The scale problem can also be overcome if digesters are planned at a central location. In this case, the feedstocks have to be transported by truck or by a special transport line for concentrates.

27.3.4 The potential for nutrient recovery

To recover organic cake from the raw or digested concentrate by conventional drying beds, these may be used alone or combined with a solar drying process (greenhouse) (Mathioudakis *et al.* 2009) or a constructed wetland (Stefanakis *et al.* 2009) for dewatering. Sludge drying bags represent another interesting alternative for concentrating the particulate matter. Aerobic composting enables the production of a soil conditioner. Substantial pathogen control is simultaneously ensured. Small-scale pyrolysis/gasification units can produce a soil conditioner with recovery of the energy (capacity of 10 kW electricity, Gekgasifier 2012).

Apart from the direct reuse of the nutrients present in the concentrator effluent (after advanced treatment), they can be concentrated into a liquid (concentration by a factor of five is possible by using reverse osmosis, van Houtte and Verbauwhede 2008) or processed to make solid fertilizer (e.g., by zeolite adsorption or struvite precipitation, Doyle and Parsons 2002). At the level of a village or small-scale community, recycling of the nutrients often does not warrant specific legislative measures and these products will normally find valuable applications, for example, in the cultivation of non-edible plants, which minimizes potential problems in terms of the micropollutants or pathogens they may contain.

27.4 APPROACHES AT HOUSEHOLD LEVEL

The organics present in mixed wastewater can also be pre-concentrated at household level. This section analyzes different fractionation strategies for small-scale private applications. If the solids are separated very close to the production site, hydrolysis and liquefaction are minor and a maximum of the solid COD can be recovered.

Mechanical separators such as the Aquatron faecal separator (Vinnerås and Jönsson 2002), the Rottebehälter pre-composting bag (Gajurel *et al.* 2003), or a combination of these, may be of interest. The solid product needs to be carefully handled, and appropriate waste management schemes (e.g., storage facilities, separate containers and transportation by truck) are crucial to ensure proper hygienic conditions and odour control. In rural regions, the collected solids can be composted on site and reused as soil conditioner.

Clarification tanks are another option widely used throughout the world. They can separate COD and suspended solids (SS), and reduce faecal coliforms (Rich 2008). Advanced multi-compartment septic tanks are efficient for the enhanced separation of wastewater solids. Filters appropriately designed to maximize solid capitation can also be installed (e.g., Richard 2002). The sludge accumulated in the first compartment needs to be regularly discharged and transported to a processing facility. It can be removed by a conventional vacuum tank truck. However, septic tanks produce diffuse methane emissions which have to be controlled because of their environmental burden. Moreover, the energy contained in the volatilized methane is lost.

A coagulant mixture can be dosed as a chemically enhanced clarification step or even directly inside the toilet water tank, for maximum separation of suspended matter and phosphorus. At high coagulant doses, the digestibility of the concentrates can be minimized. For a single family (design flow rate of $1\ m^3{\cdot}d^{-1}$), a holding tank of 1–2 m^3 volume suffices (minimum design holding time = 24 h). The cost of a conventional septic tank (up to 5 m^3) is in the order of 2,000–3,000 EUR, including installation.

Contact aeration (or bioflocculation) is also an interesting option for pre-concentrating wastewater organics at household level. A rapid activated sludge system, designed at low HRT (in the order of 0.5–1.0 h) may be used; the overall volume for a single family is 0.20 m^3 for the bioreactor, plus 0.10 m^3 for the clarifier. This type of process with sludge transfer to the septic tank allows water reuse in Japanese households (Gaulke 2006). Most of the organics are transferred (mainly by adsorption) from the liquid to the solid phase. For a single family, a compact, pre-fabricated aerobic treatment module costs 3,000–5,000 EUR including installation. The energy input is approximately 1 $kWh \cdot m^{-3}$ of water treated. The overall costs are in the order of 0.5–1.0 $EUR \cdot m^{-3}$ of water treated.

27.5 THE FUTURE OF MEMBRANE FILTRATION

In the future, it is likely that membrane technology will be used to recover water directly from the septic tank. Thus, dissolved nutrients will be extracted via the filtrate and recycled by irrigation, at least during the summer months. In a further step, direct membrane filtration can be installed upfront of new or existing sewage treatment plants (Figure 27.2). This permits maximum pre-concentration of the organics for a waste-to-energy strategy and one-step water recovery.

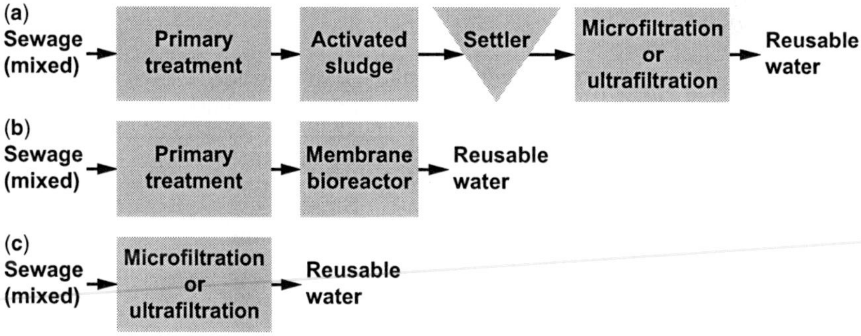

Figure 27.2 Evolution of membrane technology for water recovery from sewage. The process can be applied (a) in the tertiary phase, (b) the secondary phase, (c) the primary phase.

Pre-concentration using an advanced form of membrane filter is motivated primarily by the expectation that this new type of water treatment will be applicable at comparable cost to the conventional activated sludge treatment (Verstraete *et al.* 2009). Unlike the latter, however, it offers plenty of extra scope for the recovery of water and nutrients and carries a lower risk in terms of pathogens.

It is currently impractical to pre-concentrate raw sewage continuously by membrane filtration. A concentration factor of ten was achieved by using shear-enhanced crossflow UF, but the energy input was in the order of 5–10 $kWh \cdot m^{-3}$ (Diamantis *et al.* 2009). Membrane filtration operated on secondary effluent requires an energy input of 0.5–1.0 $kWh \cdot m^{-3}$ of water treated.

Previous studies of sewage filtration revealed the complete removal of suspended solids and a COD decrease below secondary effluent disposal standards (i.e., 100–120 $mg \cdot L^{-1}$), often lower than 20 $mg \cdot L^{-1}$ (Gan and Allen 1999, Bourgeous *et al.* 2005, Bendick *et al.* 2004). At household level, where separate collection of low polluted greywater is possible, direct membrane filtration (MF or UF) can also be applied for greywater reuse. This is already done in Japanese buildings, the recovered water being used to flush toilets (Gaulke 2006, Ogoshi *et al.* 2001). The costs of such systems are in the order of 0.30–0.40 $EUR \cdot m^{-3}$ of water treated (Durham *et al.* 2001, van Houtte and Verbauwhede 2008).

Despite the current difficulties for the long-term application of this approach due to fouling, the rapid evolution currently experienced by membrane filtration and the new advantages offered by nanofilters and novel membrane materials make it likely that it will make progress in the near future and become of major importance.

27.6 SUMMARY

Water may be recovered from sewage without using energy-intensive conventional aerobic biological processes aimed at total disintegration without recycling, even in

Table 27.1 Balance scorecard of the fractionation of wastewater relative to the conventional approach (sewerage and activated sludge treatment).

	Conventional[1]	Fractionation approach		
		City[2]	Small community[3]	Household[3]
Sewerage	In full	In full	In part	No
Treatment	In full	Cradle-to-cradle design	In part	In part
Major recovery		Energy, Water, Nutrients	Energy, Water	Water
Major strength	Reliability	Sustainability	Sustainability	Sustainability
Major challenge	Societal costs	Regulatory aspects of reuse	Public acceptability	Personal investment and care

References: [1] van Haandel and van der Lubbe (2007), [2] Verstraete and Vlaeminck (2011), [3] This work.

small communities or households. The concept of concentrating the organics from mixed wastewater requires appropriate handling of the concentrates (faecal matter), especially as regards the control of diffuse methane emissions and hygienic issues.

For small community-size wastewater treatment plants, the choice of sewage pre-concentration is driven by the availability of an anaerobic digester in the vicinity to minimize the overall carbon footprint. We definitely need more anaerobic digestion plants for this purpose. By implementing pre-concentration technology, the capital and operational expenses for the subsequent aerobic treatment can be reduced or completely eliminated (e.g., by implementing a natural treatment system). Table 27.1 attempts to summarize the advantages and disadvantages of the fractionation-based approach relative to conventional activated sludge (CAS).

ACKNOWLEDGEMENT

This work was supported by the Multidisciplinary Research Partnership Ghent Bio-Economy.

REFERENCES

Akanyeti I., Temmink H., Remy M. and Zwijnenburg A. (2010). Feasibility of bioflocculation in a high-loaded membrane bioreactor for improved energy recovery from sewage. *Water Science and Technology*, **61**, 1433–1439.

Antakyali D., Krampe J. and Steinmetz H. (2008). Practical application of wastewater reuse in tourist regions. *Water Science and Technology*, **57**, 2051–2057.

Bendick J. A., Modise C. M., Miller C. J., Neufeld R. D. and Vidic R. D. (2004). Application of cross-flow microfiltration for the treatment of combined sewer overflow wastewater. *Journal of Environmental Engineering ASCE*, **130**, 1442–1449.

Bohnke B. and Diering B. (1986). Method for treating sewage in sewage treatment installations having an adsorption stage. US Patent 4568462.

Bourgeous K. N., Darby J. L. and Tchobanoglous G. (2005). Ultrafiltration of wastewater: effects of particles, mode of operation, and backwash effectiveness. *Water Research*, **35**, 77–90.

Brombach H., Weiss G. and Fuchs S. (2005). A new database on urban runoff pollution: comparison of separate and combined sewer systems. *Water Science and Technology*, **51**, 119–128.

Carballa M., Omil F. and Lema J. M. (2005). Removal of cosmetic ingredients and pharmaceuticals on sewage primary treatment. *Water Research*, **39**, 4790–4796.

Conn K. E., Barber L. B., Brown G. K. and Siegrist R. L. (2006). Occurrence and fate of organic contaminants during onsite wastewater treatment. *Environmental Science and Technology*, **40**, 7358–7366.

Culp R. L., Wesner G. M. and Culp G. L. (1978). Handbook of advanced wastewater treatment. 2nd edn, van Nostrand Reinhold Company, New York, USA.

De Clippeleir H., Yan X., Verstraete W. and Vlaeminck S. E. (2011). OLAND is feasible to treat sewage-like nitrogen concentrations at low hydraulic residence times. *Applied Microbiology and Biotechnology*, **90**, 1537–1545.

Durham B., Bourbigot M. and Pankratz T. (2001). Membranes as pretreatment to desalination in wastewater reuse: operating experience in the municipal and industrial sectors. *Desalination*, **138**, 83–90.

Diamantis V., van de Caveye P., Jingxing M., Carballa M. and Verstraete W. (2009). Maximum up-concentration of domestic wastewater by a novel shear enhanced cross flow ultrafiltration. 8th Aachen Conference Water and Membranes, Aachen, Germany, 27–28 October 2009.

Diamantis V. I., Antoniou I., Athanasoulia E., Melidis P. and Aivasidis A. (2010a). Recovery of reusable water from sewage using aerated flat-sheet membranes. *Water Science and Technology*, **62**, 2769–2775.

Diamantis V. I., Leontaridou A., Antoniou A., Melidis P. and Aivasidis A. (2010b). Water recovery from sewers: sewer water abstraction plant (SWAP) technology for decentralized application (in Greek), Third Conference on Small Decentralized Water and Wastewater Treatment Plants, Skiathos, Greece, 14–16 May 2010.

Diamantis V. I., Verstraete W., Vlaeminck S. E., Melidis P. and Aivasidis A. (2011). Efficiency and sustainability of urban wastewater treatment with maximum separation of the solid and liquid fraction. In: Comprehensive Biotechnology, Moo-Young *et al.* (eds.), 2nd edn, Elsevier, Amsterdam, The Netherlands.

Doyle J. D. and Parsons S. A. (2002). Struvite formation, control and recovery. *Water Research*, **36**, 3925–3940.

Even-Ezra I., Beliavski M., Tarre S., Dosoretz C. and Green M. (2010). Chemical versus biological pretreatment for membrane filtration of domestic wastewater. *Desalination*, **272**, 85–89.

Gajurel D. R., Li Z. and Otterpohl R. (2003). Investigation of the effectiveness of source control sanitation concepts including pre-treatment with Rottebehaelter. *Water Science and Technology*, **48**, 111–118.

Gan Q. and Allen S. J. (1999). Crossflow microfiltration of a primary sewage effluent-solids retention efficiency and flux enhancement. *Journal of Chemical Technology and Biotechnology*, **74**, 693–699.

Gaulke L. S. (2006). On-site wastewater treatment and reuses in Japan. *Proceedings of the Institution of Civil Engineers – Water Management*, **159**, 103–109.

Gekgasifier (2012). Gasifier Experimenters Kit. http://gekgasifier.com (accessed 05 April 2012)

Haruvy N. (1997). Agricultural reuse of wastewater: nation-wide cost-benefit analysis. *Agriculture Ecosystems and Environment*, **66**, 113–119.

Koivunen J. and Heinonen-Tanski H. (2008). Dissolved air flotation (DAF) for primary and tertiary treatment of municipal wastewater. *Environmental Technology*, **29**, 101–109.

Mathioudakis V. L., Kapagiannidis A. G., Athanasoulia E., Diamantis V. I., Melidis P. and Aivasidis A. (2009). Extended dewatering of sewage sludge in solar drying plants. *Desalination*, **248**, 733–739.

Mels A. R., Rulkens W. H., van der Meer A. K., van Nieuwenhuijzen A. F. and Klapwijk A. (2001). Flotation with polyelectrolytes as a first step of a more sustainable wastewater treatment process. *Water Science and Technology*, **43**, 83–90.

Mendoza-Espinosa L. and Stephenson (1999). A review of biological aerated filters (BAFs) for wastewater treatment. *Environmental Engineering Science*, **16**, 201–216.

Müller E. A. and Kobel B. (2004). Energy-analysis at utilities in Nordrhein-Westfalen representing 30 millions PE – energy-benchmarking and savings potential (in German). *Korrespondenz Abwasser*, **51**, 625–631.

Odegaard H. (2001). The use of dissolve air flotation in municipal wastewater treatment. *Water Science and Technology*, **43**, 75–81.

Ogoshi M., Suzuki Y. and Asano T. (2001). Water reuse in Japan. *Water Science and Technology*, **43**, 17–23.

Onesios K. M., Yu J. T. and Bouwer E. J. (2009). Biodegradation and removal of pharmaceuticals and personal care products in treatment systems: a review. *Biodegradation*, **20**, 441–466.

Pointon K. and Langan M. (2002). Distributed power generation using biogas fuelled microturbines. Report, Advantica Technologies Ltd, accessible at: http://www.berr.gov.uk/files/file14934.pdf (accessed 05 April 2012).

Richard J. (2002). Double pass septic tank outlet filter. US patent 6447680 B1.

Smith J. A. and Carliell-Marquet C. M. (2008). The digestibility of iron dosed activated sludge. *Bioresource Technology*, **99**, 8585–8592.

Stefanakis A. I., Akratos C. S., Melidis P. and Tsihrintzis V. A. (2009). Surplus activated sludge dewatering in pilot-scale sludge drying reed beds. *Journal of Hazardous Materials*, **172**, 1122–1130.

Suarez S., Lema J. M. and Omil F. (2009). Pre-treatment of hospital wastewater by coagulation-flocculation and flotation. *Bioresource Technology*, **100**, 2138–2146.

Sundaravadivel M. and Vingeswaran S. (2001). Constructed wetlands for wastewater treatment. *Critical Reviews Environmental Science Technology*, **31**, 351–409.

Tchobanoglous G., Burton F. and Stensel H. D. (2003). Wastewater engineering: treatment and reuse, 4th edn, Metcalf and Eddy Inc., McGraw Hill, New York, USA.

van Dijk P., Menkveld W., Schellekens D., Broeders E., Renkens G., Postmes L. and de Jonge J. (2011). *Praktijkonderzoek naar vergaande zuivering van riooloverstrortwater en effluent op de RWZI Eindhoven* (Practical research in relation to advanced treatment of water from sewage overflows and from the sewage treatment plant at Eindhoven). *WT-Afvalwater*, **11**, 143–151.

van Haandel A. and van der Lubbe J. (2007). Handbook Biological Wastewater Treatment. Design and optimization of activated sludge systems. Quist Publishing, Leidschendam, The Netherlands.

van Houtte E. and And Verbauwhede J. (2008). Operational experience with indirect potable reuse at the Flemish Coast. *Desalination*, **218**, 198–207.

Verstraete W. and Vlaeminck S. E. (2011). ZeroWasteWater: short-cycling of wastewater resources for sustainable cities of the future. *International Journal of Sustainable Development and World Ecology*, **18**, 253–264.

Verstraete W., van de Caveye P. and Diamantis V. I. (2009). Maximum use of resources present in domestic “used” water. *Bioresource Technology*, **100**, 5537–5545.

Vinnerås B. and Jönsson H. (2002). The performance and potential of faecal separation and urine diversion to recycle plant nutrients in household wastewater. *Bioresource Technology*, **84**, 275–282.

Wett B., Buchauer K. and Fimmi C. (2007). Energy self-sufficiency as a feasible concept for wastewater treatment. Proc. IWA Leading Edge Technology Conference, Singapore, 03–06 June 2007, pp. 21–24.

Part IV

The international experience

Chapter 28

The Swedish experience with source separation

B. Vinnerås and H. Jönsson

28.1 THE EARLY 1990s – INTRODUCTION OF UD

Environmental concerns grew in Sweden during the 1980s and intensified around 1990. The Swedish Environmental Code, introduced in 1999, included the ambition to create closed loops for substances. The general concerns about the environment and resources were also reflected in the building of eco-villages, which started in Sweden at the end of the 1980s and boomed during the first half of the 1990s (Norbeck 1998). One important aspect of these eco-villages was on-site treatment of wastewater to promote recycling of plant nutrients. To achieve this, several of the first eco-villages, such as Tuggeliten and Solbyn, used composting toilets. However, urine diversion, with and without flushing, was installed in many of these eco-villages, starting from Åkesta in 1990. The villages organized the reuse of urine directly, usually with help from local farmers. The municipal authorities, however, were not involved.

Urine diversion (UD) systems spread and were also installed in several eco-houses built by housing companies looking for an environmental profile, such as Miljöhuset in Hallsberg and Ekoporten in Norrköping, as well as at the Universeum science centre in Gothenburg. Here too, the local authorities were not involved in using the collected urine as fertilizer. The urine was often not used at all, as no-one took responsibility for organising a reuse system. This meant the loss of potential environmental benefits such as lower energy consumption and eutrophication (Jönsson 2002).

Many of the systems, mainly those where the urine was not recycled, were converted to conventional flush systems after 5–10 years. One important reason for this was the lack of incentive to provide the extra maintenance required by the diversion toilets and the urine pipe system when the urine was not used as a fertilizer. Those systems that actually use the urine as a fertilizer, including in most of the Swedish eco-villages, have lasted far better and most are still in use.

28.2 LATE 1990s TO PRESENT – ON-SITE SANITATION

Sweden has approximately 700,000 on-site sanitation systems, about 60% of them serving residential homes, while the others serve holiday homes. Studies indicate that about 40% of these systems are deficient, with poor treatment and high emissions. In fact, the on-site systems, which serve about 10% of Swedish households, contribute 12% of Swedish anthropogenic emissions of phosphorus to water, compared with a 20% contribution from the other 90% of households connected to municipal wastewater plants (SEPA 2008). To tackle this problem, many municipal authorities recommend UD, especially for holiday homes, and several municipalities require source separation, urine diversion or blackwater separation to obtain a building permit for an on-site system. One of the pioneers was the municipality of Tanum (see below) and several others have followed, for example, Västervik, Norrköping, Söderköping and Norrtälje. The environmental benefits represent a major driving force (Tidåker *et al.* 2007).

The institutional framework for urine recycling has recently been greatly improved. In 2006, the Swedish EPA published new advice on interpreting the Environmental Code with respect to on-site sanitation (NFS 2006). This explicitly stresses that nutrient recycling should be considered when choosing a system. Moreover, the government recently proposed a bylaw that will specify the legal conditions for using urine as a fertilizer. The Swedish institutional infrastructure is consequently becoming more welcoming to urine diversion and source separation in general.

Kvarnström *et al.* (2006) estimated that there were at least 120,000 UD systems in Sweden in 2006, and that the vast majority of these involved dry faecal handling. Many of them do not recycle urine. The main reason for urine diversion is that it improves the handling of dry faecal matter, as the smell is greatly reduced when faeces are not mixed with urine, volumes are smaller, there are less flies and the composting process works better (Jönsson and Vinnerås 2007).

At an increasing number of holiday homes, the urine is collected and used as a fertilizer on-site. A common system is to simply collect the urine in a plastic container (5–20 litre volume), and then apply it as fertilizer when it is full. Following the recommendations of the Swedish EPA, an increasing number of municipalities are developing recycling schemes for source-separated urine as part of their promotion of UD. They do this because source separation has proved to be an efficient and economic means of decreasing eutrophying emissions from on-site systems (Kärrman *et al.* 2005). Moreover, the Swedish EPA has determined that source-separated urine is household waste, just like latrine waste, and consequently the responsibility of the municipal authority. Recycling schemes are especially common where urine diversion is recommended for residential housing, and there were at least 15,000 dual-flush UD toilets in Sweden in 2006, most installed in regular residences (Kvarnström *et al.* 2006). In these systems, the urine is usually collected in a tank. The recommendation is that this tank should be large enough for one collection per year to be sufficient: 500 $L \cdot p^{-1} \cdot a^{-1}$ can normally be expected. In some municipalities, one collection is also free of charge, while any additional collections are subject to a charge. In a number of municipalities, households can obtain permission to use the urine as a fertilizer on their plots. This usually requires a certain plot size in relation to the size of the household in order to minimize the risk of excessive nutrient accumulation and inconvenience to neighbours. A serious challenge to new UD systems in residential housing is that only a few types of porcelain UD toilets are currently (2010) on the market and these have not been tested. In spite of this, the trend towards improved acceptance of source separation for on-site systems seems clear, although it is still fairly slow.

The number of blackwater systems in Sweden is several tens of thousands, mainly in densely populated and environmentally sensitive rural areas. In the past, the blackwater was transported long distances by tanker to a central treatment plant, but new systems are now being developed. These usually use vacuum toilets and a tank that only needs to be emptied once a year. The blackwater is transported to a storage tank on a farm, preferably in the neighbourhood, for sanitization by long-term storage or by ammonia treatment (Nordin 2010), and then reused as a fertilizer. The fertilized crop is often short-rotation forest (energy forest). The underlying driving forces are economics, the work environment and other environmental factors. The economic benefits accrue from shorter transport distances, lower fuel consumption and increased machine capacity.

A recycling sanitation system involves many stakeholders, and it is important that the drivers and restrictions of each stakeholder are understood (Jönsson *et al.* 2010). Farmers are a new key stakeholder in the system and as important as households (see Jönsson and Vinnerås 2013). It is also important for an arena to exist where the different stakeholders of the system can meet and communicate, since the systems are new and their potential for improvement large. This arena is just as essential for maintaining a shared stakeholder vision of the system, which is very important for achieving the goal. The modern Swedish on-site sanitation systems are of three main types (individual, locally driven by the municipality and municipal), as presented in the case studies below.

Table 28.1 Overview of selected Swedish pilot projects.

Project according to main type	Municipalities	Main driver	References	Special technical focus
Individual households				
Individual residents; apartments and separate houses >10 houses	>30	Environment	Kvarnström *et al.* 2006.	Urine diversion (UD), with more than 10 houses per municipality or housing complex (in total 15,000 houses)
Schools K-12	11	Environmental, and education	Kvarnström *et al.* 2006.	Urine diversion in all toilets in the buildings
Information centres, training stations, university buildings	7	Information and education	Kvarnström *et al.* 2006.	Urine diversion in all toilets in the buildings
Municipal driven urine/blackwater systems				
Wet compost treatment (blackwater)	2	Environment	http://www.jti.se/uploads/jti/RKA38-DE.pdf	Closed septic tanks, septic tanks. Wet co-compost with waste or manure.
Long term storage to energy crop (blackwater)	>3	Environment, bioenergy production	http://www.bioenergiportalen.se/attachments/42/407.pdf	Closed septic tanks, septic tanks, long term storage and reuse to energy crop. Mainly short rotation forest, for example, *Salix salix*
Long-term storage to agriculture (urine and blackwater)	>5	Environment	http://www.tanum.se/	UD (dry & flushed), closed septic tanks (normal, low-flush and vacuum)
Urea treatment (blackwater)	1 (3 planned)	Economy, environment	http://www.laqua.se/downloads/Rapport_vaxtnaring_fran_avlopp_070131.pdf	Closed septic tanks. Urea treatment, reuse in agriculture for food, fodder or energy production

K-12; Kindergarten through grade 12, including all years from pre school education through 12 years of basic education.

28.3 THREE TYPICAL SWEDISH CASES STUDIES

28.3.1 Single household with local reuse of urine and faeces

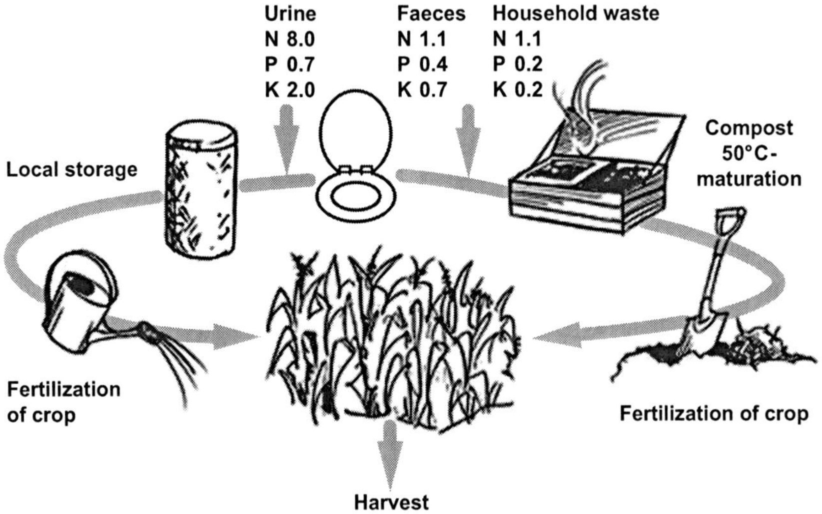

Figure 28.1 Mass flows in a single household with local reuse of faeces and urine.

The house is located in a suburb of Uppsala and has three urine-diverting toilets, two dual-flush and one dry. The urine, with some flushwater, goes to a 3 m^3 underground concrete tank, whereas greywater and faecal water from the flush toilets end up in the municipal sewage system (Figure 28.1). The faeces from the dry toilet are collected in a bucket lined with a paper bag, with leaves and newspaper in the bottom for moisture absorption. The dry faeces fraction is emptied about every two weeks by the household, a popular option for users (Jönsson and Vinnerås 2007), as it minimizes problems with overflow, smell and flies, and becomes a natural part of the bi-weekly routine. The faeces are co-composted with kitchen waste. Fresh faeces are only added (in the hot centre of the pile) if the temperature is 50°C or above as required for proper sanitization; else they are stored.

28.3.2 Local blackwater system for eco-fertilizer production

The municipality of Lund, in southern Sweden, has a large number of on-site sanitation systems, some of which are located in areas with sensitive water recipients, where the blackwater has to be collected in sealed tanks. Since 2002

"generally accepted rules of technology," resulting in education programs and publications. DWA standards have a very high practical and legal impact in Germany, and it is consequently difficult to develop new solutions such as source separation of wastewater without DWA support. The association promotes research and development, but like many others it was rather sceptical at first.

In order to speed up a discussion on source separation and promote the idea of resource-oriented sanitation, a small working group on "alternative sanitation concepts" was established in 1999. A report published in 2002 (ATV-DVWK 2002) led to controversial and sometimes even emotional discussions. However, its ideas met with increasing interest, and a large working group with more than 60 active members in six subgroups was established in 2004. The results of this fundamental work were published in book form in 2008 (DWA 2008) and presented at a conference, which was well attended. Reports in the association's journal KA (Schneider and Christian 2009) and in a special edition of the same journal (Londong 2008) made its more than 10,000 readers familiar with the topic of "new sanitation systems." Indeed, this topic was presented at many German conferences in 2009. In 2010 the public relations sub-working group produced a 16-page marketing leaflet entitled "Do we need new sanitation systems?," which was presented at IFAT in Munich. By synthesizing practical experience and setting up accepted standards, these efforts help to pave the way for the practical implementation of new, innovative technologies.

29.3 THE PILOT PLANT STORY

The main idea of source separation is to make use of wastewater, and many possible solutions have been discussed. The focus in the pilot projects has depended on the interests of the parties involved. Not all sources were used in every case. Blackwater and urine separation systems were tested, whereas the use of dry toilets has remained a niche technology in Germany. Greywater treatment and reuse were widely discussed, but the pilot projects showed that greywater reuse is uneconomical under most boundary conditions in Germany. Today, only a single new large-scale project is in sight – Hamburg Jenfelder Au – where black and greywater will be treated separately to create a fully decentralized treatment system. The project started in November 2011 with a research phase (www.kreis-jenfeld.de). The construction of houses and infrastructure will begin in spring 2012 and will be finished in 2013. Table 29.1 gives an overview of the projects running in Germany; only some of them will be described in more detail below.

29.3.1 Pilot projects with anaerobic digestion of blackwater

"Lübeck Flintenbreite" is an early pilot project that was started in 1999, with vacuum toilets and anaerobic digestion of blackwater. Greywater is treated in a

Table 29.1 Overview of selected German pilot projects.

Project	*	Driver	References	Special Focus
Urine diversion systems				
Berlin Stahnsdorf 2003–2006; 10 Apartments, 1 oper. building	A/OB	Wasser Berlin	Peter-Fröhlich *et al.* (2007, 2008), www.kompetenz-wasser.de/SCST.22.0.html	Optimizing of components
Giz-Building Eschborn, 2006-ongoing, 50 NoMix toilets	OB	Giz	www.gtz.de/en/dokumente/gtz2009-en-presentation-gtz-eschborn-building1.pdf	Demonstration
Lambertsmühle, 1999-ongoing, small museum, 1 caretaker's house	SH	Water company	Londong and Otterpohl (2001), Bastian *et al.* (2005)	Presentation, pharmaceuticals
Emscherquellenhof	OB	Water company	Teschner *et al.* (2009)	Representation
Huber office Berching, 13 NoMix toilets, 10 urinals	OB	Manufacturer	Bischof and Meuler (2004)	Finding market, demonstration
Vacuum system for blackwater				
Lübeck Flintenbreite 1999–ongoing; 117 households	SH	University	Oldenburg *et al.* (2008), Oldenburg (2010), www.susana.org	Ana.digestion, recycling to agriculture
Kaiserslautern/Oberhausen, completed 2007/2008, 8 apart.	A/OB	University	Kaufmann Alves *et al.* (2008)	Treatment, toilet, remote supervision
Hamburg Jenfelder Au (construction phase, in operation in 2013)	A/SH	Hamburg water company	Augustin and Schonlau (2008), www.kreis-jenfeld.de	Decentralizing part of a city
Dry toilets				
Hamburg Allermöhe, 36 homes, 140 people, 1985-ongoing	SH		www.susana.org/docs_ccbk/susana_download/2-56-en-susana-cs-germany-hamburg-eco-settlement-2009.pdf	Eco-village

* Single Households (SH), Apartments (A), Office Building (OB), Institution (I).

constructed wetland and discharged into a creek (Figure 29.1). The transient phase with a slowly increasing number of inhabitants was the main technical challenge. The anaerobic reactor could not be used from the outset due to extreme underloading. It took several years to reach full operation of the system.

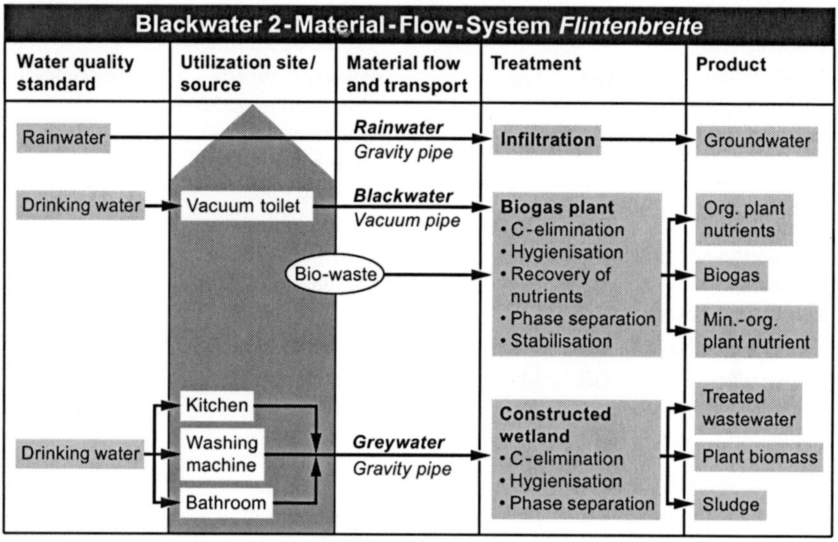

Figure 29.1 The Flintenbreite technical concept.

A very interesting aspect of the Flintenbreite project is its organization. In communities with a centralized sewerage system, homeowners may be forced by German law to connect their houses to it, but in this case the city did not impose its right to insist on a sewer connection. However, to assure a proper water supply and wastewater disposal system, a public private partnership contract with a PLC (infranova GmbH & Co KG) as a general partner had to be established. This made the company liable for developing the site. A very recent German example from Hamburg is based on the Flintenbreite experience. The idea is to implement a source separation system from 2011 for the new urban residential area of Jenfeld and the environmental park of Gut Karlshöhe. The demonstration project is intended to promote the image of the Hamburg Water company. More than 700 apartments will be connected to the separation system. The technology will use a 2-flow system. It includes on-site management of rainwater, recycling of nutrients, concentration of micropollutants in blackwater to permit cost-effective treatment, energy autonomy thanks to blackwater digestion plus the co-digestion of lipids (Augustin and Schonlau 2008).

29.3.2 Pilot projects with urine source separation

A different technological route is followed in various other German pilot projects, initiated by a publication of Larsen and Gujer (1996). Discussions with Ralf Otterpohl during my time as technical director of the Wupper Association, a major company involved in designing and operating large wastewater treatment plants, led to the small Lambertsmühle pilot project in 1999 (Londong and Otterpohl 2001). The Lambertsmühle concept involves urine source separation and full recycling of nutrients from faeces and urine (Figure 29.2).

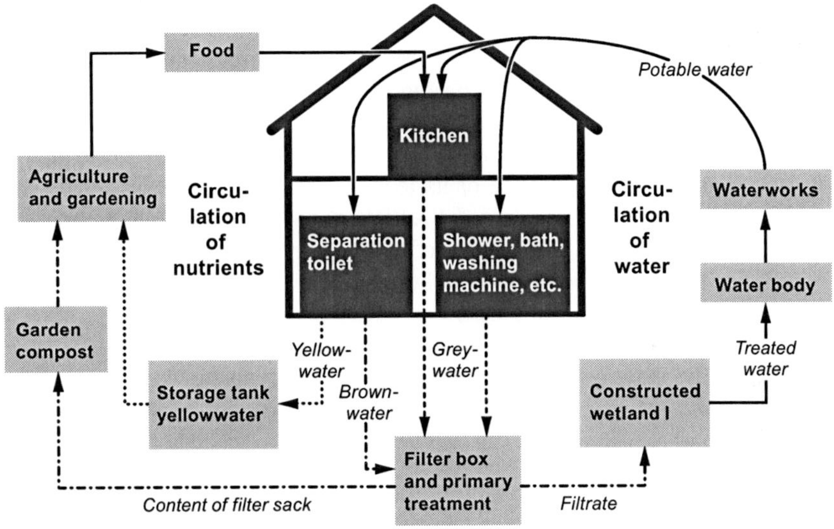

Figure 29.2 Closing the water and nutrient loop in the Lambertsmühle demonstration project (Bastian *et al.* 2005).

The main obstacle was again to obtain legal permission for the new technology. This was only possible because the Wupper Association was made responsible for the operation. The plant is still in operation with good results, although scientific support stopped long ago. Moreover, legal consent for using faeces and urine in agriculture has still not been granted in Germany.

A demonstration project conceptually similar to that of Lambertsmühle, but larger in size, is being run in the buildings of the Berliner Wasserbetriebe at the Berlin Stahnsdorf wastewater treatment plant. The project involves an office building for the operating staff with ten NoMix toilets and ten apartments for the operators and their families. Ecological and economical assessments have been focal points of investigation in addition to gaining operational experience.

An interesting finding was that the urine separation ratio was much lower than expected (35% instead of 75%). Urine losses were caused by the toilet construction, but also by urinating into greywater, which showed a higher ammonia content than expected (Peter-Fröhlich *et al.* 2008). The accompanying investigations by the Technical University of Hamburg at Harburg focused on treating the source-separated urine (Tettenborn *et al.* 2007).

If liable operators such as Berliner Wasserbetriebe (BWB) invest in new concepts, general acceptance can be increased. The Berlin-Stahnsdorf project is consequently of great importance for the development of decentralized technologies in Germany. BWB, a large partly private (RWE, Veolia) and partly public (city of Berlin) company has a commercial interest in the new paradigm of reusing wastewater for the world water market (Peter-Fröhlich *et al.* 2007).

A similarly interesting but entirely private demonstration project was set up in 2004 by a leading German manufacturer of wastewater technology, Huber SE. This company, with a worldwide scope of operations, installed source separation wastewater facilities in its new administration building for about 200 employees to demonstrate the performance of its technology. It created the marketing label of Huber DESA/R (DEcentralized SAnitation and Reuse) (Bischof and Meuler 2004). Unfortunately, only reuse of greywater and mixed sewage proved economically viable within the time horizon of a private company, so the other ideas are not currently being pursued.

NoMix-toilets and waterless urinals have been installed at several sites in Germany, for demonstration purposes and/or to generate waterless urine for research purposes (e.g., the gtz building, the Universities of Aachen, Rostock, Hamburg-Harburg, Braunschweig and Weimar).

29.4 NECESSITY FOR TECHNICAL IMPROVEMENTS AND SCIENTIFIC RESEARCH

Despite many demonstration projects of ever larger size, many detailed questions remain unanswered. A German working group in DWA has currently published a report on urgent research topics (Dockhorn *et al.* 2011). Their main findings are: Improvement of urine-separating technologies has high priority. A service friendly NoMix flush toilet (gravity flow and vacuum) and urinals for women must be developed. Challenges include incrustations of pipes and functional parts, a sound design of vacuum toilets and insufficient separation of urine with existing NoMix toilets. Transport systems also need to be optimized for separated flows, especially if pipes are used for transporting bio-waste.

Moreover, treatment technologies for separated flows need to be developed further. Although much existing equipment for wastewater treatment may be used, optimization and adaptation are necessary. Examples include technologies for elimination of micropollutants from urine and process-stable separation of faeces and flush-water. Downsizing for decentralized applications is necessary.

Dry sanitation, which could find general acceptance in Germany, is not yet on the horizon. Considerable development would seem to be needed here.

29.5 PERSPECTIVE

Visions and innovations will only be operative if they are implemented, but for this to happen, new solutions must be developed. First courses on new sanitation systems have already been introduced at many German universities following a programme developed in Weimar based on the DWA working group report.

Decision makers, consultants in the water and urban planning sectors as well as administrators have to be aware of the challenges of source separation in order to promote implementation. To overcome the difficulties involved in the implementation of new sanitation strategies is consequently a major task of the enlarged DWA working group on "new sanitation systems." A German standard on how to include such systems in design processes has been worked out and will be presented for public discussion in 2012.

The activities of the working group have aroused the interest of members of the German administration at ministerial level. The German Environmental Ministry (BMU) invited researchers, industry and administrators to an expert workshop in 2009, and chose the topic of decentralized sanitation for the annual "Sektorgespräch" (conference on topical aspects of water and waste management) held in March 2010 in order to inform ministerial staff of new solutions.

REFERENCES

ATV-DVWK (2002). Überlegungen zu einer nachhaltigen Siedlungswasserwirtschaft, Arbeitsbericht der AG GB 5.1, ATV-DVWK, Hennef 2002.

Augustin K. and Schonlau H. (2008). Der Hamburg Water Cycle am Beispiel der internationalen Bauausstellung in Hamburg. 2. Aachener Kongress dezentraler Infrastruktur. Gewässerschutz, Wasser, Abwasser GWA, 215, 10/1–10/8.

Bastian A. 14 further authors. (2005). Nährstofftrennung und –verwertung in der Abwassertechnik am Beispiel der „Lambertsmühle". Verein zur Förderung der Agrikulturchemie e.V., Institut für Pflanzenernährung, Bonn (2005), ISBN 3-937941-02-9 (final research report in German).

Bischof F. and Meuler S. (2004). DESA/R – A pilot system in HUBER Administration Building, DESA/R-Symposium, Berching, Germany, 14th July 2004.

Dockhorn T., Hillenbrand T., Langergraber G., Londong J., Niederste-Hollenberg J., Steinmetz H. and Otterpohl R. (2011). Zukünftiger Bedarf für Forschung und Entwicklung im Bereich Neuartiger Sanitärsysteme, Arbeitsbericht der DWA AG KA 1.8. *KA Abwasser Abfall*, **58**(7), 646–655.

DWA (2008). Neuartige Sanitärsysteme, Themenband des DWA-Fachausschusses KA 1 und seiner 6 Arbeitsgruppen, DWA, Hennef, Germany.

Kaufmann-Alves I., Knerr H. and Schmitt T. G. (2008). Auswirkungen der Integration neuartiger Abwasserentsorgungskonzepten bestehende Infrastruktursysteme, KA, Korrespondenz Abwasser, Abfall **10**(55), 1074–1084.

Larsen T. A. and Gujer W. (1996). Separate management of anthropogenic nutrient solutions. *Water Science and Technology*, **34**(3–4), 87–94.

Londong J. (2008). Wandel ohne Katastrophe, KA Korrespondenz Abwasser, Abfall, **10**, 1047.

Londong J. and Otterpohl R. (2001). The Lambertsmühle case: ecologically and economically efficient decentralized wastewater and resources management, 2nd IWA World Water Congress, Berlin 15–19.10.2001.

Otterpohl R., Grottker M. and Lange J. (1997). Sustainable water and waste management in urban areas. *Water Science and Technology*, **35**(9), 121–133.

Oldenburg M. (2010). Flintenbreite – Organisation und Betrieb, presentation at DWA-Symposium "NASS-Tage", Weimar 4th March 2010.

Oldenburg M., Albold A., Wendland C. and Otterpohl R. (2008). Erfahrungen aus dem Betrieb eines neuen Sanitärsystems über einen Zeitraum von acht Jahren, KA Korrespondenz Abwasser, Abfall, **10**, 1100–1005.

Peter-Fröhlich A., Pawlowski L., Bonhomme A. and Oldenburg M. (2007). EU demonstration project for separate discharge and treatment of urine, faeces and grey-water – Part I: Results. *Water Science and Technology*, **56**(5), 239–249.

Peter-Fröhlich A., Pawlowski L., Bonhomme A. and Oldenburg M. (2008). Experiences with sanitation concepts for separate discharge and treatment of grey, brown and yellow water based on an EU Demonstration Project. IWA World Water Congress and Exhibition 7–12 Sept. 2008 in Vienna, Austria (paper on conference CD).

Schneider Ch. (2009). Neuartige Sanitärsysteme, DWA Fachausschuss KA1 stellt Arbeitsergebnisse und neuen Themenband vor, KA Korrespondenz Abwasser, Abfall, **2**, 108–111.

Teschner B., Geisler S. and Drzisga G. (2009). Der Emscherquellhof – neue Techniken im historischen Umfeld. KA Korrespondenz Abwasser, Abfall, **56**(12), 1241–1246.

Tettenborn F., Behrendt J. and Otterpohl R. (2007). Resource recovery and removal of pharmaceutical residues. Treatment of separately collected urine within the EU-funded SCST project. Institute of Wastewater Management and Water Protection, Hamburg University of Technology, Hamburg, Germany, 2007.

WHG (2009). Gesetz zur Ordnung des Wasserhaushalts (Wasserhaushaltgesetz – WHG) in der Fassung vom 31.7.2009, §55 Grundsätze der Abwasserbeseitigung, Bundesgesetzblatt I p. 2585.

Chapter 30

The Netherlands: “Nieuwe Sanitatie”

D. (Bjartur) Swart and A. J. (Bert) Palsma

30.1 INTRODUCTION

Towards the end of the last century, certain groups in the Netherlands were becoming dissatisfied with the way in which human wastewater was collected and treated in their country. Among other things, they were critical of the amount of energy used, the large quantities of drinking water required for toilet flushing, the use of chemicals, the inflexibility of the system and the loss of valuable nutrients. At the time, several initiatives were launched to handle things differently, but they did not get much further than introducing composting toilets, which got a bad press, and a few projects in which greywater was not sent down the sewer system but was purified with helophyte filters instead.

In the past decade (2000–2010), however, relevant developments have gained momentum in the Netherlands. This chapter will describe their current status as well as possible future directions.

30.2 “NEW SANITATION” IN THE NETHERLANDS

The idea that our wastewater system could be designed better and more sustainably had already taken root in limited circles at the end of the last century. Yet, for various reasons, such initiatives did not really get off the ground. Several serious attempts were made, but in most cases the parties involved balked at the costs and the substantial uncertainties. Perhaps the dearth of knowledge at the time also played a role. This knowledge gap was remedied after the turn of the century, however, when STOWA[1] decided to extend its focus to include the water chain.

From 2000, STOWA has been closely involved with the development of New Sanitation in the Netherlands. Until then, wastewater treatment mainly took place within the closed world of purification and sewerage technologists. Wastewater was not an issue on the public agenda and the main aim was to get rid of it at the lowest possible cost. STOWA has taken a different approach from the outset, dedicating itself to broadening the playing field and inviting new players to contribute their ideas and assume responsibility. And it was the contribution of these new players – such as housing associations, schools, users and property developers – which placed wastewater treatment in a much wider social context. From their perspectives it was investigated which requirements that new forms of sanitation should meet, both with regard to ease of use and to achieving sustainability goals. The willingness of these parties to contribute their share to new developments has also induced many water boards to launch their own initiatives.

From the very beginning, the underlying thought was that New Sanitation should, above all, be modern and geared to future needs. In order to make the concept appealing to as wide an audience as possible, any association with terms like "ecological sanitation" or "Ecosan," which might put all but a small group of people off, was deliberately avoided.

Initially – that is in the first few years – projects were mainly limited to tentative case studies. But 2004 saw a breakthrough when two projects in which housing associations, users and municipalities cooperated, were successfully realized. These projects, a small urine separation project in Meppel and a larger blackwater project in Sneek using vacuum technology and anaerobic purification, proved that changes in the water chain were accepted by the general public after all and that the technology was applicable on a practical level. This gradually made many water boards realize that separation and concentration of wastewater streams might have significant advantages for wastewater treatment. Together with STOWA, several water boards started small-scale pilot projects for research purposes or installed urine-separating toilets in their own office buildings to set an example. At the same time, the demand for knowledge increased. In order to help fill this gap, STOWA organized information meetings and contacted research groups in Switzerland (Eawag), Berlin and Stockholm. In this way, STOWA gradually became the spider in the web of what is now called "New Sanitation" in the Netherlands.

---

[1]STOWA (Dutch acronym for the Foundation for Applied Water Research) was founded in 1971. The foundation coordinates and commissions research on behalf of a large number of local water administrations. The bodies contributing to STOWA are the 26 Water Boards, the Provinces and the Ministry of Infrastructure and the Environment.

Another important event in the same period was the creation of a "Coordinating Body" chaired by STOWA in which knowledge institutes and water boards are widely represented. This Coordinating Body focuses on stimulating the further development of knowledge as well as the implementation of new sanitation systems. To this end, it bundles knowledge and indicates which knowledge gaps impede further implementation. In a Strategy Memorandum entitled "Anders omgaan met huishoudelijk afvalwater" (A different approach to domestic wastewater; STOWA 2006), the Coordinating Body has outlined the options for further knowledge development and implementation of these technologies. The aim is to develop a wastewater system yielding the highest purification efficiency at the lowest cost, which will utilize waste products more fully than hitherto and is more flexible thanks to small-scale techniques. The Strategy Memorandum is also meant to inspire enthusiasm and stimulate stakeholders to take the right initiatives at the right time and in the right place. A different approach to wastewater entails looking for alternatives to the present "conventional" methods of treatment. In the development of new sanitation systems, the Coordinating Body focuses on techniques based on separate collection and transport of wastewater, in which both decentralized and centralized wastewater treatment are options.

The course that was set turned out to be successful. Within a few years, more than half of all water boards in the Netherlands were involved in one of the 40 research and pilot projects. Some of these projects were mainly aimed at gaining experience with new toilet systems (vacuum and urine-separating toilets), but most of them also included research into the recycling of blackwater (to biogas) or urine (to struvite). Other studies in this period focused on the removal of drug residues from urine (in Sleen), decomposition of drug residues from urine in the soil (Anderen), and urine salt deposits in drains (Zwolle). The possibility of processing human and animal (manure) wastewater streams together was also investigated, for instance in the SOURCE project. Another important initiative was the development of the Pharmafilter system, which processes hospital wastewater together with as much biodegradable material from the hospital departments as possible (e.g., biodegradable crockery and cutlery), thereby producing energy and removing harmful substances.

The research projects did not always involve the actual installation of new toilets, however. Some of them were basically literature reviews (e.g., a study of a mobile urine-processing unit). Others used urine that was already being collected, for instance from mobile urinals in the city centre of The Hague, and by the "Mothers for Mothers" scheme run by the Organon pharmaceutical company, in which pregnant women donate urine to help infertile couples. This urine was, among other things, used to find out how struvite and ammonium sulphate could be recovered from urine on a commercial basis.

The above list of projects and studies is by no means exhaustive, but gives an indication of the outstanding creativity and innovative power mobilized in the

Netherlands in a relatively short period of time. It shows that we definitely have more at our disposal than 1950s technology to solve the problems of 2050. We realize that the next few years will bring many more new developments, so projects like those carried out in Sneek, Meppel or Anderen certainly do not provide a blueprint for future solutions. However, the knowledge collected in these projects can be used as a basis to build upon.

STOWA played a significant role in the initial phase of almost all the above projects, which was undoubtedly important for the development and exchange of knowledge. But what is much more important is that local parties provided the driving force behind almost all these projects. It is due to visionary housing associations, water boards, schools and companies that so many projects could be carried out successfully in the Netherlands.

30.3 FROM RESEARCH TO IMPLEMENTATION

By 2011, New Sanitation had gradually reached a transitional phase and it was time to move from research to implementation projects. The first steps in this direction were taken by up scaling several of the most promising initiatives:

- The project in Sneek (vacuum toilets and anaerobic digestion) is now being scaled up to a residential area with 250 houses up for renovation. The concentrated blackwater will yield energy and the system will save drinking water. Phosphate will be recovered and drug residues removed. This project will be realized in 2012/2013.
- In the Reinier de Graaf Hospital in Delft, a full-scale Pharmafilter installation has been set up. The hospital departments use biodegradable plastics as much as possible (e.g., for cutlery), which are shredded after use and added to the wastewater. In the subsequent Pharmafilter process, energy and nutrients are recovered and micropollutants are removed.
- In Apeldoorn it is investigated whether wastewater from a residential area can be converted into energy at the local wastewater treatment plant, and if so, how. The business case that is currently being drawn up should make it clear whether the concept is economically viable.
- The Saniphos installation in Zutphen recovers ammonium sulphate and struvite from urine collected by Organon (see above), from a number of office buildings and the large summer festivals.
- Urine-separating toilets were installed in several office buildings. The urine that is collected will be processed in the Saniphos installation or used for new studies.

But just up scaling pilot projects to full-scale format does not seem to be enough. Their social relevance will have to be made abundantly clear. If the present developments are to continue in a country like the Netherlands, where the standard of sanitation is high throughout, we will need to find social drivers powerful enough to promote the further advance of New Sanitation, even in times of recession. Keeping in mind that 99.9% of all houses in the Netherlands are equipped with some form of sanitation, we can distinguish four important drivers for the Dutch situation:

- The marketplace: The threat of phosphate shortages on the world market and rising energy prices may make large-scale recovery of energy and nutrients from wastewater attractive to various parties – not just large multinationals (like Thermphos) but also local agricultural companies interested in using phosphate from locally collected urine if that is more cost-effective than buying it on the world market.
- Sustainability: There is a growing awareness among the general public that we need to make our world more sustainable. And it is increasingly recognized that New Sanitation can make a valuable contribution to this end. An increasing number of builders and renovators will be looking for ways to save water and recover energy and nutrients, even though the short-term costs may be higher.
- Technological developments: The Netherlands is focussing heavily on the development of water technology, partly with the aim of exporting its expertise. For this to succeed, an ample number of showcases will be needed in the country.
- Replacement investments: The existing infrastructure in the Netherlands will soon be technically obsolete and we will face the need for large reinvestments in our sewer system. In places where the present system requires substantial investments and good alternatives are available due to local circumstances, New Sanitation will get the opportunity to show that it can compete with conventional systems.

During the past few years, a lot of knowledge and experience have been gained in the separate collection and processing of urine and the recovery of energy from concentrated wastewater streams. This has not yet led to large-scale implementation, but that was only to be expected, as 99% of the Netherlands is already equipped with sewerage systems. Because of the investments already made in the present infrastructure, a transition to a new water chain will unavoidably take a long time. Furthermore, many building projects have been put on hold due to the economic crisis – among them unfortunately several projects that were seriously considering New Sanitation.

In order to gain insight into the development prospects in the short term, STOWA published a Prospects Memorandum in 2010 (STOWA 2010). It outlines where the main opportunities for a transition in the water chain lie in the coming decade in light of current social developments. Opportunities occur particularly where prospects exist to join projects that are already in the pipeline and where interests converge. This can happen in sparsely populated areas with houses that are currently connected to expensive and high-maintenance pressure sewer systems, where

New Sanitation may be a more economical alternative. It can also happen in agricultural areas where farmers are interested in manure digestion and recovery of nutrients. Or it can happen on new housing locations producing so much wastewater that new investments in conventional WWTPs are necessary. And finally, it can happen in all areas where sustainability is recognized as an important factor.

The Prospects Memorandum especially shows parties not involved in wastewater treatment on a daily basis that New Sanitation is a very attractive option. And this is of the utmost importance, for it is only when wastewater is no longer viewed as something that we need to get rid of at the lowest possible cost, but rather as a valuable resource that society cannot afford to lose, that New Sanitation will really gain a foothold. This means that we need to work on raising public awareness and support in the Netherlands as well as abroad.

Table 30.1 Overview of selected Dutch pilot projects.

Project	*	Main driver	References
Urine Diversion Systems (Numbers in brackets refer to NoMix toilets installed)			
Delft – IHE 2003, waterless urinals	OB	IHE	
Leiden – Offices Water Board, 2003 (2)	OB	Water Board	
Meppel – Ambachtshuys, day-care centre, 2003 – 2007 (3)	SH	Housing Ass.	Swart (2006)
Meppel – Offices Waterboard, 2005 – present (20)	OB	Water Board	
Nieuwegein – Offices Research Institute, 2005 – present (6)	OB	Private Res.	
Anderen – Assisted living centre, 2005 – 2010, 12 apart. (20)	A	Housing Ass.	Kujawa-Roeleveld *et al.* (2009), De Wit and Vergouwen (2010)
Sleen – De Schoel, apartments for elderly, 2005 – 2011 (25)	A	Water Board	Meulenkamp (2012)
Zwolle – Windesheim, University, 2006 – present (125)	I	University	
Doetinchem – Offices Water Board, 2006 – present (21)	OB	Water Board	
Enschede – Bonhoeffer, high school, 2008 – 2009 (2)	I	Water Board	

(*Continued*)

Table 30.1 Overview of selected Dutch pilot projects (*Continued*).

Project	*	Main driver	References
Leeuwarden – Ark, centre for architecture, 2008 (2)	OB	Owner	
Assen – Office Province of Drenthe, 2010 – present (60)	OB	Province	
Heino – Natuurlijk Huus, education centre 2009 – present (2)	SH	Owner	
Vacuum System for Blackwater (Numbers in brackets refer to vacuum toilets installed)			
Sneek – Lemmerweg Oost, 32 houses, 2003 – 2010 (64)	SH	Manufacturer	
Sneek – Noorderhoek, 230 houses, 2010 – present, until 2011 75 toilets realized/350 to be built	SH	Manufacturer	
Wageningen – NIOO, 2011 (25)	OB	NIOO	www.nioo.knaw.nl/nieuwbouw.php
Venlo – Floriade Villa Flora, in operation in 2012 (40)	OB	Manufacturer	www.desah.nl/projecten/floriade-2012-venlo/
Ameland – C2C (1)	SH	Province	www.desah.nl/projecten/c2c-vakantiewoning-ameland/
Sneek – Landustrie, 2010 (8)	OB	Manufacturer	
Park 2020, 2012 (3)	OB	Manufacturer	

* Single Households (SH), Apartments (A), Office Building (OB), Institution (I)
Source: Most projects can be found from this webpage: http://nieuwesanitatie.stowa.nl/Projecten/index.aspx?pId=1340

REFERENCES

Kujawa-Roeleveld K., Kragić Kok D., Schuman E., Jacobi J., Grotenhuis T. and Mels A. (2009). Sorption and biodegradation of pharmaceutical compounds in soil, report Wageningen University & LeAF, Wageningen, The Netherlands.

Meulenkamp R. J. A. (2012). Lokale zuivering van brongescheiden urine (Decentral treatment of source separated urine), prnr 287539, W&E-1034567-RM/hh, Grontmij, De Bilt, The Netherlands.

STOWA (2006). Anders omgaan met huishoudelijk afvalwater (A different approach to domestic waste water), Stowa-report 18/2006, Utrecht, The Netherlands.

STOWA (2010). Perspectieven Nota (Prospects Memorandum), STOWA-report 13/2010, Amersfoort, The Netherlands.

Swart D. (2006). Ervaringen met het Nieuwe Plassen in Meppel. (Experiences with New Sanitation in Meppel), report prnr 172632, Grontmij, Drachten, The Netherlands.

De Wit J. and Vergouwen A. A. (2010). Nuttige inzet urine als meststof voor energiegewassen (The use of urine to grow energy crops), Report prnr. 284641, ref. Nr 13/99100215/JdW, Grontmij, Houten, The Netherlands.

Chapter 31

Source control and source separation: the Swiss experience

Markus Boller

31.1 INTRODUCTION

In recent years, several buildings have been constructed in Switzerland based on decentralized concepts of fully or partly self-sufficient energy and water management. These first attempts covered rainwater and snowmelt harvesting, reuse of greywater, brownwater and blackwater, urine separation and recovery of nutrients from urine. They demonstrate that in-house technology is now ready to accommodate dramatic changes in urban water cycles to achieve higher levels of environmental sustainability. As a rule, the technologies tested in six key buildings usually performed successfully and could be operated satisfactorily over long periods. Some conditions necessary for constructing such systems and possible reasons for project failures are discussed.

31.2 DRIVERS FOR CHANGE IN SWITZERLAND

After half a century of water pollution control in Switzerland based on major efforts to implement end-of-pipe solutions, conclusions can be drawn about the efficiency of existing urban drainage systems with respect to ecological and economic factors. The quality of the receiving waters has improved enormously thanks to constant advances in public and industrial wastewater treatment, but we now realize that even this near-perfect system does not fully meet the environmental requirements. Residual micro- and nanopollutants in various forms, excess nutrients, heavy metals and other pollutants as well as the sludge quality for agricultural use cannot be controlled with current treatment technologies. Membrane filtration, activated carbon adsorption and ozonation are considered as further end-of-pipe expansions. The question thus arises whether pollution control might be achieved more efficiently and economically by controlling the water pollutants at the source. This perspective served as a vector in the last decade of our research to look for examples of the successful introduction of source-control measures to current urban water systems and to visualize successful approaches to the stakeholders involved.

There are various ways of changing current end-of-pipe systems towards more sustainable source control concepts. The easiest way to reduce hazards is to prohibit or limit the use of certain substances, products and materials. However, this is often not politically feasible. In Switzerland for instance, it took more than ten years of debate to prohibit phosphates in washing powders. In 1986, Switzerland took the lead in Europe in prohibiting phosphates in textile detergents by law. In subsequent years, successful measures at source cut phosphate loads from domestic wastewaters by 60%, and together with appropriate end-of-pipe measures successfully reduced eutrophication in Swiss surface waters (Siegrist and Boller 1999). Another example of source control by prohibition with a beneficial impact on water and sediment quality was the ban on lead in petrol in 1988.

It will hardly be possible to introduce source control of recently discussed pollutants such as nutrients and certain micropollutants without changing the current wastewater drainage systems, which still have 65% combined sewers. A first attempt to change the system was to introduce on-site infiltration of stormwater and separate sewer systems for all new or renovated buildings and roads by law in 1991. This enabled source control to be studied for a limited number of hazards originating from construction materials and motor traffic. The concept is based primarily on reducing pollutants from roofs and facades by either providing a catalogue of more ecological alternative materials or offering incentives to manufacturers to change their product composition on the basis of competition with other more ecological products on the market (Boller 2004). A typical example of the first case is the replacement of the widely used Cu and Zn sheets by more sustainable metallic materials such as steel and aluminium (KBOB 2001) or by non-metallic minerals. In past years, many architects changed their original idea of using copper on building surfaces of more than 200 m^2 due to the Swiss regulations on stormwater management (VSA 2000). An example of the second case is the use of a new bitumen isolation sheet on flat green and gravel roofs containing 90% less of the biocide Mecoprop than the widely used alternative (Burkhardt *et al.* 2008). On the other hand, new technical barrier systems were introduced for the efficient on-site removal of water hazards contained in the surface runoff from roofs, roads and highways. Special adsorber systems have been developed which are now widely used, especially for the treatment of runoff from Cu and Zn roofs and of road runoff (Boller *et al.* 2007, Steiner *et al.* 2007).

Source control in stormwater management is an example of a transition phase in which the new system is gradually introduced. Another system change with more far-reaching consequences has not yet been accepted as a state-of-the-art technology, namely the introduction of in-house installations to control water and waste flows, including water and nutrient reuse systems. Several pilot and full-scale projects were realized in recent years in Switzerland to test decentralized water and wastewater technologies in detached houses. The introduction of NoMix toilets in particular has been studied under several full-scale conditions (Lienert and Larsen 2007). Projects combining decentralized energy and water management show most promise with respect to sustainability improvement. Some examples will be presented below.

31.3 CASE STUDIES IN SWITZERLAND

Several projects have started which should demonstrate that decentralized energy and water management concepts are ready to be transferred into practice. They all implement new elements in the energy and water management of detached houses designed to substantially increase environmental sustainability.

Basically, all these projects involve the reuse of decentralized greywater and/or the use of rainwater. Potable water is taken from the public supply or in high-altitude mountain resorts from snow melt. Only in the "Self" project is potable water produced from roof water by membrane treatment. In three cases, NoMix toilets and waterless urinals were installed, allowing for urine separation, storage and processing. In the following chapter, these projects are divided into three groups with 1) on-site wastewater treatment and reuse, 2) separate collection and processing of urine, and 3) small-scale autarkic material and water cycles.

31.3.1 On-site wastewater treatment and reuse

31.3.1.1 Cableway station Zermatt

In 2005, the highest altitude wastewater treatment plant in Europe started operation at a cable car station in the Zermatt ski zone. The plant treats the wastewater of up to 500 toilet flushes per day (4 $L{\cdot}flush^{-1}$) during the high winter season, producing a maximum of 2 $m^3{\cdot}d^{-1}$ of wastewater. With an average content of 130 $mgNH_4\text{-}N{\cdot}L^{-1}$, the nutrient composition is dominated by nitrogen components. The treatment plant consists mainly of three reactors (Boehler *et al.* 2007). Firstly an equalisation tank where urea hydrolysis, denitrification and bio-P elimination are induced by the recycle stream from the second unit. This is a membrane biological reactor (MBR) operated alternatively in aerobic and anoxic mode, assuring practically complete denitrification. The third chamber is a holding tank for the effluent, used to flush the toilets. This three chamber system (Figure 31.1) is an essential core element of all these concepts, including the treatment of black or brownwater and wastewater reuse. It is also applied in the Aquamin, Monte Rosa, and "Self" projects discussed in more detail below.

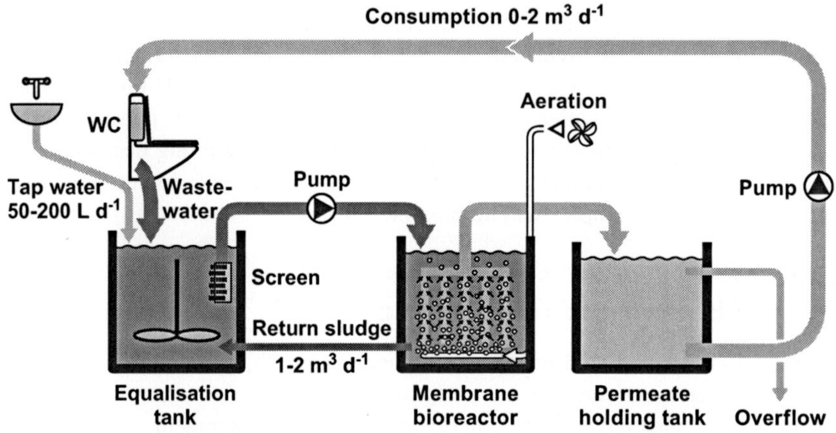

Figure 31.1 Small-scale MBR process scheme for wastewater reuse applied at the Zermatt-Hohtälli cableway station and other on-site treatment projects in Switzerland.

The main lessons learned from this type of reactor is that full recycling of wastewater leads to the accumulation of dissolved salts and non-biodegradable organic matter. Within one winter season, oversaturation of salts led to precipitation in pipes. The organics induced strong colorization of the permeate. Powdered activated carbon (PAC) was consequently dosed to the MBR reactor at a concentration of $100\,\mathrm{mg{\cdot}L^{-1}}$ for decolorization. After some preliminary adaptations, 80% COD removal, 100% N removal and 65–80% P removal was achieved in this small treatment plant. Satisfactory operation by local staff is now possible. Because of its small size, energy consumption is relatively high at about $8\,\mathrm{kWh{\cdot}m^{-3}}$. In view of this positive experience, two more plants of this type were installed at other cableway stations in the same area.

31.3.1.2 Aquamin detached house

The purpose of the Aquamin project was to demonstrate the availability of the in-house water technology required for decentralized treatment options. It involves a conventional supply of potable water for all uses with hygienic requirements, and rainwater as lower-grade water for the washing machine. Other processes are the separate collection of urine and faeces in NoMix toilets, MBR treatment of all wastewater except urine, struvite precipitation for phosphorus recovery from urine, activated carbon treatment of the permeate and reuse for toilet flushing and irrigation in the garden. Plus sludge dewatering in filter bags and infiltration of the permeate overflow into an infiltration pond. Detailed information on this project is found in Abegglen (2008) and Abegglen *et al.* (2008). The project ran successfully for three years, during which it demonstrated its technical feasibility including the stepwise improvement of several treatment steps and equipment changes. It was then stopped, mainly due to the excessive commitment by the homeowner and relatively high costs.

31.3.2 Urine separation and processing

31.3.2.1 Office building Forum Chriesbach at Eawag

The Eawag headquarters were constructed as a "zero energy building" for 150 office workplaces including a new water supply and disposal concept. This large building was opened in 2006 (www.Forum.Chriesbach). It uses no conventional heating, but the water concept comprises a conventional water supply for drinking, hand washing and the canteen. The rest of the water is supplied from the green flat roof covered with a calcite-free substrate for extensive plant growth to maintain low hardness in the roof runoff (avoiding staining tanks and toilets). The roof water is stored in an open tank with a capacity of 80 m^3 as part of the landscape, treated via textile cartridge filters and used for flushing toilets. The urine from waterless urinals and NoMix WCs is collected, while the brownwater is discharged to the public sewer system. A dry fertilizer is produced from the urine not needed for research purposes (www.eawag.ch/vuna). The 37 NoMix toilets and seven waterless urinals produce on average of 90–110 $L_{urine}\cdot d^{-1}$. Substantial amounts of N (about 50%) are lost as NH_3 via the ventilation pipe on the roof, so the urine storage has to be improved in order to benefit from the full nutrient value.

31.3.2.2 Urine processing at the Liestal public library

A urine processing plant was installed close to the public library of the town of Liestal from where the urine was collected via NoMix toilets. It was stored and transported to the processing plant, which consisted mainly of an electrodialysis and an ozonation unit. The 3.9 m^2 electrodialysis stack allowed the nutrients N and P contained in the urine to be almost completely separated and used for liquid fertilizer. Ozonation guaranteed a virtually micropollutant-free product. The plant achieved high and stable performance with concentration factors in the electrodialysis stage of between 2.7 and 3.5. More than 90% of the investigated pharmaceutical products and estrogens were removed (Pronk *et al.* 2006; Pronk *et al.* 2007). It was shown that the proposed technical process designed to produce a marketable fertilizer containing up to 12 $gN\cdot L^{-1}$, 0.6 $gP\cdot L^{-1}$, and 5.6 $gK\cdot L^{-1}$ could be operated under full-scale conditions. The "Urevit" product was tested on field maize crops and compared with other approved fertilizers in Switzerland, showing almost equal performance to a commercial ammonium nitrate fertilizer (Boller 2007). Taking into account the energy needed for the

services, (2) Deferring infrastructure upgrades, (3) Environmental protection, (4) Showcasing sustainability, (5) Water conservation, (6) Enhancement of local amenity, (7) Technology showcase.

Sustainability is often interpreted by the innovator/developer as reducing the reliance on external (mains) water sources by using a local treatment solution to avoid "long" pumping distances to/from a central treatment plant. Avoided energy use almost ranks as highly as water self sufficiency. Nutrient recycling objectives are usually captured by local irrigation of crops, pasture or amenity horticulture. Deferring infrastructure upgrades is an emerging issue for water/sewerage utilities as there is a strong motivation in avoiding the construction of new or upgraded treatment plants to prevent exceedance of the licenced discharge limits of N and P. Similarly increasing densification in the inner areas of most cities has let to a few examples of limited water mains and sewer capacity driving decentralized sewage applications (Sharma *et al.* 2007).

32.3 OVERVIEW OF DECENTRALIZED SYSTEMS

In most case studies presented here, decentralized water supply and effluent recycling are intermixed. We focus our text on effluent treatment and recycling components. See also Table 32.1 and Tjandraatmadja *et al.* (2009).

32.3.1 Cluster Scale Developments

There are an increasing number of urban developments in Australia where the developer has chosen to install their own "small scale" sewerage system because of regulatory resistance to domestic on site systems, and/or because they wish to demonstrate an environmentally sustainable technology. Examples include Capo de Monte, Payne Road, Sunrise at 1770 and The Ecovillage at Currumbin to name but a few. All of these examples were developed by the private sector with scientists often "tagging on" to quantify their performance and provide objective evidence to separate *green facts* from *greenwash*.

Payne Road (Gardner *et al.* 2006), the first such study, focused on individual and communal rainwater tanks, off peak discharge of sewage to council mains, and on-site use of greywater for a 22 lot development. The greywater was of particular interest as it was treated by the Biolytix vermiculture treatment method (www.biolytix.com) to produce high nutrient effluent for sub surface irrigation. Whilst the treatment system was reliable, the high phosphorus and sodicity load to the 200m^2 grassed irrigation area indicated that long term (e.g., $\geq$10 years) sustainable irrigation would likely need further treatment (Beal *et al.* 2008a).

The next scale was the community based sewerage system of Capo di Monte (45 lots) and the Ecovillage at Currumbin (144 lots with 100 connected to sewers). Both

Table 32.1 Overview of selected Australian pilot projects.

Project/type	*	Main driver	References	Main technical aspects
Sustainable house				
Michael Mobbs House	SH	Showcase	http://www.abc.net.au/science/planet/house/default.htm	Produces its own power, uses rainwater and reuses sewage
Limited access to reticulated sewerage/water				
Capo di Monte 46 lots	SH	Water scarcity	http://www.capodimonte.com.au/downloads/UDIA%20capo%20sustainability%20award%20submission.pdf	Local wastewater recycling and water supply from rainwater
Payne Road The Gap (Brisbane, Queensland), 22 lots	SH	Limited sewer system capacity	Gardner *et al.* 2006	Individual rainwater tanks with communal tank/mains back up, sewage off peak pumping to council main, greywater reuse
Aurora Development Victoria, 8500 lots	SH	Effluent disposal limitation	http://yourdevelopment.org/casestudy/view/id/13	Local wastewater recycling through dual pipes
Septic Tank Effluent Drainage System in South Australia	SH	Public health	http://www.lga.sa.gov.au/webdata/resources/files/STEDS_Review_Report_Vol_1_LGA_2002_pdf1.pdf. (Palmer *et al.* 1999)	Septic tanks, common effluent drain, lagoon, public park irrigation
Sewer mining				
Sydney Olympic Park New South Wales	A OB	Innovative design	http://www.sydneyolympicpark.com.au/education_and_learning/environment/water/wrams	Non-potable water drived from sewer mining/stormwater reuse
Greywater reuse				
Inkerman Oasis, Melbourne, 245 app.	A	Sustainable development	http://www.yourhome.gov.au/technical/fs75.html	Greywater for toilet flushing and outdoor irrigation

(*Continued*)

Table 32.1 Overview of selected Australian pilot projects (*Continued*).

Project/type	*	Main driver	References	Main technical aspects
Sustainable development and innovative design				
The Currumbin Ecovillage 144 lots	SH	Showcase: Innovative Design	http://theecovillage.com.au/site/index.php/village/index/43/. (Hood *et al* 2010)	Sustainable housing design, rainwater for potable supply and local effluent recycling.
Woodford Folk Festival in South East Queensland	I	Innovative Design	http://www.woodfordfolkfestival.com/main/index.php?cID=2184&menuID=474&apply=&webpage=friends_of_woodford_2009	Recycling of effluent in the complex and for irrigation
Sensitive Receiving Environment				
Sunrise at 1770 (Queensland); 172 lots, 650 ha development,	SH	Effluent disposal limitations	http://www.sunriseat1770.com.au/conservation.html	Pressurised sewer system, local WWTP, effluent reuse at household rainwater for potable use
Rouse Hill (New South Wales) 36000 lots	SH	Effluent disposal limitations	http://www.rhtc.com.au/content.aspx?urlkey=environment	Local WWTP, reclamation plant, dual pipe for Class A + effluent reuse at households
Showcasing sustainability examples				
Council House 2 (CH2) (Victoria) 12500 m^2, 9 story building	OB	Innovative design	http://www.melbourne.vic.gov.au/Environment/CH2/Pages/CH2Ourgreenbuilding.aspx	Energy conservation, self sufficiency in water supply and minimising export of sewage.
Mawson Lakes (South Australia), 8500 lots 4000 lots	SH A OB	Sustainability	http://www.sawater.com.au/SAWater/WhatsNew/MajorProjects/mawson_lakes.htm	Aquifer stormwater storage/ recovery and dual reticulation for recycled effluent and stormwater

* Single Households (SH), Apartments (A), Office Building (OB), Institution (I).

collected combined domestic wastewater and treated it to class A standard at a central facility for reticulation back to households for toilet flushing and external use. The Capo development used immersed membrane bioreactor technology followed by UV and chlorine disinfection, whilst the Currumbin Ecovillage used a communal septic tank followed by an Orenco recirculating fabric filter (http://www.orenco.com/), then UV and chlorine disinfection. Water quality from both systems was quite suitable microbiologically and aesthetically for the intended end uses. As these cluster scale systems were of considerable interest to scale up and replicate in more traditional greenfield developments in Queensland, their robustness to shock loads and greenhouse gas (GHG) footprint were of particular interest. Consequently detailed studies were undertaken and results (Chong *et al.* 2011) showed high system resilience to load variations, but large differences in specific energy consumption (kwh/m^3). This difference tended to even out when greenhouse gas emissions (i.e. methane and nitrous oxide) were taken into account resulting in a GHG footprint of about 7 $kg_{CO2equiv.}{\cdot}m^{-3}$ for both systems, which is much larger than that for a centralized, tertiary sewage treatment and water reclamation plant (Lane *et al.* 2011).

Another type of cluster scale system is the STED/STEP (Septic Tank Effluent Drainage/Pumped) systems which was pioneered in South Australia in 1962 and "reimported" from the USA in the 1970's to provide cost effective sewage services for rural and peri urban communities. The essential features of the scheme are individual household septic tanks which discharge by gravity or pumping into shallow, small bore ($\leq$100 mm diameter) sewers constructed without traditional man holes, with subsequent treatment in a communal facility, usually oxidation lagoons. Over 70% of the effluent is reused for irrigation of agriculture and public open space, with additional filtration and disinfection required for the latter end use (LGASA 2003). There are more than 160 such schemes operating in South Australia, generating about 7000ML$\cdot a^{-1}$ of sewage from over 130,000 people in communities ranging from about 100 EP to 10,000 EP, with the majority in the 500–3000 EP range (LGASA 2003). Their construction cost per allotment is about 1/3 that of conventional sewerage because of simplification in the reticulation and treatment engineering (Palmer *et al.* 1999). Operational costs are similar to conventional sewerage due in part to the cost of sludge removal from the individual septic tanks every 4 years.

We find it surprising that STED/STEPs have not become more popular in other states of Australia. A notable exception is Kinglake West, a rural community north of Melbourne which was largely destroyed in the 2009 bushfire. Here Yarra Valley Water, the local water utility, decided to test a range of innovations for this septic trench serviced community (105 lots) which had previously been earmarked for a sewerage retrofit program. The innovations (see Figure 32.1) included greywater treatment for potable substitution, urine-separating toilets, and a STEP sewerage system (Narangala *et al.* 2010).

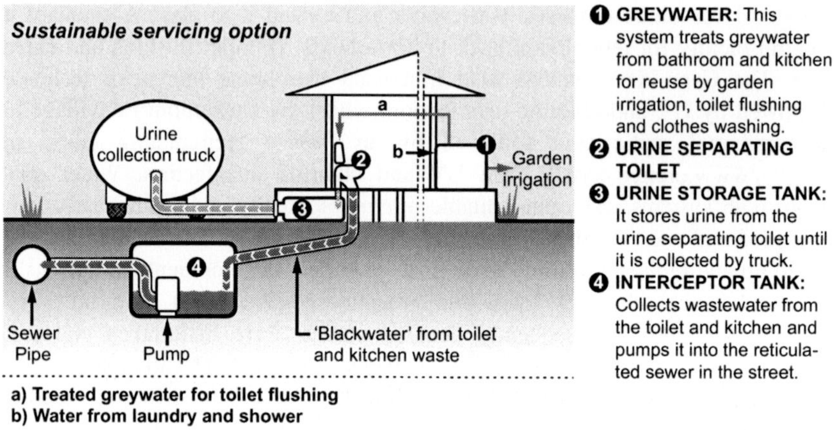

Figure 32.1 Kinglake, Victoria: Greywater reuse system, urine-separating toilets with storage tanks, and a blackwater STEP pumping system (Narangala *et al.* 2010).

32.3.2 Urine-Separating Toilets

The nutrient arguments in favour of urine separation are well known (Larsen and Gujer 1996) but there was essentially no experience with this technology in Australia before a 10 house trial was established the Ecovillage at Currumbin (Beal *et al.* 2008b). Gustavsberg toilets were installed with individual 300 L tanks, which were emptied every 4–6 weeks. After initial serious odour issues, and minor toilet use and cleaning issues were solved, the scheme ran smoothly for over 2 years, with high user acceptance (see e.g., Barton *et al.* 2011). The second stage of the project aimed to reuse the urine for open space and fruit tree irrigation, but research funding ceased before approval for reuse (a microbiological safety issue) was obtained from regulatory agencies.

Kinglake is a more sophisticated example of urine-separating toilets where Wostman EcoFlush USTs have been installed in about 20 dwellings. Urine is collected in a 1000 L vessel and then removed via vacuum truck into individually labelled 1000 L polyethylene "caged" tanks on pallets. After sanitation by extended storage is confirmed by *E.coli* count on each tank, the urine solution is reused on turf pasture (Graeme Julier, pers com.)

32.4 CONCLUSIONS

The original driver for decentralized systems in Australia was the provision of on site sanitation systems in non-sewered urban and peri-urban communities. To a large extent this lack of access to reticulated facilities is still a major driver today, but with the technical evolution into cluster scale (>20 EP) facilities that benefit from centralized management and beneficial reuse such as irrigation.

The real test of decentralized philosophy occurs when developers have the option to choose between a decentralized and centralized system. Apart from the drivers of *green star* building rating (http://www.gbca.org.au/green-star/) and other marketing incentives, we think that cluster scale systems will probably only thrive in situations where access to reticulated sewerage is too expensive. However as externalities are increasingly included into the economic assessment by government regulators (e.g., Productivity Commission 2011), alternatives that reduce demand on natural resources, reduce nutrient discharge, and reduce GHG footprint will be increasingly adopted. It is important that sound biophysical, operational and social acceptance information on alternative technologies is available when the main stream market moves to these "local solutions." We believe that scientifically rigorous documentation of those case studies, built by the innovation pioneers, will be a critical part of the process of informed choice.

REFERENCES

Barton R., Tonks M. and Gardner T. (2011). Installation, operation and acceptance of urine separating toilets in detached dwellings in SEQ. Presented at Integrated Water Management 2011 IWA Conference, Perth. February 2–5, 2011.

Beal C. D., Gardner E. A., Christiansen C. and Beavers P. (2005). A review of on-site wastewater management practices in south-east Queensland. *Journal of the Australian Water Association*, **32**(4), 69–72.

Beal C., Hood B., Gardner E., Lane J. and Christiansen C. (2008a). Energy and water metabolism of a sustainable subdivision in south east Queensland: A reality check. *Enviro'08*. Melbourne Exhibition and Convention Centre, 5–7 May 2008.

Beal C., Gardner T., Ahmed W., Walton C. and Hamlyn-Harris D. (2008b). Urine-Separation and Reuse Trial. AWA Water (Australia) **35**(1), 46–49.

Chong M. N., Ho A., Gardner T., Sharma A. and Hood B. (2011). Assessing decentralised wastewater treatment technologies: Correlating technology selection to system robustness, energy consumption and fugitive GHG emission. International Conference on Integrated Water Management, February 2–5, 2011, Perth, Australia.

Diaper C., Sharma A. and Tjandraatmadja G. (2008). Decentralised water and wastewater systems. In: Transitions: Pathways Towards Sustainable Urban Development in Australia, P. Newton (ed.), CSIRO Australia Publishing and Springer, The Netherlands, pp. 479–493.

Gardner E., Millar G., Christiansen C., Vieritz A. and Chapman H. (2006). Energy and Water Use at a WSUD Subdivision in Brisbane, Australia. *Aust. Journal of Water Resources*, **10**, 283–291.

Hood B., Gardner T., Barton R., Gardiner R., Beal C., Hyde R. and Walton C. (2010). The Ecovillage At Currumbin – A Model For Decentralised Development. *AWA OZWATER' 10 Achieving Water Security*, 8–10 March 2010, Brisbane, Australia.

Lane J., DeHaas D. and Lant P. (2011). Life Cycle Assessment of the Gold Coast Urban Water System. Urban Water Security Research Alliance Technical Report No. 52.

Larsen T. A. and Gujer W. (1996). Separate management of anthropogenic nutrient solutions (human urine). *Water Science and Technology*, **34**(3–4), 87–94.

LGASA (2003). Local Government Assoc. of South Australia. Review of Septic Tank Effluent Disposal Schemes (STEDS) program in South Australia. http://www.lga.sa.gov.au/webdata/resources/files/Review_of_STEDS_in_SA__(11).pdf (accessed 7 February 2012).

Narangala R., Pamminger F. and Knight K. (2010). Kinglake West decentralised urban water solution-turning a wee problem into a solution. *AWA OZWATER '10 Achieving Water Security*, 8–10 March 2010, Brisbane, Australia.

Palmer N., Lightbody P., Fallowfield H. and Harvey B. (1999). Australia's most successful alternative to sewerage: South Australia's Septic Tank Effluent Disposal Schemes. Proceedings Australian Water and Wastewater Association 18th Federal Convention, 11–14 April 1999, Adelaide, Australia, pp. 9. ISBN 0-0908255-47-0.

Productivity Commission (2011). Australia's Urban Water Sector. Productivity Commission Inquiry Report No 55. August 2011.

Sharma A., Grant A., Tjandraatmadja G., Grant T., Opray L. and Pamminger F. (2007). Towards sustainable water and sewerage servicing options for a multi-story infill development, In: Water Management Challenges in Global Change, B. Ulanicki *et al.* (eds), Taylor and Francis, London, pp. 507–513.

Tjandraatmadja G., Cook S., Sharma A., Diaper C., Grant A., Toifl M., Barron O., Burn S. and Gregory A. (2008). ICON Water Sensitive developments. CSIRO report prepared for Joint Steering Committee for water Sensitive Cities. www.nwc.gov.au/resources/documents/CSIRO_Icons_Report.pdf

Tjandraatmadja G., Cook S., Ho A. and Sharma A. (2009). Drivers for decentralised systems in SEQ, Urban Water Security Research Alliance Technical Report no.13.

Chapter 33

Source separation in middle- and low-income countries

Christoph Lüthi and Arne Panesar

33.1 INTRODUCTION

This chapter argues that source-separating sanitation systems still have a limited, but potentially growing role in middle- and low-income countries. It notes that the drivers and constraints for introducing source separation on a larger scale differ dramatically between industrialized and middle- or low-income countries. It gives an overview of current trends and developments in these systems in the latter countries and presents two cases from South Africa and China, where their dissemination has been scaled up in recent years.

In OECD countries like Switzerland, Sweden, Germany or Japan, environmental concerns and limited resources of materials such as high-grade, non-renewable phosphorus are often mentioned as future drivers for a system change from capital, energy and resource intensive sanitation systems to more sustainable and closed-loop ones. However, the realities of most middle- and low-income countries are often dramatically different:

- Poor cities without sunk investments (e.g., great majority of African and poor South Asian cities) cannot afford to build and properly maintain end-of-pipe systems.
- Extreme poverty leads to situations where products from source-separating sanitation systems are not only a competitive source of fertilizer, energy or irrigation water for poor farmers, but more or less the *only* one available.
- Lack of water resources often rules out high-tech engineered systems. The implementation of end-of-pipe systems that use water to transport waste

often contributes to water stress in countries where climate change is already expected to worsen a difficult situation still further.

- Past experience has demonstrated further common bottlenecks for the sustainable functioning of large end-of-pipe systems in middle- and low-income countries, such as the lack of a stable energy supply (electricity, fuel) and spare parts, limited skills to operate such systems, weak sector governance, and so on.

For these reasons, the vast majority of households will continue to be served by some form of on-site sanitation for the foreseeable future. These on-site systems may be well-serviced septic tanks but will often be some rudimentary and poorly constructed pit latrine or cesspit. This picture is common in most cities throughout the global South. However, this deplorable situation also presents a golden opportunity to leapfrog into the future as recent initiatives in China and South Africa are beginning to do (see case studies below). Source-separating sanitation systems may thus play a key role in allowing a fundamental system change in previously under-served areas.

33.2 DRIVERS FOR SOURCE SEPARATION IN MIDDLE- AND LOW-INCOME COUNTRIES

The main drivers for source-separating systems in middle- and low-income countries and selected cases are:

(i) The stable and cheaper availability of energy through biogas production from blackwater and animal manure
(ii) The supply of competitive and safe fertilizers and soil-conditioners from products of urine-diversion and composting toilets
(iii) The reliable availability of (nutrient rich) irrigation water for example, from constructed wetlands or greywater reuse systems
(iv) Less complex treatment, hygienization and transport requirements if flow streams are source-separated (e.g., reduced volume of blackwater; Lüthi *et al.* 2011).

In the section below, we explore these points in more detail and give two examples of how these "drivers" are providing cutting-edge solutions for the global South.

A growing number of domestic biogas units now combine anaerobic treatment of animal manure with sanitation provision and household blackwater treatment. These anaerobic reactors produce digested sludge appropriate for use as liquid fertilizer in agriculture and renewable energy in the form of biogas. China is leading the way, with an estimated 40 million units of domestic anaerobic digesters currently in operation, to be used by 80 million households (300 million people) by 2020 (NDRC 2007). The system has also proved successful in other Asian countries such as India, Nepal and Thailand.

33.3 EXAMPLES OF SCALED-UP DISSEMINATION OF SOURCE SEPARATION IN AFRICA AND ASIA

This part presents two successful examples of source separation at scale from Africa and Asia:

- The celebrated Basic Water and Sanitation Programme (BWSP) in eThekwini, South Africa, and
- The development NGO *PLAN, China's* successful introduction of urine-diverting (UD) toilets at scale in 500 villages and 200 schools.

33.3.1 Scaling-up urine diverting toilets in peri-urban areas of eThekwini, South Africa

Type of project: Scaling-up urine-diversion dry toilets (UDDTs) to peri-urban areas of eThekwini Metropolitan Municipality
Project period: 2004 – ongoing
Project scale: 90,000 households outside the "waterborne sewage line"

eThekwini Metropolitan Municipality is the third largest metropolitan area in South Africa and is one of the 11 districts of KwaZulu-Natal province. The municipal area is located on the country's eastern seaboard and covers approximately 2295 km^2. Settlement patterns in its peri-urban interface are characterized by relatively large plots (200 to 1000 m^2). Several key factors are changing settlement patterns and land use in the area: a decrease of agricultural land due to urbanization, a shift of agricultural production from maize to high-value vegetable crops and an increased dependency on wage income for peri-urban households.

The eThekwini Municipal Water Services Department was looking for cost-effective alternatives to waterborne sewage for the vast peri-urban settlements of its metropolitan area which were unlikely to be connected to the city's sewerage in the short to medium term. Poor households within the "waterborne sewage line" are provided with flush toilets connected to the municipal sewer system, while low-cost houses situated outside this line are provided with urine-diversion dry toilets (UDDTs).

A national subsidy scheme called the Municipal Infrastructure Grant (MIG) was applied to subsidise high quality urine-diversion dry toilets for these households. The subsidy was administered by the municipality's Basic Water and Sanitation Programme (BWSP), whose main purpose is to assist the poor to gain access to basic infrastructure at household level. To date, over 90,000 UDDTs have been installed in eThekwini's peri-urban settlement areas. This represents almost 10% of all sanitation facilities in the metropolitan area. The standard urine diversion dry toilets feature double vaults with a brick and mortar base and superstructure with a plaster finish and a precast concrete slab floor. Ventilation pipes are provided to control flies and odours. The minimum storage period of faecal matter is about one year to ensure its safe handling when emptying the toilet vaults.

33.3.1.1 Costs and economics

Although double-vault urine diversion units (Figure 33.1) cost less to maintain, they are considerably more expensive than the VIP options: € 940 as opposed to € 725. However, they incur considerably lower operational costs because they need no external service providers for emptying (WSP-Africa 2009). This is the main reason why traditional VIPs have ceased to be provided as a sanitation option in peri-urban eThekwini since 2004, the cost of emptying a full VIP being about R1,900 (€ 166 in 2008 prices).

Figure 33.1 Images from left to right: peri-urban settlement with two UDD toilets on the right; a typical double-vault urine diversion dry toilet; inside view. © Duncan Mara

It is of interest to note that these urine diverting toilets were not introduced for the purpose of reusing excreta or urine, but to provide a viable low-maintenance and lower cost on-site technology. Initially, neither the dried faeces nor the urine were collected, but buried on site and infiltrated into a soak-away respectively (Atkins and hydrophil 2009). It should be noted that the municipality does not currently promote excreta reuse due to health concerns. Now that urine diverting toilets are so widespread, and given the changing land use patterns noted above, future nutrient recovery is set to become a reality in eThekwini. Furthermore, the municipality is interested in improving the management and treatment of urine. A four year research project carried out by Eawag, the University of KwaZulu Natal and eThekwini Municipality, and funded by the Bill & Melinda Gates Foundation, will study the technical, managerial and economic feasibility of fertilizer production from source-separated urine. This will offer the potential of turning urine into a marketable commodity and a potential source of revenue for the poor (www.eawag.ch/vuna).

Since UDDT technology was new for most users, social marketing techniques were applied to familiarise people with the UDD toilets and address their needs: they pointed out that this marketable product has no smell, is safe, comfortable and offers privacy. Acceptance among first time users is high, as most peri-urban households previously used unimproved pit latrines. Since September 2008, UDDTs have been officially recognised as a sustainable toilet option for South Africa by the Minister of Water Affairs.

33.3.2 Community-led water and environmental sanitation improvement in Shaanxi, China

Type of project: Scaling-up three types of sustainable sanitation systems
Project period: 2005 – 2009
Project scale: 27,000 sanitation systems

The second case study presents the WES (Water and Environmental Sanitation) programme of PLAN China, an international NGO working towards sustainable water and sanitation in rural and peri-urban areas. The programme began in 2005 and aimed to cover 500 communities and 200 schools in Shaanxi province.

Three types of sanitation options were presented to the communities in the programme area: (i) urine diverting dehydration toilets, (ii) biogas sanitation systems and (iii) twin-pit series latrines (double vault toilets). The communities were informed about the costs and benefits of all three types as well as the financial support available from PLAN China. The community as a whole then selected one of these technology options.

After a successful pilot phase in one village, the construction of these systems in larger numbers began in July 2005. The experience gained from the 14,847 toilets constructed in 2006/2007 shows widespread acceptance of the new sanitation technology. The main reasons for this acceptance by the targeted communities are:

- The pro-active promotion and individual subsidy by PLAN China,
- The low price in comparison to other toilet systems (e.g., biogas, flush latrines and twin pit latrines),
- Cultural attitudes in rural China which favour resource recovery.

33.3.2.1 Costs and economics

A standard urine diverting toilet (turn-key installation) in Shaanxi cost RMB 750 (€ 80) in 2007. However, the use of local materials for the superstructure has since brought the cost down to RMB 400 (€ 44). Where the entire toilet is built inside the house (Figure 33.2), it can cost as little as RMB 300 (€ 32). A cost breakdown for a standard indoor UD toilet may be found in Kumar (2008).

Figure 33.2 Typical (high-cost) situation of an in-house UDD toilet in the PLAN Shaanxi programme. A total of more than 2 million UDDTs were built in China between 2000 and 2010 (NDRC 2007). © Ph. Feiereisen

A helpful feature is that human excreta are traditionally used as fertilizer for crops and vegetables in China. The toilet users are generally small farmers owning an average plot of <0.4 ha. Most of them use land near their home for vegetable farming and some fruit orchards. The annual income of a household of 4–5 is in the range of € 400–600. Farming and fruit orchards are the primary source of income in the programme area. The dried faeces are removed from the vault once or twice a year, depending on the filling rate. They are applied to the field before planting or between two planting periods. The average fertilizer needed by a household of five for these farming activities is 150 kg per year. The use of composted faeces and urine from a UD toilet can meet around 15% of the household's fertilizer need – worth $40 according to 2006 market prices for commercial fertilizer – and hence increase the household's income by 5-10% (Feiereisen and Wu 2009).

China today features more than 2 million UDD toilets and more than 40 million biogas sanitation systems for households in villages and peri-urban situations (NDRC 2007).

33.3.2.2 Evaluation

A mid-term evaluation carried out in 2009 by PLAN China showed that not all the installed UD toilets are in constant use: 43% of 219 households surveyed use them exclusively, 13% use a combination of the new toilets and the old latrines, whereas 44% stated that they do not use the new UD toilet regularly. The fact that the old pit latrines have not yet been removed may be a contributing factor. The report lists the main reasons for the limited interest: the decisions for the new toilets were not made by the house owners individually, but by the village committee; some users had not participated in the training and therefore did not know how to operate the UDD toilets correctly, and some UDD toilets were located next to the kitchen and the residual odours prevented them from being used.

To achieve a higher acceptance rate among existing users and gain new ones, PLAN China has now increased its training and post-installation supervision activities. The project is regularly monitored at three levels – community, programme units and county office(s). The county offices are now planning to conduct follow-up research on community participation and the reuse of urine and faeces in agriculture.

33.4 CONCLUSION AND OUTLOOK

Over the last decade, an increasing number of pilot and demonstration projects implementing source separation in sanitation systems have been run world-wide. A sector review providing information on some 310 projects has been brought together by the GTZ sector programme "sustainable sanitation – ecosan" (GTZ 2010). Approximately 90 of these projects serve more than 1000 users in middle- or low-income countries (25 serve more than 10,000 users) – so source-separating systems are increasingly gaining ground there.

There are success stories as well as failures, and we can learn from both. However, implementation of these projects at scale has contributed to the further development of a whole range of source separation technologies and to the improvement of operation and management skills and reuse options. As the two examples from China and South Africa show, there can be a variety of drivers for introducing new source-separating technologies. A key lesson learned is that technological innovations require targeted subsidies coupled with a social marketing phase for market entry; they must be carefully introduced and accompanied by long-term and targeted awareness-raising and training interventions to guarantee widespread success. So far, source-separating technologies show most promise in rural or peri-urban communities where agriculture is the dominant source of income.

REFERENCES

Atkins and hydrophil (2009). Draft Case Study Report for the Financial and Economic Analysis of Ecological Sanitation in Sub-Saharan Africa. Unpublished report by M. Wilkinson.

Feiereisen P. (2010). Personal communication.

Feiereisen P. and Wu T. (2009). Surveying the long-term feasibility and acceptance of different sustainable sanitation systems in Shaanxi and Yunnan. GTZ Eschborn. http://www2.gtz.de/Dokumente/oe44/ecosan/en-long-term-feasibility-sustainable-sanitation-systems-2009.pdf (accessed 20 December 2010).

GTZ (2010). Worldwide list of documented ecosan projects by various organisations. http://www.gtz.de/en/themen/umwelt-infrastruktur/wasser/30631.htm (accessed 20 December 2010).

Kumar (2008). Community-led water and ecosan Programme, Shaanxi Province, China. SuSana case study of sustainable sanitation project, Eschborn, Germany.

Lüthi C., Panesar A. and Schütze T. (eds.) (2011). Sustainable Sanitation in Cities – a framework for action. Papyroz Publishing, The Netherlands.

NDRC (National Development and Reform Commission) (2007). Medium and Long-term Development Plan for Renewable Energy in China. Abbreviated Version translated from Chinese, Peking, PR China.

WSP-Africa (Water and Sanitation Program). (2009). Study for Financial and Economic Analysis of Ecological Sanitation in Sub-Saharan Africa. WSP Nairobi, Kenya.

Part V

The paradigm shift

Chapter 34

Why question the prevailing paradigm of wastewater management?

M Bruce Beck

34.1 INTRODUCTION

From the perspective of Earth Systems Analysis, we might view the metabolism of the city as that of Figure 34.1 (Beck 2011). Under this gross simplification, the city is caricatured as taking in water and food on its "upside," while emitting a single stream of wastewater on its companion "downside" (Figure 34.1a). Driven by the self-evidently global concerns of sustainability and climate change, we may ask: where and how can we recover and recycle resources from this metabolism; and, if we can, what might be the implications for, say, greenhouse gas emissions? Passing through the sequence of Figures 34.1a–34.1d, we observe that source separation (with wet sanitation) in Figure 34.1c should succeed in pulling apart the schematic water and nutrient cycles in which the city's metabolism participates. The inference is that this would be desirable for resource recovery (including water) and for rectifying some of the distortions in global material cycles wrought by the city. Source separation with dry sanitation (in Figure 34.1d) extends the prospect of reducing the city's water metabolism to nearly zero, if extrapolated to its logical end point (albeit with some artistic license). It is suggestive of enhancing the recovery of resources from the city's metabolism, in particular, resources other than water. Understood literally, Figure 34.1d would imply a very high degree of recycling and/or urban agriculture.

At this strategic and simplified level, one salient point is striking: all structural change is within or on the downside of the city. In other words, the topology and engineering of the infrastructure for bringing water and food into the city remain essentially unchanged. More specifically, Figure 34.1a reflects the configuration of the prevailing paradigm of wastewater management and infrastructure, that is, for cities largely of the so-called Global North. We have it because of the overwhelming sanitary and technological success of the Water Closet (WC).

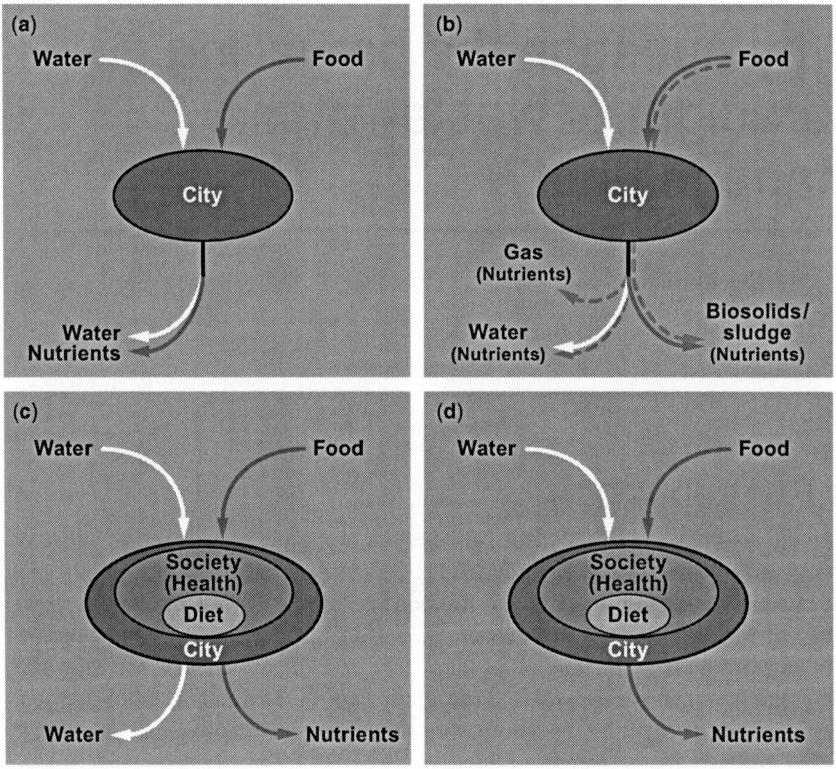

Figure 34.1 Schematic of the city, its daily water and daily bread, its metabolism, and its water and nutrient return infrastructures: (a) current water-based wastewater infrastructure of cities of the Global North; (b) the current paradigm of (a) bent to some other purpose, for example, the recovery of resources from the solids (sludge) stream, with yet resource losses to the water and atmospheric media; (c) future vision of uncoupled water and nutrient return infrastructures, *including* a change of lifestyle (see text); (d) the logical limit of a maximally eco-efficient city water metabolism, that is, with a dry sanitation system.

Given the equally strategic long view over time, history records that the structure of the metabolism of the city of Paris (between 1790 and 1970) passed from Figure 34.1d, through Figure 34.1c, to end up with the paradigm of the Global North, that is, Figure 34.1a (Barles 2007a, b). Just prior to the First World War (in 1913), 40% of the dietary nitrogen (N) in the biological residuals of the city's human (as opposed to horse) population, was being recycled in fertilizers applied to the agricultural lands of Paris's surrounds. As elsewhere today, there is doubtless great interest in Paris in seeking to achieve nutrient, energy and water recovery from its wastewater infrastructure, according to the arrangement of Figure 34.1b – in other words, from a centralized, source-mixed configuration of the system.

The conceptual and schematic difference in proceeding from Figure 34.1a to Figure 34.1c (or even 34.1d), as opposed to striving for the situation in Figure 34.1b, is what this book has been all about.

The present chapter will begin by re-phrasing this core challenge in modest but significant ways. Mindful of what was famously expressed by Otterpohl *et al.* (1999) – to the effect that, once the public health of citizens has been secured, the purpose of the urban wastewater infrastructure is to keep the soil fertile – the original motivation of the present discussion (in the early 1990s) was to answer the question of how a "perfect fertilizer" might be recovered within the current paradigm of wastewater infrastructure (Beck 2011). Invoking source separation as a prime component of the means to do so came only later, inspired by the original work of Larsen and Gujer (1997). This latter notion has been central, however, in a case study of the city of Atlanta (within the Upper Chattahoochee watershed in the south-eastern USA), itself a principal focus of the current research of the International Network on Cities as Forces for Good in the Environment (Beck *et al.* 2010, 2011; www.cfgnet.org).

The nature of today's point of departure matters greatly, of course, to the way in which one can respond to the challenges of questioning the current paradigm. The opportunities, constraints, advantages and disadvantages of source separation and decentralization will present themselves very differently, according to whether one is departing from the initial conditions of Figure 34.1a (broadly those of cities in the Global North) as opposed to those of Figure 34.1d (generally of the Global South).

Future technological options and innovations may promise much in respect of restoring, sustaining, even enhancing future (inter-generational) environmental well-being. But what features of their economic feasibility will spark any transformation away from unsustainability in the metabolism of the city; and what aspects of their social legitimacy will achieve ignition from that spark (not extinguish it)? These questions will be addressed in closing the chapter.

34.2 IMAGINING THE CITY AS A FORCE FOR GOOD IN ITS ENVIRONMENT

With hindsight, the previous relationship between the city of Paris and its surrounds (of the Seine watershed) could well be viewed as a mutually beneficial symbiosis. Put simply, securing public health in the city, so that city dwellers may participate in the obviously successful economy of the city (Glaeser 2011; Dobbs *et al.* 2011), has historically been bought at the expense of water pollution and severing of the city-watershed symbiosis (Barles 2007a, b; Neset *et al.* 2008).

Appreciation of any environmental benefits deriving from this symbiosis of former times seems to have faded from public memory across the 20th century. Cities came to be viewed as the industrial "bads" of modernity, planted destructively in fragile environments (Odum 1989). And in due course, environmental engineers also became the subject of considerable disapproval, even though they themselves felt that they were *self-evidently* doing good for the environment – witness Niemcynowicz (1993), who indeed had few words of praise for the prevailing paradigm of wastewater management.

In 2006/7 the US National Academy of Engineering convened and hosted a blue-ribbon panel to identify (with the public at large) what might be the "Grand Challenges for Engineering in the 21st Century." An essay on "Cities as Forces for Good in the Environment" was prepared as a background document for the panel. It enquired (Crutzen *et al.* 2007; see also Beck 2011):

> How can the built infrastructure of the city be re-engineered to restore the natural capital and ecosystem services of the nature that inhabited the land before the city arrived there, in "geological time"?
> How can this infrastructure be re-engineered to enable the city to act as a force for good, to compensate deliberately and positively for the ills of the rest of Man's interventions in Nature?
> How can cities of the Global South avoid adopting the same technological trajectory as those of the Global North? Can they, as it were, "leap-frog" the Global North by forgoing the entire human-waste-into-the-water-cycle phase, thereby ending up one step ahead?

These questions cast the discourse of this book in a somewhat wider frame: that of the ever-changing relationship between the city and its surrounding environment, at the core of which is the evolving nature of urban wastewater management. They oblige us to consider and assess changes beyond the water sector, in the energy, food, waste-handling, and forestry sectors, for example (Beck and Villarroel Walker 2011). They anticipate future re-configurations of the city and its infrastructure capable even of deliberately *enhancing* the ecosystem services of the watershed (Beck 2011; Force 2011).

34.3 SOURCE SEPARATION AND DECENTRALIZATION

From this wider perspective of what matters to the environment and the provision of ecosystem services, it has clearly been the intent of this book to enquire whether the force of logic – driving us towards the choices of source separation and decentralization – is compelling.

If one were to start from scratch (*de novo*), metaphorically from a blank sheet of paper for Figure 34.1, as in developing a satellite city in the vicinity of Beijing, China, Dong *et al.* (2012) would be quite bold in their advocacy of decentralization. "Design of a sustainable city has changed the traditional centralized urban wastewater system towards a decentralized or clustering one," they say. Where and how collection networks and materials-processing facilities might be placed across the future urban landscape can be determined as a function of the proximate, spatially-contingent requirements for recovery of water, nutrients and energy (Dong *et al.* 2012).

Alternatively, if one were to start from the arrangement of dry sanitation in Figure 34.1d, Issue 23 (April 2010) of the *Urban Agriculture Magazine* of the Resources Centres on Urban Agriculture & Food Security (RUAF) is revealing. It bears the title: *The Role of Urban Agriculture in Sustainable Urban Nutrient Management* (www.ruaf.org; accessed 22 January 2011). One of its articles (Drechsel and Erni 2010) is concerned with "Analysing the Nexus of Sanitation and Agriculture at Municipal Scale." Another, that of Dagerskog *et al.* (2010), opens with these words: "Since March 2009, there has been a 'human fertiliser' market in Ouagadougou, the capital of Burkina Faso. Human urine and dried faeces are collected and taken to eco-stations, where they are sold to farmers after adequate storage."

In Ouagadougou, source separation is achieved (in principle) with a so-called Urine-Diverting Dry Toilet (UDDT). Decentralization is manifest in the fact that this trial – in sparking a transition with market incentives in the absence of governmental "pressure" – was implemented in four out of the thirty administrative districts of the city, within which urban farming was already taking place. Distance to the market place is crucial to getting the economics right. Tight, closed-loop recycling of nutrient resources within the city (Ouagadougou) is therefore quite apparent. Trespassing into the developmental stage of "human-waste-into-the-water cycle," that is, Figure 34.1a, has thus far been avoided. Configuration of the city's infrastructure is broadly that of Figure 34.1d, albeit with grey water still an issue of concern (this was not illuminated above in Figure 34.1, in the interests there of clarity and simplicity).

Under the prevailing paradigm of wastewater infrastructure in the Global North (Figure 34.1a), neither of the foregoing design and evolutionary "freedoms" of Ouagadougou or the Beijing satellite city are open to us. Intuition and practice (Force 2011) suggest that proceeding from Figure 34.1a to Figure 34.1b, as opposed to either of Figures 34.1c or 34.1d, should be the easier adaptation, both economically and socially. Structural change in the means of sewage conveyance, and/or in the households and office blocks further upstream, is not implied. It is confined to innovations and changes in the unit processes deployed in the centralized wastewater treatment plant. In the foresight studies of Atlanta, proceeding from Figure 34.1a to the source-separated configuration of Figure 34.1c can be shown to be quite successful – in principle – both in terms of nutrient resource recovery and the reconstruction and enhancement of ecosystem services (Beck *et al.* 2010). In practice, however, it is not hard to imagine such (radical) change would potentially be quite disruptive socially. It would also be very difficult to justify on economic grounds, *not* because the change cannot be justified, but because the precise framing of the problem is not straightforward, hence the economic data for its assessment are currently difficult to assemble.

34.4 SOCIAL SCIENCE AND ECONOMICS

When the discussion is of paradigms and of their questioning, engineers are well aware of the fact that novel schools of thought in their subject and an abundance of potential technological innovations may not be the essential engines of paradigmatic change. For the social and economic structures that come with the paradigm both maintain the hegemony of its orthodoxy and allow it to resist change, in order that it may endure, that is, preserve itself.

Writing on *Uncertainty and Quality in Science for Policy*, Funtowicz and Ravetz (1990) introduced what they called a research-pedigree matrix. As a field of enquiry matures, they argued, colleague consensus passes from "no opinion," to "embryonic field," "competing schools," to "all but rebels" and "all but cranks." The present book is an expression of an emerging, broad school of thought that (clearly) seeks to question the prevailing paradigm. In the words of Funtowicz and Ravetz, it has elements of "rebelliousness" about it. Yet one mature, broad school of thought – in our view – is not to replace or displace any other existing school of thought, but to co-exist most fruitfully with it. Somewhat in contrast to Funtowicz and Ravetz, it can be argued that constructive, disputatious debate amongst *plural*, vigorous schools of thought should prevail (Gyawali 2001; Dixit 2002; Beck *et al.* 2011; Beck 2011).

There is much of relevance to the present discussion in Thompson's (2011) historical account of how London's housing stock from the 19th and 18th centuries survived the prevailing "centralized," hierarchical planning paradigm of the mid-20th century – and survived to prosper to stunning effect, through a combination of decentralized, local entrepreneurship and similarly localized egalitarian-minded activists (Beck *et al.* 2012).

But what, at bottom, might provide the incentive of the economic spark, to ignite the transition to some other less unsustainable *modus operandi*? Maurer (2013) observes that economic analyses of the transitions amongst any of the structural arrangements of Figure 34.1 are sparse, rudimentary and uncertain. He himself concentrates largely on the *costs* of possible transitions.

Halting eutrophication of water bodies surely does not come cheaply these days – at least not for the associated strategies for eliminating nutrients from the "waste" residuals in cities. From other on-going work on the Atlanta-Chattahoochee case study, it is apparent that ridding the wastewater of Metro Atlanta of a further 50 tonnes of phosphorus annually, beyond current performance levels, might cost around \$0.3 M–0.5 M each year (based on Jiang *et al.* 2005). In contrast, when set against operating and capital works costs of appropriate system re-configurations (based on the figures of Dockhorn 2009), an annual net income of \$4 M might be possible. This would derive from the economic *benefits* attaching to the 1700 tonnes of recoverable phosphorus and 16,600 tonnes of nitrogen, when sold into the fertilizer market. In addition, there should be savings in operating costs of about \$1.2-2.0M from not having to eliminate the recovered phosphorus and nitrogen as pollutants.

Instead of water pollution control being some never-ending, policy-driven drain on the public purse, an entrepreneurial benefit could seemingly be conjured up (Beck and Villarroel Walker 2011).

34.5 CONCLUSIONS

Why indeed should we question the prevailing paradigm of wastewater management?

In this chapter, we have embedded this question within the somewhat broader enquiry into the possible forms and purposes of the urban wastewater infrastructure in serving the development of a less unsustainable interaction between the city and the environment of its surrounding watershed. What contribution, for example, might the city make to restoring and enhancing the business of the ecosystems services sector? Asking the question, as such, does not presume that source separation and decentralization *have to be* parts of the answer.

However, without a deeper awareness of the practical successes and theoretical potential of source separation and decentralization – in fact, their potential to break the mould of conventional thinking – our appreciation of the social, economic and environmental opportunities for strategic change in the nature of the city-watershed relationship will be markedly impoverished (Beck *et al.* 2012).

We are all doubtless aware of the aphorism "*Think Globally, Act Locally*." In opening this chapter, three challenges from the essay by Crutzen *et al.* (2007) on Cities were quoted. There is a fourth such challenge:

> How can the engineering of city infrastructure be deployed expressly so that those at the bottom of the pyramid of dignified human development may be brought to a level where they care to engage in a debate over such a grand challenge for this century – of cities as forces for good – beyond their desperate needs of survival for just today and tomorrow?

At their core, source separation and decentralization drive our attention to focus on the most intimate and personal facets of the lives of each of us as individuals – *very* local indeed, one might say. What then might engender empathy – in the host of the so very many individuals in the world still awaiting the sanitary and health benefits of their own urine-separating toilet or urine-diverting dry toilet – for the big and the personally remote issues of sustainability and climate change? Quite pragmatically and prosaically, how might engineers have an ethical role in responding to this challenge? For this would be a case of (Beck 2011):

> "[Engineers] *Acting Locally*, [as deliberate stimulus to community yearning for] *Thinking Globally*"

Now the goal would be to promote posterior debate through prior action. In turn, of course, further (posterior) action should emanate from what would by then have become that prior debate.

From the myriad highly localized actions already taken and observed upon in so many of the chapters of this book, we should look forward to what ought now to follow as the framing of the debate that is about to be had, not least to engage constructively with the upholders of the paradigm this chapter has sought to question.

REFERENCES

Barles S. (2007a). Feeding the city: food consumption and flow of nitrogen, Paris, 1801–1914. *Science of the Total Environment*, **375**, 48–58.

Barles S. (2007b). Urban metabolism and river systems: an historical perspective — Paris and the Seine, 1790–1970. *Hydrology and Earth System Sciences*, **11**, 1757–1769.

Beck M. B. (2011). Cities as Forces for Good in the Environment: Sustainability in the Water Sector. Warnell School of Forestry and Natural Resources, University of Georgia, Athens, Georgia (ISBN: 978-1-61584-248-4), pp. xx + 165 (online as http://cfgnet.org/archives/587; accessed 17 April 2012).

Beck M. B. and Villarroel Walker R. (2011). Global water crisis: a joined-up view from the city. *Surveys And Perspectives Integrating ENvironment & Society*, [Online], **4**.1, Online since 27 December 2011. Url: http://sapiens.revues.org/1187.

Beck M. B., Jiang F., Shi F., Villarroel Walker R., Osidele O. O., Lin Z., Demir I. and Hall J. W. (2010). Re-engineering cities as forces for good in the environment. *Proceedings of the Institution of Civil Engineers, Engineering Sustainability*, **163**(ES1), 31–46.

Beck M. B., Thompson M., Ney S., Gyawali D. and Jeffrey P. (2011). On governance for re-engineering city infrastructure. *Proceedings of the Institution of Civil Engineers, Engineering Sustainability*, **164**(ES2), 129–142.

Beck M. B., Villarroel Walker R. and Thompson M. (2012). Changing the metabolism of coupled human-built-natural systems. In: Earth Systems Engineering 2012: A Technical Symposium on Systems Engineering for Sustainable Adaptation to Global Change, R. J. Dawson, C. L. Walsh, and C. G. Kilsby (eds), Centre for Earth Systems Engineering Research, Newcastle University, UK, pp. 11–32.

Crutzen P. J., Beck M. B. and Thompson M. (2007). Cities. Essay (posted at website www.engineeringchallenges.org), Blue Ribbon Panel on Grand Challenges for Engineering, US National Academy of Engineering (see also *Options* (Winter 2007), International Institute for Applied Systems Analysis, Laxenburg, Austria, p 8).

Dagerskog L., Coulibaly C. and Ouandaoga I. (2010). The emerging market of treated human excreta in Ouagadougou. *Urban Agriculture Magazine*, **23**, 45–48 (April) (online at www.ruaf.org).

Dixit A. (2002). Basic Water Science. Nepal Water Conservation Foundation, Kathmandu.

Dobbs R., Smit S., Remes J., Manyika J., Roxburgh C. and Restrepo A. (2011). Urban world: mapping the economic power of cities". Report, McKinsey Global Institute. http://www.mckinsey.com/Insights/MGI/Research/Urbanization/Urban_world (accessed 17 April 2012).

Dockhorn T. (2009). About the economy of phosphorus recovery. In: Proceedings International Conference on Nutrient Recovery from Wastewater Streams, IWA Publishing, London, pp. 145–158.

Dong X., Zeng S. and Chen J. (2012). A spatial multi-objective optimization model for sustainable urban wastewater system layout planning. Accepted for publication in *Water Science and Technology*, **66**(2), 267–274.

Drechsel P. and Erni M. (2010). Analysing the nexus of sanitation and agriculture at municipal scale. *Urban Agriculture Magazine*, **23**, 11–12 (April) (online at www.ruaf.org).

Force J. (2011). The ultimate recycling. Treatment Plant Operator, 12–17 (September).

Funtowicz S. O. and Ravetz J. R. (1990). Uncertainty and Quality in Science for Policy. Kluwer, Dordrecht, The Netherlands, pp. 229.

Glaeser E. (2011). Triumph of the City. How Our Greatest Invention Makes Us Smarter, Greener, Healthier, and Happier. Penguin, New York.

Gyawali D. (2001). Rivers, Technology and Society. Himal Books, Kathmandu, Nepal.

Jiang F., Beck M. B., Cummings R. G., Rowles K. and Russell D. (2005). Estimation of Costs of Phosphorus Removal in Wastewater Treatment Facilities: Adaptation of Existing Facilities. Water Policy Working Paper # 2005–011, Andrew Young School of Policy Studies, Georgia State University, Atlanta, Georgia.

Larsen T. A. and Gujer W. (1997). The concept of sustainable urban water management. *Water Science and Technology*, **35**(9), 3–10.

Maurer M. (2013). Full costs, (dis-)economies of scale and the price of uncertainty. In: *Source Separation and Decentralization for Wastewater Management*, T. A. Larsen, K. M. Udert and J. Lienert (eds.), IWA Publishing, London, UK, pp. 85–100.

Neset T-S. S., Bader H-P., Scheidegger R. and Lohm U. (2008). The flow of phosphorus in food production and consumption — Linköping, Sweden, 1870–2000. *Science of the Total Environment*, **396**, 111–120.

Niemczynowicz J. (1993). New aspects of sewerage and water technology. *Ambio.*, **22**(7), 449–455.

Odum E. P. (1989). Ecology and Our Endangered Life-support Systems. Sinauer Associates, Sunderland, Massachusetts.

Otterpohl R., Albold A. and Oldenburg M. (1999). Source control in urban sanitation and waste management: ten systems with reuse of resources. *Water Science and Technology*, **39**(5), 153–160.

Thompson M. (2011). Material Flows and Moral Positions. Insight, Cities as Forces for Good (CFG) Network (online as http://cfgnet.org/archives/531).

Chapter 35

How to spur innovation?

Peter A. Wilderer

35.1 WHY INNOVATION?

The authors who contributed to this book scrutinize our well established, technically well developed and operationally well functioning concept of central urban water supply and sanitation. The term "central" refers to the technical installations pooled within an urban area in distinct locations, the waterworks and the wastewater treatment plant, respectively. Water is used just once and then discarded. For a number of reasons the authors recommend a different approach characterized by expressions such as "source separation," "recovery" and "reuse." The technical treatment installations should get located close to the origin of the used water. To distinguish this innovative solution from the classical concept the term "decentralization" has been chosen.

What is the motivation which keeps us asking for a change of paradigm of urban water management? To answer this question we ought to consider four major categories of drivers for such a shift of urban water supply and sanitation systems:

(1) Excessive growth of urban population, worldwide.
(2) Globalization of the life style prevailing in the industrial countries.
(3) Global warming and climate change.
(4) Our obligation to contribute to the sustainable development of economy, ecology and the human society.

Concerning the latter category of drivers it is fair to assume that the inherited concept of urban water supply and sanitation does not satisfactorily meet the commonly accepted sustainability criteria. Main points of critique are:

- Use of large portions of fresh water for transportation of pollutants, and as means to avoid sediment build-up in sewers.

- Mixing and dilution which makes wastewater treatment and recovery of usable water and other valuable materials a difficult and expensive task.
- Discarding instead of recovering and reusing valuable materials and energy.

It is worth mentioning that our way of treating water and wastewater is not an invention of the "throw-away" economy which developed in the industrialized countries over the past decades. It is just a copy of the methods which were introduced thousands of years ago already in the ancient city cultures. The Romans were particularly good in transporting water, partly over long distances, into cities, and transporting the used water including stormwater out of the city limits (Figure 35.1). This is exactly what we exercise today. In modern times we just added to this "single pass concept" technology to purify the incoming water, and technology to remove pollutants from the outgoing wastewater which then are either dumped somewhere or incinerated. This old and subsequently upgraded concept fulfils only partly sustainability criteria because water, materials and energy are to a large part dissipated.

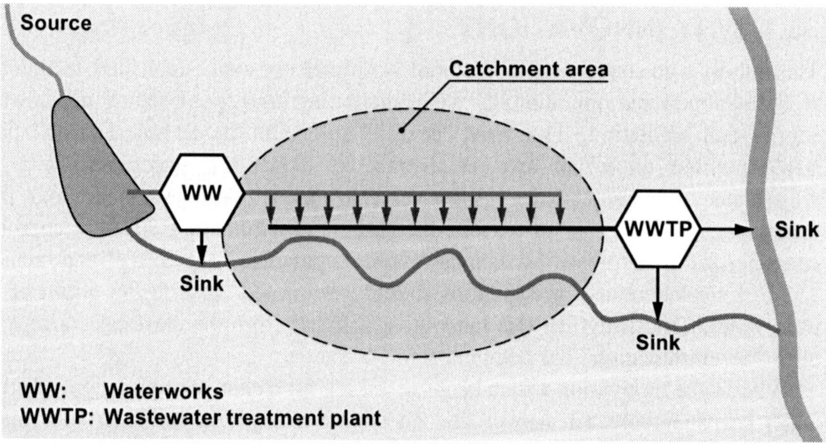

Figure 35.1 Our inherited method of using water in urban areas: Water is used just once and discarded thereafter.

As long as the population size and urban water consumption remains moderate the traditional water concept is somewhat acceptable. In recent years, the situation has significantly changed, however. Whereas 2000 years ago only about 200 million people lived on Earth we are now about 7 billion, all of them relying on water in sufficient quantity and quality. In the high income countries flushing toilets, showers, washing machines and dishwashers have become basic commodities, and it is foreseeable that people in the medium and low income countries will insist on using such commodities as well. And finally, we should realize that urban water management practices as well as water and wastewater technology were developed in areas where water is rather abundant. Most of the people on Earth, however, live in areas where the availability of water is low, naturally and as a consequence of climate change and pollution.

In summary, we are facing a tremendous increase of the demand for water, worldwide and particularly in the rapidly growing urban areas. At the same time and at the local scale, the capacity and quality of the fresh water resources is insufficient, and often shrinking because of over-abstraction and pollution. Innovation in the field of both, water management and water technology is necessary to overcome the discrepancy between availability and demand of raw water of reasonably high quality. Innovative solutions are required to serve people appropriately, and to avoid societal and economic instabilities to emerge. Innovation in the water sector is not only justified for humanitarian reasons. It is also in the interest of national and supra-national security.

35.2 THEORY OF INNOVATION

Innovation is to be understood as the cognitive expression of evolution. It is a process initiated by human intuition, by vision of future benefits, by the desire to overcome unsatisfactory conditions and not to forget: by the desire of entrepreneurs to gain advantages on the market and subsequent monetary profit.

Joseph Schumpeter (1883–1950) was the first to have analyzed the importance of innovation to economic development (Schumpeter 1934). The theory of innovation he suggested is based on the assumption that the overcoming of factors such as uncertainty, interaction and change of market economies requires a "process of creative destruction" mediated by innovation (Schumpeter 1950). Products are abandoned – are subject to destruction of their market value – as soon as innovations materialize.

Basic requirements of innovation are creative thinking, experience and scientific knowledge. The first step on the road to innovation is marked by the formulation of a clever idea, the invention and the filing of a patent.

At this point it is worth mentioning that an invention does not constitute innovation. Innovation materializes not until successful market introduction (von Hauff and Joerg 2010; see also Truffer *et al.* 2013).

In economic science the innovation process is subdivided into three stages (Schumpeter 1939):

- Invention:
  The discovery of a new production process or product
- Innovation:
  The successfulness of market introduction
- Diffusion:
  Successful distribution of the innovation among users over time

In other words, the innovation process begins with a thorough investigation of known and unknowns followed by scientific research, piloting, up-scaling, full-scale trials, market introduction and distribution.

Market introduction and diffusion appears to be the most difficult part of this sequence of processes, particularly when a paradigmatic change is in focus. Nobody expressed this better than Niccolo Machiavelli (1469–1527), when he wrote in his treatise "The Prince:" "*Remember that nothing is more difficult to achieve, more risky in realization and more uncertain concerning success than to introduce a new paradigm, since the innovator has as an enemy those who were successful under the old, and who get only lukewarm support by those who may profit from the new paradigm.*"

Innovation is not restricted to products. On the contrary, we have to discriminate between product, process and service innovation (Hagedoorn 1996). In this context, service innovation enables substantial improvements and changes of management and organizational practices, particularly with respect to consumption of resources (eco-innovation). Process innovation enables enhancement of productivity. Product innovation is aimed to satisfy tangible customer benefits providing a positive feedback on production and selling price (tangible producer benefit).

35.3 SUSTAINABLE INNOVATION

Sustainable innovation is a new paradigm in the study of innovation. It merges the sustainability axiom with Schumpeter's concept of innovation. According to recent studies of von Hauff and Joerg (2010) sustainable development requires innovative services, processes and products with respect not only to the economic but also to the societal and the ecological dimension.

The aim of sustainable development is to minimize or completely avoid innovation risk or undesirable developments. The fact that innovation also carries ecologic risk was largely ignored until recently. The potential for ecologic risk occurs when technologies are developed that threaten the environment, now and in the future. Risk may also occur indirectly when new technologies generate higher productivity but lead to an additional pressure on the environment (Hauff and Joerg 2010).

From the viewpoint of the traditional, entirely market oriented economy an invention is considered positive when commercially successful. Over the past years, however, we had to learn that such a single-edged evaluation of invention is dangerous. It neglects possible side-effects with respect to public health, ecosystem functions and stability of societal systems. With the introduction of the concept of sustainability innovation, the aim is to minimize such limitations. The intention is not to change the direction of innovation but to introduce a holistic assessment of innovation. A product, a process or a service may qualify as innovative only when with respect to economy, ecology and society overarching benefits have been successfully demonstrated.

In this book, a major attempt is made towards sustainable innovation in the field of wastewater management – or to take it one step further: to sustainable innovation of water management and water related processes and products serving the needs of people to receive water in sufficient quantity and quality, to have access to decent sanitation and public health measures while not compromising the long-term functioning of the economic, societal and ecological system characteristics for the very location to be served.

35.4 SUSTAINABLE WASTEWATER MANAGEMENT

The question is now whether or not the technical and management methods summarized in this book already qualify for sustainable innovation. Answering this question is not a trivial exercise for the author of this chapter. As being a strong proponent of source separation and decentralization from early on I cannot avoid to argue *pro domo* to a large extent. I will try to take as much as possible a neutral position by throwing some light on the recent "historical" development of the process of innovation in the field of water supply and sanitation.

The basic incentive to consider source control measures in wastewater management was given by Tove Larsen and Willi Gujer in the nineties of the past century. The papers published by the authors (Larsen and Gujer 1996, 1997) were an eye-opener for many, and led to consternation of others. In simple terms the authors pointed out that the problem we are facing in urban water management stems from mixing and dilution.

In this context it is worthwhile to notice that in the field of solid waste management we learned that separate collection of paper, glass, metals and plastics permits recovery of valuable materials for re-introduction into the cycle of materials. Thus, consumption of precious resources gets reduced as well as the demand for landfill volume and costs. Despite early warnings the population accepted separate collection very quickly, worldwide – at least in the high and medium income countries.

Apparently the concept of separate collection and recovery works on the solid waste side. Why should it not work on the wastewater side? The benefits are obvious as described in several chapters of this book. Plant nutrients can be recovered and used as fertilizers and costs for nitrification, denitrification and phosphorus removal in central wastewater treatment plant can be significantly reduced. Greywater can be treated on site and reused for various non-potable purposes while the demand for abstraction of fresh water from water sources gets minimized. Slurries containing faeces and kitchen waste can be readily used as an energy source.

When such arguments were brought up the response of the mainstream community was rather negative and in line with the Machiavelli statement. In the meantime, the scientific basis of source separation and decentralization has been solidified, however. Major contributions to science and technology were achieved in the framework of the integrated project "Novaquatis" executed by researchers of the Swiss Federal Institute of Aquatic Science and Technology (Eawag) and presented in the form of a great number of publication and a summarizing report (Larsen and Lienert 2007). The initial idea turned into products which are successfully marketed. Separation toilets are now on the market based on suggestion made by Larsen *et al.* (2001) and by several others. Separation and recovery processes are practised at a variety of full-scale installation. At an increasing rate, end-users accepted the new technology as discussed by Lienert (2013). It appears that the new concept of source separation and decentralization and the derived services, processes and products qualify for "sustainable innovation."

At the same time Tove Larsen and Willi Gujer presented their ideas to two staff members of a company, who visited my laboratory in Garching near Munich (Germany). The company, specialized in military weapon production and destruction, had won a bid for a project which included building residential areas for Russian soldiers leaving the Eastern part of Germany. The engineers of the company wondered why houses are built at the end of the 20th century using a "*stone-age approach*" as they put it. The question was: Which construction and building methods might be applied if we would have to invent the building process from scratch using the entire arsenal of scientific knowledge available today?

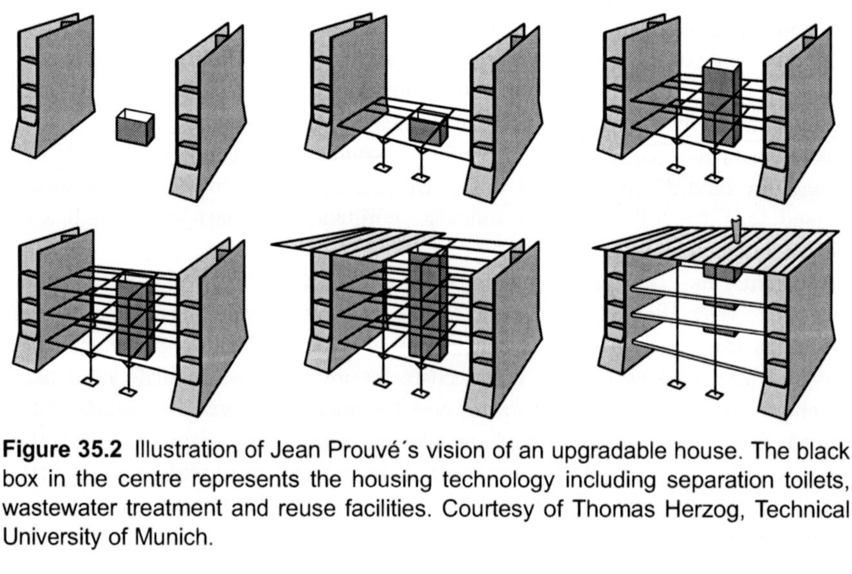

Figure 35.2 Illustration of Jean Prouvé´s vision of an upgradable house. The black box in the centre represents the housing technology including separation toilets, wastewater treatment and reuse facilities. Courtesy of Thomas Herzog, Technical University of Munich.

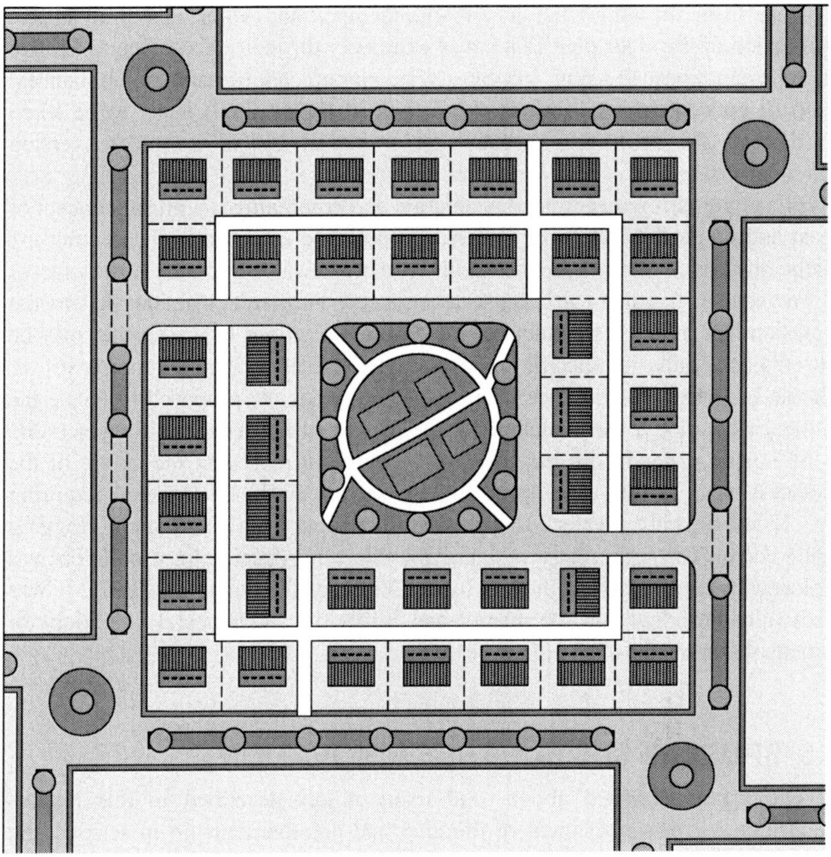

Figure 35.3 Design of a housing estate. The wastewater treatment facilities are located in the basement of a central courtyard from where the purified water is returned to the individual houses for non-potable use. Courtesy: Arnold Consult, Kissing (Germany).

During a brain-storming event attended by faculty of the departments of architecture and civil engineering at the Technical University of Munich (TUM), Thomas Herzog, professor of the department of architecture, showed us a sketch (Figure 35.2) made by the French architect, Jean Prouvé (1901–1984). He proposed designing a house so that it can be modified in a stepwise process according to the changing demands of the tenants, and the technical progress of housing technology. To be able to replace "old for new" without disturbing the tenant it appears to be advantageous to avoid attributing alien functions to the components of a house: the supporting structure, ceilings and walls, facade and roof and housing technology. For instance, the function of a wall is to separate one room from the other, but implementing pipes and wires in a wall makes modification of the floor plan of a house extremely difficult. According to Prouvé housing technology (heating, cooling, telecommunication, sanitary appliances) should be concentrated in a block (black box in Figure 35.2) ready to be taken out (through the roof, for instance) and replaced by an up-to-date version when appropriate.

While taking this concept in consideration and combining it with the concept of Larsen and Gujer the task force at TUM proposed to integrate in the central box installations which allow source separation of wastewater fractions (urine, faeces and greywater), treatment and reuse technology. Further evaluations led to the conclusion that in-house treatment of the different fractions of wastewater may be cost effective only in specific cases (single residences, hotel complexes or high-rise buildings, for instance). For residential areas we proposed to locate the treatment technology, the container for the recovered nutrients and the reservoirs for the treated water in the basement of a courtyard placed in the centre of the residential area. In this case, steep gravity sewers could be installed requiring only little flushing water to prevent sediment build-up and clogging (Figure 35.3). The technology required for this semi-decentralized solution was developed in cooperation with the Huber Company (Huber *et al.* 2007). It was successfully introduced in the market and fulfils the fundamental conditions of "sustainable innovation," therefore.

35.5 SPURING SUSTAINABLE INNOVATION

The two cases described above – and many others described in this book – demonstrate the power of creative thinking and entrepreneurship in science and technology. But in line with what Machiavelli wrote it has been and still is difficult to overcome the resistance of those who in monetary terms are successful under the paradigm which we inherited from the ancient city cultures, and which was further developed over the past 150 years.

The answer to the question how to spur innovation can be derived from the experience made by the authors of this book. The secret of success can be summarized in three key-words: flexibility, good science and persistence. Sustainable innovation materializes not only when the market accepts the new product, process or service. As important are honesty, persuasiveness and dedication of the innovators, and their readiness to act responsible with respect to humanity and nature. The call is for a truly and overarching participative approach.

REFERENCES

Hagedoorn J. (1996). Innovation and entrepreneurship: Schumpeter revisited. *Industrial and Corporate Change*, **5**(3), 883–869.

Von Hauff M. and Joerg A. (2010). Innovation im Kontext nachhaltiger Entwicklung (Innovation in the context of sustainable development). In: Nachhaltige Entwicklung: Das neue Paradigma in der Oekonomie, H. Hagemann and M. von Hauff (eds.), Metropolis Verlag, Marburg, Germany, pp. 185–212.

Huber H., Wilderer P. and Paris S. (eds.), (2007). Water Supply and Sanitation for All. WEMS 14: Obligation of the water professionals for our common future. IWA–Publishing, London, UK. ISBN: 9781843395140.

Larsen T. A. and Gujer W. (1996). Separate management of anthropogenic nutrient solutions (human urine). *Water Sci. Technol.*, **34**(3–4), 87–94.

Larsen T. A. and Gujer W. (1997). The concept of sustainable urban water management. *Water Sci. Technol.*, **35**(9), 3–10.

Larsen T. A. and Lienert J. (2007). Novaquatis final report. NoMix – A new approach to urban water management. Eawag, Duebendorf, Switzerland.

Larsen T. A., Peters I., Alder A., Eggen R., Maurer M. and Muncke J. (2001). Re-engineering the toilet for sustainable wastewater management. *Environ. Sci. Technol.*, **35**(9), A192–A197.

Lienert J. (2013). High acceptance of source-separating technologies, but…. In: *Source Separation and Decentralization for Wastewater Management*, T. A. Larsen, K. M. Udert and J. Lienert (eds.), IWA Publishing, London, UK, pp. 193–207.

Schumpeter J. A. (1911). Theorie der wirtschaftlichen Entwicklung (The Theory of Economic Development). Berlin, Germany.

Schumpeter J. A. (1934). The theory of economic development. Harvard University Press, Cambridge, Mass. (originally published in German in 1911 by Duncker and Humblot, Munich, Germany).

Schumpeter J. A. (1939). Business cycles: A Theoretical, Historical and Statistical Analysis of the Capitalist Process. McGraw–Hill, New York, USA.

Schumpeter J. A. (1950). Capitalism, Socialism and Democracy. 3rd edition, Allen and Unwin Publ., London, UK.

Truffer B., Binz C., Gebauer H. and Störmer E. (2013). Market success of on-site treatment: a systemic innovation problem. In: *Source Separation and Decentralization for Wastewater Management*, T. A. Larsen, K. M. Udert and J. Lienert (eds.), IWA Publishing, London, UK, pp. 209–223.

Index

N

O

P